Springer Natural Hazards

The Springer Natural Hazards series seeks to publish a broad portfolio of scientific books, aiming at researchers, students, and everyone interested in Natural Hazard research. The series includes peer-reviewed monographs, edited volumes, textbooks, and conference proceedings. It covers all categories of hazards such as atmospheric/climatological/oceanographic hazards, storms, tsunamis, floods, avalanches, landslides, erosion, earthquakes, volcanoes, and welcomes book proposals on topics like risk assessment, risk management, and mitigation of hazards, and related subjects.

Apolonia Diana Sherly da Costa

Building Resilience for Flood Disaster in Malaka-Timor, Indonesia

Risk Perception from Human Geography

Springer

Apolonia Diana Sherly da Costa
Department of Southeast Asian Studies
University of Bonn
Bonn, Nordrhein-Westfalen, Germany

Graduate Program of Environmental
and Urban Studies
Faculty of Science and Environment
Technology
Soegijapranata Catholic University
Semarang, Central Java, Indonesia

Present Address:
Institute of Landscape Ecology
University of Münster
Münster, Nordrhein-Westfalen, Germany

ISSN 2365-0656 ISSN 2365-0664 (electronic)
Springer Natural Hazards
ISBN 978-3-031-63271-6 ISBN 978-3-031-63272-3 (eBook)
https://doi.org/10.1007/978-3-031-63272-3

This Springer imprint is published by the registered company Springer Nature Switzerland AG
The registered company address is: Gewerbestrasse 11, 6330 Cham, Switzerland

For Sophia, Augusto, Tony, Ria, Una, Angel, Iky, Noi, Jibi, Rei, and Luís

Preface

Building resilience for natural disasters is a complex, comprehensive, integrative field and knowledge that needs to be known and understood in a familiar and interdisciplinary manner from time to time. This topic is of major interest among academics with mono-disciplinary and interdisciplinary scientific expertise and focus on both natural sciences and social sciences. The most important fundamental reason for understanding this study well and critically is to increase the knowledge of academics in understanding the concept and context of flood resilience when faced with academics who carry out fieldwork in natural disaster areas in a country affected by natural hazards and how local or urban rural communities coping with the natural phenomenon using their resilience assets.

Hence, this book is here to describe the situation of natural disaster areas in Eastern Indonesia with the resilience of women's indigenous land communities (mother/matrilineal land) in Malaka-Timor, and is intended for academics such as professors, lecturers, researchers, scientists, and fellows at universities. I am oriented toward types of local community resilience assets and flood disasters from an interdisciplinary perspective based on empirical research in social geography. I present the results of empirical research findings into a contextual participatory qualitative and quantitative interdisciplinary explanation from Malaka-Timor, Indonesia. So, I believe that the understanding of horizons and discussions related to spatial conflicts, flood mitigations, resiliencies, and local communities' coping in this book can be used as an example and reference as well as a fundamental horizon of knowledge for academics in studying and exploring more critically through empirical and non-empirical researches related to natural disaster cases and the affected communities' coping system in various corners of the country both in south hemisphere countries such as Malaka-Timor, Indonesia, other Indonesia's regions and other south hemisphere countries, and in north hemisphere countries.

This book is the result of my unpublished Ph.D.'s fieldwork/thesis at the Institute for Geography, Friedrich Schiller University of Jena, Thuringia, Germany 2018 (https://suche.thulb.uni-jena.de/Record/1024840379?sid=46755508) which was prepared with a design focus on sustainable livelihood development model regarding social-ecological integrity, resilience assets, and flood disaster risk managament in Malaka-Timor, Indonesia, and more detailed and critical flood vulnerability conclusions of the book. This in-depth insights, motivations, and knowledges in finalizing this book have been prepared by me during my research stay conducting a progressive-collaborative and multi-partnership senior postdoctoral project (research professorship stage) from October 2023 until March 2024 through the full-financial supports of the Catholic Academic Exchange Service/ *Katholischer Akademischer Ausländerdienst* (KAAD) fund, Bonn Germany at the Institute of Landscape Ecology (ILÖK), Faculty of Geo Science University of Münster, supported by the Department of Southeast Asian Studies University of Bonn Germany. Long before that, my critical thinking for inter-disciplinary disaster studies' horizon to present my scientific works e.g., this book was driven by the spirit of writing after having quite a significant research experiences during and after the completion of my doctoral study at the Friedrich Schiller University of Jena, Thuringia, Germany, also financed by KAAD Bonn, Germany, and continuing a valuable experience as a visiting faculty for three months (April 2018–June 2018) at the Friedrich Schiller University of Jena for the postdoctoral level with short research activities, supported by the German Academic Exchange Service (DAAD) Bonn, Germany. The spirit of research to produce this book, another scientific books and other on-going research papers that benefit the wider community and the academic world was continued in western European and southern European countries within some various experiences achieved, including an intensive junior postdoctoral research experiences (postdoctoral research fellowship positions) two years in West Europe i.e., Austria, and South Europe i.e., Portugal, and also some visiting postdoctoral experiences, industry sectors, freelance job, and teaching experiences and duties both onsite and online in Portugal. The first year of my junior postdoctoral research fellow/position in Austria was completed at the Institute for Geography and Regional Science University of Graz Austria by winning the Agency for Education and Internationalization of Austrian Federal Ministry of Education, Science and Research/*Bundesministerium für Bildung, Wissenschaft und Forschung* (BMBWF), *Österreichischer Austauschdienst* (OeAD) fund, Vienna Austria (September 2019–August 2020). While, the second year of my junior postdoctoral research fellow/ position was conducted at the Department of Environment and Planning University of Aveiro, Portugal by winning the most prestigious Portuguese Private-Organization Fund, *Fundação Oriente* Lisbon, Portugal (September 2020–September 2021). I consistently continued my dedication in research and development at the postdoctoral stage as a visiting postdoctoral fellow in Research and Development of the Public University of Porto. I joined as a collaborative member in two years at the Center for Geography Study and Spatial Planning, University of Porto, Portugal (October 2021–September 2023) while working full-time as an Advisor from home in Vila Nova de Gaia, Porto in the Telecommunication, Technology, and Hospitality Industrial Sector, American Company, SITEL-Foundever, in Lisbon, Portugal (February 2023–September 2023), and teaching two subjects i.e., Ecology and Environmental Problems and Philosophy of Urban Space for the Executive Master Students in Graduate Program of Environment and Urban Studies at Soegijapranata Catholic University Indonesia in Semarang, Central Java

Province via cyber-teaching (2021–2023) and also working part-time as a Freelance University-Supervisor and University-Teacher for writing and supervising the University-Teachers and Bachelor, Master, and Ph.D Students' reports, thesis, dissertation, and scientific works from a variety of scientific backgrounds (interdisciplinary, mainly economic, business, climate change, natural disasters, and sustainable development studies/projects) from different nations via Teacher On (2021–Present). These scientific experiences and jobs were undertaken before doing this postdoctoral again at the senior level in North Rhine-Westphalia, Germany for the postdoctoral project unit that I initiated and led in the frame of the KAAD funding program which has been supported by the University of Münster, University of Bonn, and additional support from the Institute of Rescue Engineering and Civil Protection of the Technische Hochschule University of Applied Science Cologne, Germany. Several research papers were also done based on the research work unit of my leadership as a Project Lead that I led and partnered or collaborated well in each of these universities. The motivation for publishing this book also stems from my reflections on my own independent journey alone for research and experience from country to country in western and southern Europe during and after my Ph.D in Jena including a dedicated five-year Junior and Senior Postdoctoral Fellowships which has resulted in more than 73 international and regional awards, prize, recognitions, appreciations, winning nominee, and scholarships and 15 local awards in humanitarian, economic/business resilience, female studies, arts and environment, climate change/natural disaster, and social response, commencing on 23rd September, 2024. This book is the first benchmark in my scientific work in academia that will continue at other stages of more progressive research on interdisciplinary disaster science in other countries while still upholding the value of understanding humanity study, risk management study, communication study, applied-participatory study (Participatory Geographical Information Systems/pGIS), social-integrated ecology study, education for recovery study, philosophy of rural research, resilience of people understanding, and environmental ethics study.

Meanwhile, this book presents several focuses on flood resilience asset studies with maps from Focus Group Discussions or FGDs and Stakeholder Meetings or SMEs, in addition to in-depth interviews conducted with local communities (July 2016–February 2017) in Malaka-Timor, Indonesia. To simplify the understanding of flood resilience assets owned by communities affected by seasonal flooding there, I apply the concept of resilience from the perspective of the local wisdom of the community which they have believed in for generations and have implemented until now. The function of geoinformatics data in the form of mapping flood resilience assets from participatory discussions is carried out to reflect the collective quality of the community in trying to understand and guide them to determine spatial resilience assets that need to be anticipated and cared for, and significant challenges that need to be overcome.

In this book, Chap. 1 presents an overview of the sustainability, resilience, and floods sub-fields, used in critical analysis with the support of previous authors and researchers in Chap. 2. Chapters 3 and 4 present scientific methods in-depth interviews, sampling, and participatory FGDs and SMEs, regarding research tools and procedures and the reasons for conducting the research. Chapters 5–7 explain the focus of findings from in-depth interviews with 320 respondents in 16 villages, i.e., Besikama, Lasaen, Umatoos, Fafoe, Rabasahaerain, Rabasahain, Sikun, Oanmane, Motaain, Umaloor, Raimataus, Motaulun, Naas, Maktihan, Lofoun, and Rabasa as the most affected regions by the flood.

Chapters 8 and 9 emphasize the 9 results of flood resilience assets with explanations of adjustment and spatial thinking of local communities, from participatory results of FGDs and SMEs, i.e., Human Resilience Adjustment (HRA), Nature Resilience Adjustment (NRA), Physical Resilience Adjustment (PRA), Social Resilience Adjustment (SRA), Economic Resilience Adjustment (ERA), Cultural Resilience Adjustment (CRA), Regional Resilience Adjustment (RRA), Law or Legal Resilience Adjustment (LRA), and Political or Policy Resilience Adjustment (PolRA). Chapter 10 closes with conclusions and recommendations from the entire content of this book.

Münster/Bonn, Germany
October 2023

Apolonia Diana Sherly da Costa

Acknowledgements

This book originates from dissertation research work (unpublished thesis) at the Friedrich Schiller University of Jena. I did doctoral fieldwork for 7 months (July 2016–February 2017) in Malaka-Timor, Indonesia. My Ph.D. started at the end of Spring 2015 and finished at the end of Autumn 2017 and graduated on 7 February 2018 in Jena and financed by the Catholic Academic Exchange Service/ Katholischer Akademischer Ausländerdienst (KAAD) Bonn, Germany. This holistic and comprehensive book project is presented in an interdisciplinary study of social ecology and social geography, floods and resilience as well as community participation in development sustainability studies, initiated and intensified its work by myself during my progressive-collaborative of seniorship postdoctoral project as a Guest Researcher/Scientist (research professor stage) at the Institute of Landscape Ecology (ILÖK), Faculty of Geo Science University of Münster, collaborating with/ supported by Department of Southeast Asian Studies of the University of Bonn, and additional support of the Institute for Rescue Engineering and Civil Protection of the Technische Hochschule University of Applied Science Cologne, North Rhine-Westphalia, Germany, within the Postdoctoral Framework Funding of KAAD Bonn. During writing, moral support was provided by fellow professors, lecturers, and students at the Executive Graduate Program of Environment and Urban Studies, Faculty of Science and Environmental Technology, Soegijapranata Catholic University, Semarang, Central Java Province, Indonesia. I would like to express my gratitude and respect to the professor, all students, fellow of geographers and environmentalists and local stakeholders (Kak Rony, Bapa Ande, and Kak Sius), and interviewees from the disaster and environmental program "Resilience for Sustainable Development-Based Environment or RESUDENVI Community" in Malaka-Timor, Indonesia, which I founded and directed there, because without their contribution and support this book can't be finished.

In completing this book, it requires patience and thoroughness in writing, and I should be grateful to ILÖK, Geo Science Faculty, Münster and Tillmann for providing access and the opportunity to experience working on my progressive project to be carried out very impressively. The progressive project at this stage in Münster is running smoothly because of a lot of moral support from the significance of local and global achievements I achieved in my international academic career. Therefore, I have to be honest that one of the many supports present to encourage me in completing the works of my progressive project i.e., this book comes from a valuable form of appreciation both moral and achievement scientifically for me from Academic Business Current Data (ABCD) Indexing, an International Journal Monitoring Platform who evaluated and certified a Certificate as a Reviewer with a recognition and designation as Professor (see https://abcdindex.com/reviewer/certificate.php?id=prof-dr-rer-nat-apolonia-diana-sherly-da-costa-000-511-813 and https://abcdindex.com/reviewer/prof-dr-rer-nat-apolonia-diana-sherly-da-costa-000-511-813) for my dedication and contribution to be part of the Reviewer/Editorial Members Team for reviewing 15 international scientific papers in 15 international journal publishers, more than 60 international and local achievements and awards, good international publications both individual and collaborative, mainly indexed-scopus publication(s) at the postdoctoral researcher qualification (junior and senior) stage, evaluated for six consecutive years (2018–2024), after completing my Ph.D in Jena, Germany on 7th February, 2018.

One of the progressive project's outcomes i.e., this book that is within the framework of the research programme that I conceived, initiated, and led, in Münster and Bonn was ended successfully due to the tangible and intangible support of many people who might not all be acknowledged here. These many people are my colleagues, peers, co-workers, and friends during my research stays in Austria, Germany, Indonesia, and Portugal (2019–2023) who ever involved in motivating the completion of the writing of this book because of their constructive criticism which ever helped the writing of my book to be presented very well.

I am also delighted to express my deepest gratitude to the Springer staff, Ashok, Femina, and their colleagues, and especially Simon, who were always ready to provide technical and substantive support to enable this book to be completed.

Contents

Abbreviations

°C	Degree Celsius
ADPC	Asian Disaster Preparedness Centre
APFM	Associated Programme on Flood Management
AsRIA	Association for Sustainable and Responsible Investment in Asia
AWI	Australian Wool Innovation
BAKORNAS	*Badan Koordinasi Nasional*
	National Coordinating Board
BALWAS II	*Balai Wilayah Sungai Nusa Tenggara*
	The Regional River Center of *Nusa Tenggara* II
BMKG	*Badan Metereologi, Klimatologi, dan Geofisika Indonesia*
	Meteorology, Climatology, and Geophysical Body, Indonesia
BRI	*Bank Rakyat Indonesia*
	Indonesian Local Bank
CARRI	Community and Regional Resilience Initiative
CBDRM	Community-Based Disaster Risk Management
CBDRR	Community-Based Disaster Risk Reduction
CEO	Chief Executive Officer
CRA	Cultural Resilience Adjustment
CRED	Centre for Research on the Epidemiology of Disasters
CRS	Catholic Relief Service
DAS	*Daerah Aliran Sungai Watershed*
DecGG	Decentralization as Contribution to Good Governance
DPR	*Dewan Perwakilan Rakyat*
	The House of Representatives of the Republic of Indonesia
DRR	Disaster Risk Reduction
ENTP	*East Nusa Tenggara Province*
EOC	Emergency Operations Center
ERA	Economic Resilience Adjustment
EU	European Union
EWS	Early Warning System

FAO and ILO	Food and Agriculture Organization and International Labor Organization
FEMA	Federal Emergency Management Agency
FGD	Focus Group Discussion
GEMA	The Georgia Emergency Management
GFDRR	Global Facility for Disaster Risk Reduction and Recovery
GIS	Geographic Information System
GO	Government Organization
GPS	Global Positioning System
Ha	Hectare
HFA	Hyogo Framework for Action
HLMD	High Level meeting on Disaster and Development
HRA	Human Resilience Adjustment
IBM	International Business Machines
ICLEI	International Council for Local Environmental Initiatives
IDEP	Illicit Discharge Elimination Program
IDR	Indonesian Rupiah Rates
IDX	Indonesian Stock Exchange
IFAD	International Fund for Agriculture Development
IFC	International Finance Corporation
IFRCS	International Federation for Red Cross and Red Crescent
IIASA	International Institute for Applied Systems Analyses
INCASA	Indonesian National Committee for Applied Systems Analyses
ISDR	International Strategy for Disaster Reduction
km^2	kilometer (square)
LRA	Law or Legal Resilience Adjustment
m	meter
MFF	Mangroves for the Future
MLA	Meat and Livestock Australia
mm	millimeter
MOF	Ministry of Finance Singapore
MPBI	*Masyarakat Penanggulangan Bencana Indonesia* Indonesian Society for Disaster Management
MPR	*Majelis Perwakilan Rakyat* People's Consultative Assembly of Republic of Indonesia
NBFIs	Non-Bank Financial Intermediaries
NGO	Non-Government Organization
NMO	National Membership Organization
NRA	Natural Resilience Adjustment
NRM	Natural Resources Management
NTT	*Nusa Tenggara Timur* East Nusa Tenggara
OCHA	UN Office for the Coordination of Humanitarian Affairs
OJK	*Otoritas Jaksa Keuangan* Financial Services Authority

PAHO	The Pan American Health Organization
PASSA	Participatory Approach to Safe Shelter Awareness
PBBM	*Panduan Penanggulangan Bencana Berbasis Masyarakat* Community-Based Disaster Management Guideline
PDAM	*Perusahaan Daerah Air Minum* Indonesian Regional Water Utility Company
PDAST	*Panduan Daerah Aliran Sungai Terpadu* Integrated Planning of Watershed Guideline
PERSEUS	Policy-Oriented Marine Environmental Research in the Southern European Seas
pGIS	Participatory Geographic Information System
PGS	Petroleum Geo-Services
PHRA	Peoples Harm Reduction Alliances
PKBM	*Pusat Kegiatan Belajar Masyarakat* Institution Centre of Community Activity
PMPB	*Perkumpulan Masyarakat Penanganan Bencana* Community Association of Disaster Management
POSYANDU	*Pos Pelayanan Terpadu* The Centre of Integrated Service
PP	*Peraturan Pemerintah* Government Regulation
PRA	Political or Policy Resilience Adjustment
PRISE	Pathways to Resilience in Semi-Arid Economies
PSDA	*Pengelolaan Sumber Daya Air* Water resource Management
PUSKESMAS	*Pusat Kesehatan Masyarakat* The Health Centre for Community
RBI	*Rupa Bumi Indonesia* Topographic Map of Indonesia
RJPN	*Rencana Pembangunan Jangka Panjang* Long Term Development Planning
Rp	*Rupiah* (Indonesian Currency)
RRA	Religion Resilience Adjustment
RT	*Rukun Tetangga* Sub-area of Neighborhood
RW	*Rukun Warga* Sub-area of Village
SDPI	Sustainable Development Policy Institute
SISIP	Strategy Intellectual Series of Implemented Policy
SLA	Sustainable Livelihood Assessment
SLF	Sustainable Livelihood Framework
SMEs	Stakeholder Meetings
SOCAT	Social Capital Investment Tool
SRA	Social Resilience Adjustment
SWP	Six Regional of Management Unit

TANGO	The Association of Non-Governmental Organizations
TGS	*Tumpeng Gizi SeimbangTumpeng-Tumpeng-Balanced-Nutrition*
TILZ	Tearfund International Learning Zone
TTS	*Timor Tengah Selatan* South Central of Timor
TTU	*Timor Tengah Utara* North Central of Timor
UMCOR	United Methodist Committee on Relief
UMVIM	United Methodist Volunteers in Mission
UN	United Nations
UNDP	United Nations Development Programme
UNEP	United Nations Environment Programme
UNESCO	United Nations, Educational, Scientific and Cultural Organization
UNHCR	United Nations High Commissioner for Refugees
UNISDR	United Nations International Office for Disaster Risk Reduction
US $	United State of America's Currency (Dollar)
USAID	United States Agency for International Development
USGS	United States Geological Survey
VAOD	Volunteer Organizations Active in Disasters
WC	Water Closet
WMO	World Meteorological Organization
WRAP	Western Regional Air Partnership

Chapter 1
Introduction

From century to century, civilization's adaptation of resilience in the context of natural disasters, i.e., floods, changes its coping capacities/mechanism and emphasizes the fact that not everyone lives and experiences the phenomenon of flood disasters in remote villages in global south and global north countries. Such one of the southern hemisphere countries, an inland and south-border communities and a female land (a sacred-mother land or so-called *Tanah sakral mama adat* in Bahasa Indonesia or *Rai lulik loomota lalawar imbei uma ferik* in Tetum-Malaka language) i.e., Malaka-Timor Indonesia, borders the Timor-Leste region is an example that our planet is still not doing well. People who live and adapt to flood disaster situations and who have their own risk perception in Malaka-Timor are adhered to and implemented their local wisdom in defining and understanding what resilience is, what flooding is, what coping capacity is, and what sustainability is. Thus, the perceptions and perspectives of the global community, both western communities and north hemisphere countries, and eastern communities from other Asian regions and from other south hemisphere countries, will vary and differ regarding the exact contextual definition of what flooding is with the trajectory of understanding of sustainability, resilience, flooding, and coping capacity. Understanding the correct definition of the types of resilience, flood disasters and sustainability in the integrated empirical and discussion's study in this book also provides a series of complex integrative analyses. Does community resilience contextually in Timor Indonesia consist of/have a series of implementations of what type of resilience and how to overcome it? Is the empirical study of resilience, which is the ability of communities who experienced the flood hazards to adapt, the same as the contextual definition of coping capacity? To differentiate our fundamental understanding of resilience and coping capacity as well as sustainability from resilience elements, the concept of sustainability is on the path of understanding both resilience and coping capacity for adaptation efforts of the affected communities to cope with flood disasters and build flood resilience.

A. D. S. da Costa, *Building Resilience for Flood Disaster in Malaka-Timor, Indonesia*, Springer Natural Hazards, https://doi.org/10.1007/978-3-031-63272-3_1

In this part of the book, I present definitions and contextual understanding based on empirical studies with the community through in-depth interviews with 320 respondents and Focus Group Discussions and Stakeholder Meetings with 37 stakeholders from various sustainable development sectors (see Chaps. 3, 4, 8, and 9), as well as a conceptual understanding of resilience, coping capacity, flood disasters, and sustainability based on the conceptions of previous authors. Resilience is the beginning and main key to the analysis and understanding of this entire book which integrates the discussion in a holistic and complex manner into an integrated concept at the level/ space of resilience policy, i.e., sustainability (long-term sustainability of community resilience to flood disaster adaptation), where flood vulnerability is introduced according to understanding of the local adaptation of community resilience in Timor. The intersection of resilience, flood, and sustainability analysis is projected into the connectivity of the "Overview of Human Geography as the Umbrella for Building Community Resilience for Flood Disaster in the Early Twenty-First Century" discussion implied in Sect. 1.1, as an introduction for us to understand the discussion in the next subsection in this book.

1.1 Overview of Human Geography as the Umbrella for Building Community Resilience for Flood Disaster in the Early Twenty-First Century

Floods and sustainability development are two of the fifteen global challenges in the early twenty-first century, as mentioned by The Millennium Project (2009: 1); see Sect. 1.2: "Floods, Resilience, and Sustainable Development in Malaka-Timor Indonesia: Global Challenge Perspectives". These two key issues could not be handled appropriately when the flood leads to damage to human assets. Communities need to operate their resilience assets to manage flood disasters. In aligning with resilience for flood disaster, human geography provides the guiding perspective for this study. It is understood as a human subjective perspective to operate in social, economic, cultural, natural, and physical environments. Furthermore, it might particularly point at a range of human abilities (i.e., efforts and assets: actions) to run daily activities while being vulnerable and affected by harmful and/or disturbing events (flood disaster). Human abilities refer to human strengths and it implies resilience. Regarding resilience, to operate within certain types of environments, people run their daily activities to drive their assets of resilience to handle flood disaster. On this study, human geography embraces human daily activities that are vulnerable due to an unstable environment (i.e., vulnerability concept: intersection between natural, social, economic, and physical factors). Human geography functions to balance or adjust to a gap caused by a vulnerable environment. A vulnerable environment may cause a risk and thus requires resilience (i.e., resilience to cope with risk of flood). Vulnerability is defined as the quality or state of being exposed to the possibility of being attacked or harmed, either physically or emotionally, by the flood. Thus, this study discusses the means of human geography as the umbrella concept—for the key elements of this study—which usually sees predominantly spatial causalities yet specifies about the perspective and/or experience of the subjective action: building community and its resilience facing flood disaster. The idea of an action-centered geographical approach refers to Werlen (1992), (see Sect. 1.3: "Key Elements of Flood Resilience and Human Geography", point 3.1, Human Geography is presented in this Chap. 1).

As an umbrella for this study, human geography also closes a gap caused by vulnerability. Vulnerability arises from the intersection of human systems, the built environment, and the natural environment (Cutter et al. 2008: 3); see Chap. 2, Sect. 2.3. This study asserts that community resilience should be viewed from the perspective of human geography.

Resilience has a meaning, in coping with external social, political, and environmental stresses or disturbances that impact the operation of human systems; resilience implies managing to cope with undesirable events that hit human systems. Such undesirable events are termed a "disaster". *Disasters are undesirable and often sudden events causing human, material, economic and/or environmental losses, which exceed the coping capability of the affected community or society. They are caused either by natural forces or processes (known as natural disasters) or by human actions, negligence, or errors (known as anthropogenic disasters),* (Jha 2010: 1). A community copes with flood disaster with their resilience assets which enable them to minimize vulnerability. They require resilience to cope with floods. Flood is an overflow or inundation that comes from a river or a body of water and causes or threatens damage (Holmes et al. 2008: 1). In dealing with flooding, coping capacity is one of the crucial terms in resilience. *Coping capacity can be described as the manner in which people and organizations use existing resources to achieve various beneficial ends during unusual, abnormal, and adverse conditions of a disaster event or process. The strengthening of coping capacities usually builds resilience to withstand the effects of natural and other hazards* (United Nations University, UNU 2004: 4).

This book analyzes a community's sustainable livelihood development through five resources or so-called *pentagon resources* (i.e., human capital, natural capital, physical capital, social capital, and economic capital) and a community's nine resources (i.e., human capital, natural capital, physical capital, social capital, economic capital, cultural capital, religion capital, law capital or legal, and political capital or policy); see Chap. 6, "Building Community Resilience through Survey and Interviews in Malaka-Timor, Indonesia", a sustainable livelihood framework (SLF) from the Food and Agriculture Organization (FAO) & International Labor Organization (ILO) (2009: 11).

Building community resilience ultimately requires an integrated approach to building community capital that will enhance the capacity of communities for collective action in the areas of disaster risk reduction, conflict mitigation, social protection, natural resource management, and the management of public goods and services. Households may achieve some level of resilience on their own but will be limited if resilience-enhancing policies and programming are not supported through local and regional institutions and governance systems (Frankenberger et al. 2013: 8); see Sect. 1.3: "Key Elements of Flood Resilience and Human Geography", point 1.3.4. Thus, I assumed that building community resilience is a mandatory action to cope with flood disaster.

The case of building resilience has been developed from the sum of national and international humanitarian disaster relief and disaster risk management knowledge using interviews and two participatory approaches [Focus Group Discussion (FGD) and Stakeholder Meetings (SMEs)], along with Geographic Information System (GIS) in different areas to obtain information on community resilience, flooding, and sustainable development. Such information is crucial for critical assessment and local decision-making processes. Communities have the right to participate in a decision-making process as it is they who are impacted by flood disaster. Indonesia supports the idea of implementing decision-making through public participation in order to mitigate natural disaster, in this case flood disaster, as stated by the community-based disaster risk management (CBDRM), clause 26 (1: e) in Chap. 5, "*in the disaster management laws of the Republic of Indonesia, number 24 from year 2007 it is mentioned that every person has a right to participate in disaster mitigation decision making*". In this study, community engagement is seen as predominantly human actions that pay attention to space and nature. It is a so-called reciprocal method of social geography, which links people to a community as the participatory actor to managerially integrate each of their perspectives or experiences of a flood disaster (see Sect. 1.3: "Key Elements of Flood Resilience and Human Geography", point 1.3.17, details of discussions in Chap. 4 and research methodologies and the outcome of the application of FGD and SME that, I lead the discussions with the stakeholders is presented in Chap. 9).

This chapter, the objective of this book, explains the daily activities of the community in utilizing its *pentagon resources* or *nine livelihood resources* and provides information on how the risk of flooding could change human perspectives and how to improve human experience about community resilience assets *(pentagon resources* and/or *nine livelihood resources*). More details about the objectives of the research study on this book is presented in Chap. 3 "Research Methodology", Sect. 4.2.6 "Research Method of Collecting the Data", Table 4.3 "Research Method".

This book explores the question of how human geography may act actively in approaching spatial problems (flood disaster) as matters of community resilience. It highlights community resilience assets needed to achieve sustainable livelihood development, as mentioned by the FAO and ILO. Measuring community resilience adjustment to cope with flooding and focusing on the community's risk perception can be of significant interest to the local community itself. Furthermore, researchers and practitioners from local, regional, national (Indonesia), and international institutions should also consider the community's independent efforts to improve resilience actions to handle flood disaster. In addition, this research study that I studied and invented can act as a model based on the community experience in Indonesia that can be used for research in different places across the globe.

1.2 Floods, Resilience, and Sustainable Development in Malaka-Timor Indonesia: Global Challenge Perspectives

Nowadays, global challenges have become the latest issues that are changing the paradigm of society in the quest for a sustainable livelihood development solution. Global challenges are defined as any major trend, shock, or development that has the potential for serious global impacts and thus to create humanitarian needs and change the environments in which humanitarian actors will operate in coming years (Gelsdorf 2010: 4). However, facing global challenges a new idea of society has shaped focusing on resiliency systems which need to be implemented.

The Millennium Project (2009: 1) mentions the 15 key issues of global challenges in the early twenty-first century where sustainable development and climate change are the first key issues that require prior handling. An understanding of sustainable development could be assessed through building community resilience. It is defined as an action process which includes community endeavors in nurturing, improving, valuing, and maintaining their assets or capitals for long-term actions. An indication of long-term action is referred to as sustainable livelihood development. It reveals that a community is well-sustained in their assets or capitals in building their resiliency system. However, sustainable development cannot be handled appropriately as long as climate change is a cause of destroying human assets or capitals that hinder survival and planning for proper sustainable livelihood development.

Climate change increases the probability of some types of weather (Union of Concerned Scientists 2016: 1). It is influenced by heavy rain and flooding. Flooding which has occurred in the southern part of the borderland area between the 16 villages (i.e., *Rabasa, Rabasahain, Umatoos, Fafoe, Sikun, Lasaen, Besikama, Umaloor, Naas, Maktihan, Motaulun, Rabasaherain, Motaan, Loofoun, Raimataus, and Oanmane*) in the West Malaka Subdistrict of Malaka District in West Timor of East Nusa Tenggara Province, Indonesia, and East Timor (Timor Leste) is one example of a recent climate event consistent with a warming planet.

The Benanain River is known as the big river flowing through the four regions in West Timor Island, Indonesia; the four regions are South Central of Timor, North Central of Timor, Belu, Malaka, and Kovalima Region, Timor Leste. These areas maintain average temperatures, but they also have a substantial amount of rain, coming in heavy downpours and creating heavy precipitation events, like flooding.

The intensity of downpours (and therefore the risk of floods) depends on how much water can be held at any given time (The Climate Reality Project 2011: 1). Flooding in the West Malaka Subdistrict of Malaka District in West Timor Island of East Nusa Tenggara Province, Indonesia, currently holds the 51st position based on the assistantship priority in the national ranking of Indonesia (*Badan Nasional Penanggulangan Bencana*—BNPB, Indonesian National Disaster Management Body 2010: 25). River flooding has impacted the community by devastating the social-economic livelihoods of the people.

A flood is an overflow or inundation that comes from a river or body of water and causes or threatens damage (Holmes et al. 2008: 1). This particular research study focuses on the downstream area of the Benanain River (i.e., West Malaka Subdistrict of Malaka District in West Timor Island of East Nusa Tenggara Province, Indonesia). Downstream flooding occurs when large amounts of rain fall over an extended period of time over a large region. Downstream floods (regional floods) may occur with lag times, continually increasing the discharge into larger streams; such floods extend over long periods of time and affect the larger streams as well as tributary streams (Nelson 2015: 1).

The West Malaka Subdistrict of Malaka District in West Timor Island of East Nusa Tenggara Province, Indonesia, experienced the most severe flood disaster in 2011. The Malaka District is a district on Timor Island, in the East Nusa Tenggara Province, Indonesia, and has been newly expanded in 2013. It was previously one of the Subdistricts of the Belu District (1965–2014). From 1939 until 2016, the West Malaka Subdistrict has been badly affected by flood disasters due to natural factors, physical factors, and the lack of effective institutional regimes in Indonesia in handling the case of flood disaster management. The natural factors are the seasonal matter, where due to every wet season or rainy days, the Benanain River has inundated most of the settlement, garden and cultivation, and paddy fields. The inundation of the area covers ± 1000 ha, with ± 1.50 m of inundation depth and the duration is ± 1 week (Kristijatno 2008: 12). The Benanain River is meandering. The river is approximately ± 73.11 km long, and the catchment area is about ± 1594 km^2, in size with ± 0.0124 riverbed slopes. The factors of institutional intervention deal with the lack of control and corporation among the other internal actors of resilience (i.e., Indonesian government) and the external actors of resilience (i.e., public and private institutions in Indonesia). The roles of these actors are crucial in decision-making phases of Indonesian policy in terms of environmental management. The case of environmental management represents the situation of the community as the main actor with regard to the situation of the flood disaster, the situation of the ex-post period of flooding, and the situation of socio-economic characteristics nearby the Benanain River watershed.

The indigenous people living in the four big regions near the Benanain River have also experienced a lot of consequences in several capitals or assets of resilience, such as natural, physical, economic, social, cultural, and religion. In each year since 2001 until 2012, and at the present year 2016 (the dramatic year of the big flood event), a dike in Sikun Village of the West Malaka Subdistrict, Malaka District in East Nusa Tenggara Province, Indonesia, has been destroyed. The local people designed a flood simulation practice and tried to strengthen their human resilience to cover the economic and other consequences for their assets. Some portraits of the local people who are facing flood disaster are presented in Fig. 1.1.

In this book, building community resilience can be implemented through Focus Group Discussion (FGD) and Stakeholder Meetings (SMEs) as the community's fundamental method to research their own capacities of resilience to deal with the risk of a flood disaster. Community resilience encompasses their existing capitals and new capitals on their resilience concept (i.e., theory) and resilience in context (i.e.,

Fig. 1.1 Flood-prone villages in the West Malaka Subdistrict of Malaka District in the West Timor Island of East Nusa Tenggara Province, Indonesia. *Source* Disaster group community in West Malaka Subdistrict, April 2, 2017

practical). In its current form, resilience theory aims to understand three fundamental themes as suggested by Gunderson and Holling (2002). The first theme of resilience theory is to consider the characteristics of stability, resilience, and change from one state to another in systems with multiple stable states. The second theme is cross-scale interactions, and the third is adaptive change and learning using the heuristic model or metaphor of the adaptive cycle (qtd. in Allison and Hobbs 2004: 2).

The concrete model or metaphor used by the community to improve their resiliency is through FGD and SMEs. These models represent community engagement in decision-making processes and results. Alvarado et al. (2020) explain that community participation in discussion could promote self-reliance and encourage participation by local people to mitigate or prevent flood losses and damages. Therefore, the critical assessment on improving community resilience with assistantship from Indonesian institutions is on target to maintain community livelihood resources to face the risk of flood disaster in the research area. The assessment of the community's risk perception (risk factors of flooding) can be analyzed with regard to flood characteristics that consist of river buffer class distribution, periodic flooding, flood duration, flood depth, and the damage caused by the flood as perceived by the community. Communities seek their own solutions to cope with flood by accessing the five resources of livelihood (i.e., human capital, natural capital, physical capital, social capital, and economic capital (pentagon resources)) and the nine resources of livelihood (i.e., human capital, natural capital, physical capital, social capital, economic capital, cultural capital, religion capital, law capital or legal, and political capital or policy) on building community resilience. These driving livelihood resources are known as an integrated part of building community resilience that affects policy solutions and provides detailed explanations of flood resilience phases (ex-ante, during, and ex-post).

Furthermore, this book also presents the analyses of disaster risk management based on community risk perception that integrates spatial analysis. It aims at strengthening the driving livelihood resources of the community by integrating and representing their spatial knowledge in FGD and SMEs. FGD and SMEs were implemented as a proper traditional method in digging information which deals with community risk perception of floods, and which helps to find solutions of resilience. However, since the essential livelihood resources are driven by the community and other internal actors of resilience (i.e., customary leaders and religious leaders), the target of achieving a solution-based policy for a resilient community is far from being sufficient. To build sufficient community resilience for flood disaster, the implementation of the FGD and SMEs is able to find the analysis of the existing and new livelihood resources (pentagon resources and nine resources) in the community's daily life that supports them in coping with a flood. The result (discovery) of analysis of this book is presented.

1.3 Key Elements of Flood Resilience and Human Geography

The focus of this book consists of key elements which represent my conception, and which are supported by previous research. The following key elements could facilitate the reader's understanding and comprehension of this study as a whole; they are clarified as follows.

1.3.1 Human Geography

Human action is always an expression of sociocultural, subjective, and material conditions (Werlen 1992: 6–7). Werlen adds that "Only subjects can act, but there is no such thing as a purely individual action…." (ibid., 6). Sociocultural conditions are the elements that either support or hinder human actions.

He further sees that "the present task confronting an alternative human geography as precisely one which takes account of action and agency: an action-oriented social geography" (ibid.).

In this book, human geography is the umbrella term for the key elements of this study which sees predominantly spatial casualties (i.e., flood disaster) from the perspective and/or experience of the subjective action.

1.3.2 Resilience

Resilience as the buffer capacity or the ability of a system to absorb perturbations, or the magnitude of disturbance that can be absorbed before a system changes its structure by changing the variables and processes that control behavior (Holling et al. 1995 qtd. in Pecillo 2016: 2).

In this book, resilience is the ability of the community's behavior to adapt, to resist on, to cope with, and to recover from something that threatened their life and livelihood. It has been viewed as a recovery system that has been employed by the individual, the community (local people of the study area), organizations, and the local government after the flood disaster has occurred.

1.3.3 Community Resilience

Community resilience is defined as a community or region's capability to prepare for, respond to, and recover from significant disturbance-driven changes while maintaining community character, cohesion, capacity, and without permanent impairment of the community's public safety and health, economic, social, and national security functions, thus, accelerating recovery (Cutter et al. 2008: 2).

In this book, community resilience can be understood as the strength of the community to face, overcome, and minimize the risk of flood disaster.

1.3.4 Building Community Resilience

Building community resilience ultimately requires an integrated approach to building community capitals that will enhance the capacity of communities for collective action in the areas of disaster risk reduction, conflict mitigation, social protection, natural resource management, and the management of public goods and services. Households may achieve some level of resilience on their own but will be limited if resilience-enhancing policies and programming are not supported through local and regional institutions and governance systems (Frankenberger et al. 2013: 8).

In this book, building community resilience can be understood as an integrated approach that has been applied both by the community within their self-assessment and self-improvement to enhance their own assets of resilience. There are eight existing capitals on resilience (i.e., human capital, natural capital, physical capital, economic capital, social capital, cultural capital, religion capital, political capital, and law or legal capital) and the feedback of the self-assessment and self-improvement by the country's government system (i.e., Indonesia).

1.3.5 Human Capital and Improving Human Capital to Deal with Flood Disaster

Human capital consists of the skills, knowledge, ability to labor, and good health that are important to the pursuit of livelihood strategies. At the individual and household levels, educational attainment and health status of members shapes their ability to absorb the negative impacts of a shock and to successfully adapt to changing social, economic, and environmental conditions. At the community level, human capital reflects the collective level of access to skills, labor, knowledge, and physical and mental health; it is also key to innovation (The Association of Non-Governmental Organizations, TANGO 2006 qtd. in Frankenberger et al. 2014:11).

Improvement of human capital to deal with flood disaster can be assessed and applied through Focus Group Discussion (FGD) and Stakeholder Meetings (SMEs) with two kinds of assessments:

1. What are the community's human capitals that create their own identity assets (uniqueness)?
2. How could they use and improve their human capital to empower their livelihood resources (i.e., the capital on resilience)?

1.3.6 Natural Capital and Education on a Community's Natural Capital to Utilize Its Existing or Its Availability for Building Their Systems of Resiliency

Natural capital is the community's natural resources: environmental stocks from which resources useful for livelihoods are derived (e.g., land, water, forest, rangeland, fisheries, wildlife, biodiversity, and environmental services) (TANGO 2006 qtd. in Frankenberger et al. 2014: 12).

Educating a community's natural capital to utilize the existing or the available natural capital for building their systems of resiliency can be assessed and applied through FGD and SMEs where several points are going to be assessed. Points of assessment are: (1) Recognizing the natural capitals of the community (what are their natural capitals?) (2) Finding out the obstacles which the community faces (what are their obstacles or barriers in utilizing their natural capitals?) (3) Answering the questions by galvanizing the non-government institutions, related institutions, and Indonesian government to design a solution framework for building their resiliency system on natural capital (How could their resiliency system be improved by potentially utilizing their natural capitals?).

1.3.7 Physical Capital and Providing and Discussing the Information on the Use of Physical Capital

Physical capital includes basic infrastructure (e.g., transportation, shelter, energy, communications, and water systems, health facilities, and markets), production equipment, and other material means that enable people to maintain safety and enhance their relative level of well-being (Mayunga 2007; Gill and Ritchie 2011 qtd. in Frankenberger et al. 2014: 13).

This book provides and discusses the information on the use of physical capital which can be assessed and applied through FGD and SMEs with two kinds of assessments: (1) Providing information for the community about maintenance and control of physical capital in terms of community infrastructure which was provided by the non-governmental institutions, community members, and other institutions; (2) Supporting their social-economic activities such as drainage and irrigation.

1.3.8 Economic Capital: Empowering the Community's Economic Capital

Economic capital denotes financial resources that people use to achieve their livelihoods. It includes savings, income, investments, and credit. The contribution of economic capital to building community resilience is straightforward in the sense that it increases the ability and capacity of individuals, groups, and communities to absorb disaster impacts and speed up the recovery process (Mayunga 2007: 7).

Empowering community economic capital can be assessed and applied through several key strategies which are assigned to community's economic assets.

1.3.9 Social Capital: Ensuring the Community's Social Capital

The concept of social capital as features of social organization such as networks, norms, and social trust that facilitate coordination and cooperation for mutual benefit (Putnam 1995 qtd. in Mayunga 2007: 7).

The ensuring of the community's social capital can be assessed and applied through FGD and SMEs on four kinds of assessments of social capital:

1. Knowing the ties within community

2. Understanding the community's social capital
3. Ensuring a community's bonding social capital
4. Understanding a community's bridging capital and
5. Linking a community's social capital to strengthen the community's social capital.

1.3.10 The Revitalization of the Community's Cultural Capital and Religious Capital

a. Cultural capital is defined as "forms of knowledge", both tangible and intangible, that have value in a given society in relation to status and power with a more practice-oriented definition of cultural capital as the rules for engaging other types of capital (human, economic, social). (Choby 2010: 1a). Cultural capital defines how people (humans) engage each other (social) and their resources (economic). Whether the culture of an organization is good or bad—cultural capital is created when values, traditions, beliefs, and language become the currency to leverage other types of capital (Choby 2010: 1b). Community or cultural resilience is the capacity of a distinct community or cultural system to absorb disturbance and reorganize while undergoing change so as to retain key elements of structure and identity that preserve its distinctness (Healey 2006: 12). Werlen (2015: 19) adds the idea of culture in the context of sustainability in the sense that culture is regarded as the central pillar of sustainable development. He explains that culture can be considered the elephant in the room for transdisciplinary research, which addresses complex societal problems and contributes to the development of more sustainable human resources.
b. In this study, religious capital is defined in two ways; firstly, it is defined as a spiritual entity and/or a faith of certain religions, which is generally also known community's spiritual forms, its trust, and its self-defense based on the believe in the creator of the universe. In this study, the spiritual entity and/or a catholic faith provides and creates a set of faith-resources (intangible and tangible things). It is horizontally formed among people as it could serve and/ or solve their own cases. Secondly, religious capital is defined as a spiritual and or a faith medium of several certain religions that are formed vertically linking the people with the creator of the universe (i.e., priests, clergy, leaders, or servants of several certain religions). It is particularly adhering to the servants of the religious institutions. In this study, the task of the servants is to intangibly and tangibly serve and/or solve the surrounding humanistic cases of people in the universe.
c. Revitalization and enlightenment of the community's cultural capital and the community's religious capital can be assessed and applied through several key strategies based on the community's cultural and religious perspectives and/ or experience of the community's actions toward their system of resiliency. It integrates resilience ideas into contextual solutions to deal with any challenges (i.e., flood disaster).

These two kinds of capital structure the analysis to form an idea of orientation or path-determination of human or community values. They are revitalized within the self-assessment toward space and nature by devolving their own illegal and legal decisions or constituents that they have made according to the customary, pastoral, or religious rules. The purpose is to resolve their essential problems.

1.3.11 The Importance of Political Capital for Building Community Resilience in the Case of Disaster Assistantship for Each Household

Political capital is inferred to be an entity which political actors possess, build up and spend. Political activity occurs among three societal actors: individuals, groups, and institutions (Casey 2003: 8).

Informing and promoting the importance of political capital for building community resilience in the form of disaster assistantship for each household in the study area can be assessed and applied by discussing the political activity that occurs among three societal actors: individuals, groups, and institutions. The purpose of discussion is to achieve the community's self-assessment and self-decision making of their own capability, as of households and individuals, in a disaster-community. Furthermore, communities can obtain their human rights, security rights, and judicial rights to aid, assist, support, and improve their eight capitals of the resiliency system after flood disaster occurrences.

1.3.12 Law or Legal Capital: Assess the Legal or Law Capital by Involving the Community as a Decision-Maker

Law or legal capital is seen and constituted as the product of regulations that is necessary to be implemented by the Indonesian government through political action or policy implementation about the case of the natural disaster (flood disaster) in the West Malaka Region of Eastern Indonesia. Additionally, international intervention is also engaged in tackling the case of humanitarian disaster relief in the world including Indonesia. The United Nations has passed their efforts through reaffirming its resolution 46/182 of 19 December 1991, which contains the guiding principles for the strengthening of the coordination of emergency humanitarian assistance of the United Nations system, as well as all its resolutions on international cooperation on humanitarian assistance in the field of natural disasters, from relief to development, and recalling the resolutions of the humanitarian segments of the substantive sessions of the Economic and Social Council (The United Nations International Strategy For Disaster Reduction, UN-ISDR 2007: 1a).

The United Nations has also tried to forefront their efforts to be inclusive across its policy frameworks where through Reaffirming further the Hyogo Declaration, the Hyogo Framework for Action 2005–2015: Building the Resilience of Nations and Communities to Disasters and the common statement of the special session on the Indian Ocean disaster: risk reduction for a safer future, as adopted by the World Conference on Disaster Reduction, held in Kobe, Hyogo, Japan, from January 18 to January 22, 2005, and bearing in mind that the implementation of the Hyogo Framework for Action is coming to an end in 2015 (UN-ISDR 2007: 1b).The law or legal capital is defined as the strong component systems of the government's regulations and implementations, roles and rules which are created and produced, monitored and controlled, executed and punished by the legislative, executive, judicial or judicative councils, and other related institutions which also implement trial politics and certain policy frameworks that are related to the humanitarian disaster relief case. To create law capital and put it into this analysis is to provide an accessibility between community participation through their self-assessment and self-evaluation in their decision making in the government legal system in tackling three cases.

1. The case of natural and physical factors of controlling, monitoring, and protecting the Benanain River Watershed and Dike System Management within the discussion on trial politics, its roles, and rules.
2. The case of human actions (i.e., human activities/intervention or man-made) that have less governmental responsibility to protect and to give penalties and sanctions in relation to the existing customary legal system and in conjunction to requesting a new legal system of land use and human actions protecting the watershed resources management in the cross-border area between the West Malaka, East Nusa Tenggara Province Indonesia and the Timor Leste Republic (since the Benanain River Watershed divides these two countries).
3. Finally, there is the case of financial or economic assistantship in terms of micro-scale insurance for several damages due to flood disaster, which is supposed to be given by the government to the affected community, not only in the phase of ex-post of aftermath flood disaster but also in the phase of ex-ante or before a flooding disaster has occurred and during flooding disaster occurrences. For its implication, a discussion and assessment on FGD and SMEs have been conducted in the research location.

1.3.13 The Community's Sustainable Livelihood Resources

Livelihoods consist of the capabilities, the assets—both material and social resources—and the activities required for a means of living. Livelihoods are sustainable when they can: cope with and recover from stresses and shocks; maintain or enhance capabilities and assets (current standard of living) (Christensen and Pozarni 2008: 5).

In this book, community sustainable livelihood resources refer to community capitals or assets. Resilience refers to a livelihood resources system which is a model of each capital or asset held by the community in addressing disturbances and adverse events and/or catastrophic events (in this case study a flood disaster). For understanding a hierarchical definition of sustainable livelihood, I put this into a process of analysis or assessment. The sustainable livelihood assessment is the major framework of resilience. Analysis starts from ex-ante or before flood disaster, continues during the flood disaster, and includes ex-post or aftermath flood disaster occurrences. In this study, the community sustainable livelihood resources can be understood as the capital or asset that is held by the community to eagerly improve and maintain both their existing capitals or assets and new capitals or assets on their resiliency systems to achieve effective, efficient, and sufficient sustainable livelihood resources in the long run.

1.3.14 Disaster

Disasters are undesirable and often sudden events causing human, material, economic and/or environmental losses, which exceed the coping capability of the affected community or society. They are caused either by natural forces/processes (known as natural disasters) or by human actions, negligence, or errors (known as anthropogenic disasters), (Jha 2010: 1).

In this book, disasters can be defined as unexpected phenomena which is not only caused by natural disaster but also by human intervention. It has led to the weaknesses of a community's socio-economic assets and other assets. Therefore, it is important to improve and maintain the community's knowledge and skills (capability to cope with the flood disaster).

1.3.15 Flood

A flood is defined as an overflow or inundation of water from a river or other bodies of water which causes the loss of life, threatens property, and environmental damage (Ministry of Natural Resources 2007: 2).

Flood as an overflow or inundation that comes from a river or a body of water and causes or threatens damage (Holmes et al. 2008: 1).

In this book, flood can be defined as an overflow or inundation of water from the Benanain River which causes loss in terms of several components such as loss of property (the content of the house), loss of the house altogether, loss of crops, loss of livestock, nurseries, cultural heritage sites, fisheries, and accessibility disruption (damaged infrastructure, i.e., road, public infrastructure, and broken dike).

1.3.16 Coping Capacity

Coping capacity can be described as the manner in which people and organizations use existing resources to achieve various beneficial ends during unusual, abnormal, and adverse conditions of a disaster event or process. The strengthening of coping capacities usually builds resilience to withstand the effects of natural and other hazards (United Nations University, UNU 2004: 4).

In this book, coping capacity in the sense of coping with floods can be defined as a capacity of both communities who experienced flood events and/or organizations who are capable of stimulating risk reduction of flood events. They are expected to cope with, adapt, and adjust to the flood phenomena in facing, enduring, and/or reducing the risk of a flood disaster; therefore, they strive to handle the mitigation and its management for better conditions in the future.

1.3.17 The Community as the Main Actor of Participatory Actions

As written down in clause 26 (1: e) in Chap. 5, in the disaster management laws of the Republic of Indonesia, number 24 from the year 2007 mentions that every person has a right to participate in disaster mitigation decision making (Community-Based Disaster Risk Management, CBDRM 2005: 39).

In this book, the community is seen as the key actor to inform, discuss, and decide about preventive and mitigation actions (regarding their resilience for the flood disaster) in the form of FGD and SMEs.

The detailed definition and explanation of FGD and SMEs are in Fig. 4.2 in Chap. 4 and the outcome of community risk perception on nine resilience adjustment maps for flood disaster through comprehensive FGD and SMEs research in Chap. 9.

In this book, community engagement is seen as predominantly human actions that pay attention to space and nature. It is so-called as the reciprocal method of social geography. The reciprocal method of social geography is linked to the people and to the community's involvement, as the participatory actor which is supposed to managerially resolve and devolve each of the perspectives and experiences related to a flood disaster.

1.4 An Outline of the Content of the Book

The outline of this content of the book provides a framework for each chapter (this chapter to Chap. 10). It aims at concisely conveying the steps of critical and theoretical analysis which has been obtained from the author's fieldwork in Malaka-Timor, Indonesia.

This chapter consists of a general overview of research that involves background research, statements on the problem in research, originality of the research, objective of the research question, the key elements of the book.

Chapter 2 consists of a theoretical analysis of human geography as a commencement of theoretical foundation for community engagement as the participatory actor, an understanding of definitions and implicative assessments of building community resilience for the risk of flooding disaster in the study area, a definition of community resilience in the theoretical framework of the human-social geography context.

Chapter 3 presents the research site.

Chapter 4 presents the research methodology.

Chapter 5 presents and discusses the results of the dissertation thesis (i.e., social economic data based on the interviews with the 320 respondents in 16 villages of the West Malaka Subdistrict, Malaka District, East Nusa Tenggara Province, West-Timor Island, Indonesia, namely Lasaen, Umatoos, Fafoe, Sikun, Besikama, Raimataus, Oanmane, Lofoun, Naas, Rabasa, Rabasahain, Rabsahaerain, Motaain, Maktihan, Umaloor, and Motaulun).

Chapter 6 answers the first objective of the research question that is to analyze building community resilience for the flood disaster and the question of what the existing and/or new capitals in the community in the Malaka-Timor, Indonesia, are.

Chapter 7 answers the second research objective that is to analyze the risk of flooding disaster impacting community livelihood (i.e., capitals or assets, known as nine resources); the chapter also answers the second research question of this study: How does the community use their capitals or assets on resilience to deal with flood risk. The results and discussion are presented in three phases of analysis.

1. The first phase of analysis is the introduction of risk factors of flood disaster impacting the building system of community resilience.
2. The second phase of analysis is the human risk perception for the flood characteristics through interview which includes periodic flooding, duration of flood, flood depth, and flood distribution.
3. The third phase of analysis includes community assets on resilience for the flood risk which contains the risk profile and boundaries, distribution of human assets, natural assets, physical assets, social assets, economic assets, cultural assets, religious assets, law or legal assets, religious assets, and political or policy assets on resilience that are impacted by flood risk in the 16 villages of the study area in the West Malaka Subdistrict of Malaka District in East Nusa Tenggara Province, Indonesia.

Chapter 8 discusses the three components of building community resilience, risks factors of flooding and sustainable livelihood development based on the community-based approach in Malaka-Timor, Indonesia. These three components represent the quality of the community-based approach in implementing their assets or capitals on resilience as a means of resources for the power to deal with flooding as to achieve long-term actions at the high level. It can be assessed structurally according to:

1. What kind of assets or capitals do they have and thus far implemented to deal with flood disaster?

2. What kind of actions do they have and how far are they implemented to deal with flood disaster? (Are they short-term actions on a so-called basic level? Are they moderate- or medium-term actions or so-called moderate level? or are they long term actions or a so-called high level?).

The indication of short level, nearly moderate, moderate level, and high level stands for a community-based approach to creating a resiliency system to achieve a sustainable livelihood development. Overall, a sustainable livelihood development is indicated by the high level of what and how a community could potentially not only smartly implement and build their resilience but also maintain their resilience for flood disaster in the West Malaka Subdistrict, Malaka District, East Nusa Tenggara Province, Indonesia.

Chapter 9 answers and discusses the third research question of the second objective of this research:

1. How to measure community resilience for the flood disaster and how to build a resilient community through measuring the sustainable livelihood development of community actions using Focus Group Discussion (FGD) and Stakeholder Meetings (SMEs) as the final or top solution for a community-based-approach in the West Malaka Subdistrict of Malaka District in East Nusa Tenggara Province, Indonesia.
2. Answering the research question and its discussion consists of three phases of discussion, namely assessment phases, community assets for sustainable development and measuring phases, measuring community resilience adjustment on sustainable actions (nine livelihood resources scoring results-community-based-approach).

Chapter 10 presents the Conclusion and Recommendation of the book. The conclusion consists of two points based on the results of the study:

1. Presenting nine assets or so-called the nine livelihood resources on building community resilience (i.e., human capital, natural capital, physical capital, social capital, economic capital, cultural capital, religion capital, law capital or legal, and political capital or policy).
2. The community risk perception about the risk factors of flooding; flood characteristics which consist of river-buffer class distribution, flood frequency, flood duration, flood depth, and damage of flooding are perceived by the community to access the nine assets or so-called the nine livelihood resources on building community resilience.

Additionally, the recommendation describes the issues of building community resilience and flood risk perception which have been described in the chapters before. The issues deal with specific solutions and measurements which are possible in future research.

1.5 Summary

This chapter consists of two parts. *Firstly*, it discusses "*An Overview of Human Geography as the Umbrella for Building Community Resilience for Flood Disaster in the early Twenty-First Century*", as presented in Sect. 1.1. This part also discusses the sequences of the study, which unfolds the understanding of critical reasons for choosing the research topic. The introduction to the study is provided in Sect. 1.2. It establishes the scope of the study from the global perspective [see global challenge and key issues in the twenty-first century, Gelsdorf (2010: 4) and The Millennium Project (2009: 1)]. The idea of the global perspective is then conveyed to the local perspective on the empirical example. This part also explains the historical point of view that is assessed in this research project of this book to explain the Malaka community's own adaptation and awareness.

Secondly, the chapter presents Key Elements of Flood Resilience and Human Geography of the book (see Sect. 1.3). This chapter inform the analysis on how we understand the key concepts of the book as a whole. In addition, this part establishes "The Outline of the Book" in Sect. 1.4, with the content framework and the overall concept of the Book from this chapter to Chap. 10.

References

Allison, E., H. and Hobbs, J., R. *Resilience, Adaptive Capacity, and the "Lock-in Trap" of the Western Australian Agricultural Region*. Murdoch University. Ecology and Society: ES Vol. 9, No 1. Australia (2004). Retrieved from https://www.ecologyandsociety.org/vol9/iss1/art3/

Alvarado MMC, Rendon ZA, Peres VDA (2020) Integrating public participation in knowledge generation processes: Evidence from citizen science initiatives in Mexico. Environ Sci Pol J 114:230–241. Elsevier. Retrieved from https://www.sciencedirect.com/science/article/abs/pii/S1462901120300575

Badan Nasional Penanggulangan Bencana-BNPB, Indonesian National Disaster Management Body, *The Affected Area of Map in Belu Province, East Nusa Tenggara*. Geo-spatial of The National Body of Disaster Management. Jakarta, Indonesia (2010)

Casey B (2003) Defining political capital: a reconsideration of Bourdieu's inter convertibility theory. University of Missouri-St. Louis, Political International Journal

Choby P (2010) The social capital: what is cultural capital, Cobalt Group (2010) Retrieved from https://pattichoby.wordpress.com/2010/06/05/what-is-cultural-capital/

Christensen I, Pozarni P (2008) Socio-economic and livelihood analysis in investment planning. Food and Agriculture Organization of The United Nations's Policy Learning Programme (2008). Retrieved from http://www.fao.org/docs/up/easypol/581/3-7-social%20analysis%20session_167en.pdf

Cutter S, Barnes L, Berry M, Burton C, Evans E, Tate E, Webb J (2008) Communityand regional resilience: perspective from hazards, disasters, and emergency management. Hazard and Vulnerability Research Institute, Department of Geography University of South Carolina, Columbia, South Carolina

Community-Based Disaster Risk Management, CBDRM, *Masyarakat Penanggulangan Bencana.*, Community Based Disaster Risk Management's Article. Kupang, East Nusa Tenggara Province, Indonesia (2005)

Food and Agriculture Organization and International Labor Organization, FAO and ILO, The Livelihood Assessment Tool-kit. Analyzing and Responding to the Impact of Disasters on the Livelihood of People. Policy and Support Branch Communication Division. Geneva and Rome (2009)

Frankenberger, T., Mueller M., Spangler T., and Alexander S. *Community Resilience: Conceptual Framework and Measurement Feed the Future Learning Agenda.* Rockville, MD: Westat (2013)

Frankenberger T, Mueller M, Spangler T, Alexander S (2014) Community resilience: conceptual framework and measurement feed the future learning agenda. Westat, Rockville

Gelsdorf K (2010) Global challenges and their impact on international humanitarian action. Policy Development and Studies Branch, PDSB. The United Nations Office for the Coordination of Humanitarian Affairs, OCHA. OCHA's Occasional Policy Briefing Series

Gunderson, L.H., & Holling, C.S. (Eds.) *Panarchy: Understanding Transformations in Human and Natural Systems.* Washington, DC: Island Press (2002)

Healey S (2006) Cultural Resilience, Identity, and Restructuring of Political Power in Bo- livia. University of Guelph, Canada

Holling D, Weber M, Wentworth M, Sullivan J, Millemann B, Griffenberg A (1995) State of the coasts: a state-by-state analysis of the vital link between healthy coasts and a healthy economy. Coast Alliance, Washington, D.C., USA

Holmes AE, Geddes JR, Colom F, Goodwin GM (2008) Mental imagery as an emotional amplifier: application to bipolar disorder. Behav Res Therapy 46(12):1251–1258, ISSN 0005-7967. Retrieved from https://doi.org/10.1016/j.brat.2008.09.005 and https://www.sciencedirect.com/science/article/pii/S0005796708001915

Jha MK (2010) Natural and anthropogenic disasters: an overview of vulnerability, preparedness and mitigation. SpringerLink, The Netherlands. ISBN: 978-90-481-2498-5

Kristijatno, *Analisis Pengendalian dan Perbaikan Sungai Kritis Haekto (Benenain) di Pulau Timor.* Balai sungai, Puslitbang Sumber Daya Air, Balitbang Pekerjaan Umum- PU. Kupang, East Nusa Tenggara, Indonesia (2008)

Mayunga, Joseph, S. *Understanding and Applying the Concept of Community Disaster Resilience: A capital-based approach. Department of Landscape Architecture and Urban Planning.* Hazard Reduction and Recovery Center, Texas, The United States of America (2007)

Ministry of Natural Resources (2007) Emergency management. Ministry of Natural Resources's Publication, Ontario-Canada

Nelson, Stephen, River Flooding. Department of Earth and Environmental Science, Natural Disaster Issues EENS 3050–6050. Tulane University (2015)

Pecillo, M. *The concept of resilience in OSH management: a review of approaches. Central Institute for Labour Protection*-National Research Institute (CIOP-PIB). Taylor & Francis's Publishing (2016). Retrieved from https://www.ncbi.nlm.nih.gov/pmc/articles/PMC4867880/

Putnam RD (1995) Bowling alone: America's declining social capital. J Democr 6(1):65–78

The association of non-governmental organizations, TANGO (2003) Socio-political analysis: critical information needs for addressing underlying causes of poverty and marginalization. Prepared for CARE. Draft

The Climate Reality Project, Why does climate change leads to more floods and droughts? The Climate Reality Project Group. Washington DC (2011). Retrieved from https://www.climaterealityproject.org/blog/why-does-climate-change-lead-more-floods-and-droughts

The Millennium Project, Fifteen key issues of global challenges in the early 21st Century. Global Future Studies and Research. Washington DC (2009). Retrieved from http://107.22.164.43/millennium/challeng.html

Union of Concerned Scientists (2016) Heavy Warming and Global Warming: Is There a Connection? Global Warming's Issues. Cambridge. Retrieved from http://www.ucsusa.org/

The United Nations International Strategy for Disaster Reduction, UN-ISDR, *Hyogo Framework for Action (HFA), 2005–2015: Building the Resilience of Nations and Communities to Disasters.* The HFA's Research Projects, in the Andean region, Central America, the Caribbean, Asia, Pacific, Africa and Europe (2007)

United Nations University (2004) UNU, Comparative glossary for core terms of disaster reduction. Institute for Environment and Human Security. Bonn, Germany

Werlen B (1992) Society, Action and Space: An Alternative Human Geography. University of Zurich, Switzerland, Routledge International Publishing, Switzerland, Department of Geography

Werlen B, Sustainability G (2015) Cultural Perspectives and Challenges for Transdisciplinary Integrated Research. Friedrich Schiller University of Jena, Germany, Springer International Publishing, Switzerland, Department of Geography

Chapter 2
Theoretical Approach in Related Reviews of Building Community Resilience in General and Specific Frameworks of the Book

Building community resilience involves a set of components in a conceptual framework that is a substantial basis for measurement regarding flood disaster management. Several previous authors and researchers have introduced and described the key indicators of community resilience through the establishment of community assets and capacities, using a mixed-method approach with quantitative and qualitative measures. Aldrich (2012) qtd. in Frankenberger et al. (2013) section iv provides an idea as to how community resilience can be set within a framework which integrates a livelihood approach and a disaster risk reduction (DRR) approach, as well as considering contributions from recent literature on the role of social capital and collective action. This book defines resilience as a set of abilities or coping systems that are acquired by people to aid their survival while undertaking activities using their coping system for handling natural disasters (i.e., floods).

This book also reflects the concept that resilience is a process rather than an outcome. This concept relates to local people using their traditional approach in responding to flood disaster and takes the view that resilience is invaluable practice as part of a process because it is possible to absorb new strengths of resilience every time a flood disaster threatens people and their coping system (i.e., assets of resilience).

Although some previous researchers have provided resilience-related concepts within the frame of reference of natural disaster management, in the assessment of this study a new harmonious approach is required. This approach includes the context of environmental characteristics which previous researchers have not yet studied; a way of integrating both conceptual and contextual studies was not available.

This book assesses where building resilience can be measured by conceptual and contextual frameworks, where there is a new discovery or development in the study of social-human geography (i.e., building community resilience in the West Malaka Subdistrict, in the West Timor Island, Indonesia). In the context of social-human geography, a traditional approach was described that was created, shaped, and run by the people based on their perceptions and observations from flood disaster (see

A. D. S. da Costa, *Building Resilience for Flood Disaster in Malaka-Timor, Indonesia*, Springer Natural Hazards, https://doi.org/10.1007/978-3-031-63272-3_2

Chap. 1, Sect. 1.3 "Key Elements of Flood Resilience and Human Geography", point one), The idea of an action-centered geographical approach refers to Werlen (1992).

In determining the framework of conceptual ideas of resilience for the assessment process of flood disaster management, the ideas generated can be assessed in the environmental context. The assessment is not limited only to the fundamental concepts of community resilience, but also refers to the formed idea of socioecological perspectives in human-social geography. The idea of socioecological perspectives as suggested by Bene et al. (2012: 27) assesses capacities that socio-ecological systems must acquire and maintain to ensure resilience. The assessment also adopts an explicit view of resilience as a process rather than a static state, with its determinants consistently changing within evolving social, economic, and environmental contexts (Frankenberger and Nelson 2013 qtd in. Frankenberger et al. 2013: 7).

Discussion of theories, concepts, processes, understanding, and assets in building community resilience is provided in this chapter.

2.1 Theoretical Approach to Community Resilience in the Context of Human-Social Geography

The conceptual framework for community resilience is a basis for measurement. To gather information on the key indicators related to the community assets, social dimensions, and capacities for collective action described in the framework, a mixed-method approach is needed that combines quantitative and qualitative measures. Given the mix of tangible and intangible assets involved and the dynamic nature of each component, measurement should entail a combination of traditional outcome measures with process measures, as well as others that capture capacity (Aldrich 2012 qtd. in Frankenberger et al. 2013, section iv). The framework integrates a livelihood approach, a disaster risk reduction (DRR) approach, and contributions from recent literature on the role of social capital and collective action. The framework also incorporates the work of Bene et al. (2012: 27) by considering key capacities that socio-ecological systems must acquire and maintain to ensure resilience. It adopts an explicit view of resilience as *a process* rather than a static state, with its determinants consistently changing within evolving social, economic, and environmental contexts (Frankenberger and Nelson 2013 qtd in. Frankenberger et al. 2013: 7). It incorporates three types of capacities. Firstly, absorptive capacity which is the ability to minimize exposure to shocks and stresses through preventative measures and appropriate coping strategies to avoid permanent, negative impacts. Secondly, adaptive capacity for making proactive and informed choices about alternative livelihood strategies based on an understanding of changing conditions. Finally, transformative capacity which are the governance mechanisms, policies/regulations, infrastructure, community networks, and formal and informal social protection mechanisms that constitute the enabling environment for systemic change.

2.2 Understanding the Global Challenge and Coping with Flood

An understanding of the global challenge can be linked into sustainable livelihood development. In tackling the global challenges, some geographers and other scientists have been trying to solve the type of the global challenges such as climate change and natural disaster by enhancing the local people's resilience strategies. (See also Gelsdorf's definition of the Global Challenge in Chap. 1 "Background of the Study"). Another perspective on global public goods, where climate change is introduced and prioritized by the International Task on Global Public Goods as one of the global events and it requires management priorities. Its long-term consequences will likely be severe (International Task on Global Public Goods section xiv). Even though the International Task on Global Public Goods pressed the issue of climate change in the context of global public goods, it is mentioned that adaptation in one of crucial issues in developing countries. Adaptation—developing countries will need assistance in reducing their vulnerability and enhancing their preparedness to respond to climate change (International Task on Global Public Goods section xv: a). Adaptation should take its first implementation in developing countries than in developed countries, reasoned by policy for adaptation to response to climate change and natural disaster are still lacking and ineffective. The research project of this book is conducted in Indonesia which is one of the developing countries, facing potential risks of climate change and natural disaster (i.e., flood disaster). (See for details in Sect. 2.6.1 prior to its discussion and analysis on its in Sect. 2.6.2).

In regard to people's adaptation to deal with the potential risks of a flood disaster, sustainable development is a path in enhancing preparedness. Potential strategies include increasing and sustaining financing to strengthen adaptation capacities; supporting the Consultative Group on International Agricultural Research in its work on tropical agricultural R&D programs; and ultimately delivering new seeds and breeds tolerant to climate variations (International Task on Global Public Goods 2006, section xv: b).

To solve the global challenges, The Global Challenge's Foundation has been tried to forefront their efforts "*to accelerate effective-counter actions against global events with the potential for large-scale unwanted effected by deepening both decision makers' and the public's insights into the risks, and also to inspire both debate and judged decision...*" (Pamlin and Amstrong 2015: 27). The local people's resilience in coping with natural disasters (i.e., floods) in the research location is an example of a resilient community, evidenced by their active role as a decision-maker in handling the risks of a flood disaster. In handling a flood disaster, the local people tried to optimize their resilience' assets as to achieve sustainable development.

An understanding of sustainable development could be assessed through building community resilience. It is defined as an action process in which community endeavor in nurturing, improving, valuing, and maintaining their assets or capitals for long-term actions. An indication of long-term action is referred to sustainable livelihood development. It reveals a well-sustained of community' assets or capitals in building their resiliency system. However, sustainable development cannot be handled appropriately if climate change is a cause to destroy human assets or capitals to survive and to plan a proper sustainable livelihood development. Climate change may cause floods. (See Union of Concerned Scientist and The Climate Reality Project idea about the relationship between climate change and floods in Chap. 1).

Flood is defined as an overflow or inundation of water from a river or other body of water which causes or threatens loss of life and property and environmental damage (Ministry of Natural Resources 2007: 2). (See this definition in Sect. 1.3 , "Key Elements of Flood Resilience and Human Geography" Chap. 1, Point 1.3.15). Floods have the greatest damage potential of all natural disasters worldwide and affect the greatest number of people. Flood is the single most destructive type of natural disaster that strikes humans and their livelihoods around the world.

On a global basis, there is evidence that the number of people affected by flood-induced economic damage is on the rise at an alarming rate. Society must move from the current paradigm of post-disaster response (The United Nations, UN 2009, section iv). According to Associated Programme on Flood Management, APFM (2008: 5), a flood is the combination of meteorological and hydrological conditioning of an area. Some floods are also caused by human-induced factors. Therefore, a flood disaster has to be seen as a combination resulting from natural and man-made factors. The kinds of flooding that APFM has classified floods into local floods, river floods, flash floods, and coastal floods. Local flooding is caused by bad drainage as a result of urban development. This flood usually occurs during the rainy season. The second type of flood, river flooding occurs when the river run-off volume exceeds local flow capacities. This flood is triggered by heavy rainfall or snow melting from upstream or tidal influence. The third flood type, the flash flood, is common in mountain and desert areas. This flood is the result of rapid accumulation and release of run-off water. Coastal floods are caused by high tides and storms. This flood usually occurs in a city that is in an estuary. In this study, the type of flood disaster is a riverine flood. Abnormally severe flooding is likely to damage infrastructure and productive capacity, as well as directly reduce output, particularly by destroying standing crops; and disrupting economic and social activity (Benson and Clay 2003: 18).

In this context of this book, I researched and discussed flooding which is occurred in the Southern part of the borderland between the sixteen villages (i.e., *Rabasa, Rabasahain, Umatoos, Fafoe, Sikun, Lasaen, Besikama, Umaloor, Naas, Maktihan, Motaulun, Rabasaherain, Motaan, Loofoun, Raimataus, and Oanmane*) in the West Malaka Subdistrict of Malaka District in West Timor of East Nusa Tenggara Province, Indonesia, and East Timor (Timor Leste) is therefore one of examples for a recent consistent with a warming planet.

Benanain River is known as the big river flowing through the four regions in West Timor Island, Indonesia; the four regions are South Central of Timor, North Central of Timor, Belu, Malaka, and Kovalima Region, Timor Leste. These areas maintain average temperatures but with more rain, coming in heavy downpours and creating heavy precipitation events, like flooding.

Furthermore, coping with floods can be effective through the involvement of local people that leads to mitigation strategies. In developing the mitigation goals as part of the city's flood mitigation plan, officials ensured efforts were locally based by involving civic leaders, business owners, and residents in the planning. The community used a multi-objective approach to understand watershed problems, consider all flooding solutions, identify community concerns, obtain expert advice, and build partnership. The mitigation strategy is part of a resiliency system where the local people can cope with flood disasters. The coping capacity is a sub-system in resilience which has several strong resources to minimize the level of sovereign disaster as to increase the strength of the continuity process of resilience. Coping capacity is a combination of all strength and resources available within a community or organization that can reduce the level of risk, or the effect of a disaster. It may include physical, institutional, social, or economic means as well as skilled personal or collective attributes such as leadership and management.

However, according to Rossi et al. (1992: 4) coping with flood is defined as all those measures, with necessary policies and strategies of implementation, which a society may apply to alleviate the consequences of flood events. This may also include—doing nothing except learning and adjusting to flood phenomena.

2.3 Vulnerability and Resilience-Exposure and Human-Social Geography: A Perspective on the Flooding Disaster

Vulnerability arises from the intersection of human systems, the built environment, and the natural environment (Cutter et al. 2008: 3). Furthermore, the most obvious factor contributing to community vulnerability is location or proximity to hazard-prone areas such as coasts, floodplains, seismic zones, potential contamination sites, and so forth. The physical exposure, described by the characteristics of the initiating event (magnitude, duration, frequency, impact, rapidity of onset), defines the physical vulnerability of places.

A concrete example is in the prone area of flooding, the West Malaka Subdistrict where communities are living in the nearby estuary in the downstream of Benanain River watershed; this community is naturally more vulnerable to flooding and has less exposure of resiliency than other regions in the upstream area. In order to know the characteristics of the prone area, a community is expected to draw their knowledge of each point of sites which are of risk for flood disaster.

The use of participatory Geographic Information Systems (pGIS) was applied by the local community in the prone area of the West Malaka Subdistrict from August to September 2012, the researcher's Master Fieldwork. The local community has gained some knowledge and understanding of the risk of natural environment, i.e., flood disaster such as delineation practical through pGIS of the border affected of each area to the most safety area and presented description and explanation of their delineation result.

Through this, the community has a quality system or capacity to recognize the most extreme situation in their area. It can be said that the community has a good capacity in vulnerability context in West Malaka Subdistrict. As supposed by Cutter et al. (2008: 2a) defined that *vulnerability* is the pre-event, inherent characteristics or qualities of systems that create the potential for harm or differential ability to recover following an event. The ability to expose the prevention action was applied by the community in the affected area of the West Malaka Subdistrict.

Furthermore, it has been stated that vulnerability is a function of the exposure (who or what is at risk) and the sensitivity of the system (the degree to which people and places can be harmed) (Cutter et al. 2008: 2b). However, the community could not handle all sensitivity of systems and the vulnerability for flood disaster, as they had less exposure to improve their social resilience and had less involvement of government interventions to overcome their case. As suggested by Cutter et al. (2008: 2c), the terms *vulnerability* and *resilience* can be applied to many systems—social, natural, economic, engineering—and at different units of analysis (individual, household, community, and region). The need to build community resilience is therefore a great job for the me to figure out the solutions by using multi-critical analyses on each element of resilience. In order to know the multi-critical analyses on each element on resilience, coping capacity, which is one provision of the resilience concept, has seen as an effective analyses approach to understand the role of community actions toward their social capacity before (ex-ante), during, and after (ex-post) flood disaster.

To support this, it is emphasized that *vulnerability* is provisionally defined as the degree to which a system is sensitive to and unable to cope with adverse impacts of global change stimuli (qtd. in Birkman 2006: 484a). Vulnerability is a function of a system's exposure to global change stimuli and its adaptive capacity, that is, its ability to cope with these stimuli (Klein 2003 qtd. in Birkman 2006: 484b). It is the characteristic of a person or a group in terms of their capacity to anticipate, cope with, resist on, and recover from the impact of a natural disaster (Blakie et al. 1994 qtd. in Kasperson and Roger 2005: 251).

This characteristic pertains to an understanding of building community resilience for the risk of flooding disaster in the study area through a lens of human geography, where participation of a community is known as the subjective orientation, or those who see space and nature, i.e., flood disaster as the natural casualties. Extending to a broader scale, advanced knowledge on effective approaches to mitigate the effects of disaster and support communities in coping with disaster consequences has been a strategy of the United Nations (UN). The UN International Decade for Natural Disaster Reduction and/with the Yokohama strategy in the International Strategy for Disaster Reduction has created a Hyogo framework for Action (HFA) 2005–2015: Building the Resilience of Nations and Communities to Disasters (The United Nations International Strategy for Disaster Reduction). These have greatly advanced knowledge on effective the effects of disasters and support communities in coping with disaster consequences.

The Hyogo framework focuses on building national and community resilience to disasters. It outlines three strategic goals: The Introduction of disaster risk reduction into planning for sustaining development at national and local levels; development and strengthening of institutions, mechanisms, and capacities to build resilience to hazards and also systematic incorporation of risk reduction approaches into the implementation of emergency preparedness, response, and recovery programs.

2.4 Disaster and the Local Critical Risk Assessment

Disasters correspond to "undesirable and often sudden events causing human, material, economic and/or environmental losses, which exceed the coping capability of the affected community or society" (Jha 2010: 1). They are caused either by natural forces/processes (known as *natural disasters*) or by human actions, negligence, or errors (known as *anthropogenic disasters*). Natural disasters are generally classified into three major groups: (i) geophysical disasters (e.g., earthquake, volcanic eruption, rockfall, landslide, avalanche, and subsidence); (ii) hydro-meteorological disasters (e.g., flood, drought, storm, extreme temperature, wildfire, and wet mass movement); and (iii) biological disasters (e.g., epidemic, insect, infestation, animal stampede) (Centre for Research on the Epidemiology of Disasters, CRED 2009 qtd. in Jha 2010: 1). In this case study, we dealt with hydro-meteorological disasters (i.e., flood).

According to Phillips (2009: 20), disasters are more than common emergencies such as traffic accidents, personal injuries, or house fires experienced by most communities. For these types of emergencies, first responders typically roll ambulances, squad cars, fire engines, and other resources. For a disaster to occur, it must have a human impact that disrupts community functioning. In a disaster, local resources may become overwhelmed, requiring assistance from other jurisdictions, the state, or the assistance of the federal government.

Disasters cause community functions to halt at least temporarily and, in some cases, for months or years. Disasters disrupt the social functioning of communities: People become displaced, schools and businesses close, hospitals bring in additional personnel and activate special emergency plans, and the critical infrastructure—power, water, transportation sectors—may be compromised (Quarantelli and Perry 2005 qtd. in Phillips 2009: 19). Emergency managers, who have spent time laying a foundation to handle large-scale events, will now step in to guide recovery by organizing and coordinating massive resources to meet widespread needs.

Disaster risk reduction (DRR) is a conceptual framework of elements considered with the purpose of minimizing vulnerabilities and disaster risks throughout a society in order to avoid or to limit the adverse impacts of hazards and to facilitate sustainable development. DRR is a cross-cutting and development issue.

The process of DRR is a complex one consisting of political, technical, participatory, and resource mobilization components. Therefore, DRR requires collective wisdom and efforts from national policy and decision-makers from various government sectors and from representatives from civil society, including academic institutions, the private sector, and the media (United Nations-International Strategy for

Disaster Reduction, UN-ISDR 2007 qtd. in Organization for Economic Cooperation And Development, OECD 2009: 11).

Disaster response refers to the provision of assistance or intervention during or immediately after a disaster to meet the needs of those affected. It is generally immediate and short-term (UN-ISDR 2004 qtd. in the United Nations Human Settlements Programme, UN-Habitat 2007: 168). The primary objective of this humanitarian assistance is to save lives, alleviate suffering, and maintain human dignity. It includes immediate post-disaster rescue and relief activities, such as the provision of food, water and sanitation, shelter, health services, and other assistance to the affected population. It also includes the protection of vulnerable people—for example, those involuntarily displaced from their homes by a hazard event or whose access to relief assistance may be affected by factors such as a disability (The Sphere Project 2004 qtd. in UN-Habitat 2007: 180).

Disaster recovery (rehabilitation and reconstruction) refers to the decisions and actions taken after a disaster with a view to restoring or improving the pre-disaster living conditions of the stricken community while encouraging and facilitating necessary adjustments to reduce disaster risk. Recovery affords an opportunity to develop and apply disaster risk reduction measures (UN-ISDR 2004 qtd. in The World Bank 2008: 3).

Disaster requires an integrated assessment of sustainable development. Disasters are increasingly recognized as a threat to sustainable development, poverty reduction, and achievement of the Millennium Development Goals and poor households are particularly vulnerable to negative shocks arising from disaster events for a number of reasons: the poor own fewer productive assets; are more likely to reside in hazardous locations and in substandard housing; and are primarily dependent on their own labor to meet their livelihood needs (The World Bank 2008, section xi).

Furthermore, a detailed risk assessment should be carried out by the community disaster committee led by a local leader (The World Bank 2008: 20). Within the context of the International Strategy for Disaster Reduction (ISDR), the Hyogo Framework for Action (HFA) 2005–2015 says that: Building the Resilience of Nations and Communities to Disasters is the key planning document that is guiding international efforts to reduce the risk and impacts of natural disasters. The HFA's five strategic priorities for action (qtd. in the World Bank 2008) are to:

1. Ensure that disaster risk reduction is a national and local priority with a strong institutional basis for implementation.
2. Identify, assess, and monitor disaster risks—and enhance early warning.
3. Use knowledge, innovation, and education to build a culture of safety and resilience at all levels.
4. Reduce the underlying risk factors.
5. Strengthen disaster preparedness for effective response at all levels.

2.5 The Notion of Resilience

Resilience, an English word derived from the Latin for springing back, or jumping back up, took on an additional preventive meaning sometime in the last century, in part because it helped to change the focus of research from pathologies to opportunities for supportive action (Fleming and Ledogar 2008: 8a). The desire to act in support of resilience leads naturally to a search for ways to help families and communities strengthen resilience in their individual members. Research has shown that, indeed, much of what seems to promote positive adaptation despite adversity does originate outside of the individual in the family, the community, society, culture, and environment (Fleming and Ledogar 2008: 8b). Furthermore, these researchers define resilience as the natural, human capacity to navigate life well. Some researchers began to distinguish *resilience* from *resiliency,* which for them, carries a connotation that it is an individual trait or characteristic. Luthar and Chichetti (2000) said that the term *resiliency* is derived from *ego resiliency*, which differs from *resilience* in that the former is a character trait and the latter is a dynamic developmental process (qtd. in Fleming and Ledogar 2008: 2). Also, resilience requires the presence of substantial risk or adversity, whereas resiliency does not.

Resilience has been most frequently defined as positive adaptation despite adversity; this definition is synthesized from definitions from various disciplines. As defined by Alwang et al. (2001), *resiliency* is the ability to resist downward pressures and to recover from a shock (qtd. in Thywissen 2010: 8). From the ecology literature, the term means: property that allows a system to absorb and use even benefit from change (McAslan 2010). Where resilience is high, it requires a major disturbance to overcome the limits to qualitative change in a system and allow it to be transformed rapidly into another condition. From the sociology literature, *resilience* is the ability to exploit opportunities, and resist and recover from negative shocks (Alwang et al. 2011 qtd. in Thywissen 2010: 9). Thus, it needs good adaptation and capability of community to face and cope with and recover from shock caused by flood disaster. Overall, resilience is positive adaptation despite adversity.

A system, community, or society that has potentially been exposed to hazards has a varying capacity to adapt by resisting or changing in order to reach and maintain an acceptable level of functioning and structure. This is determined by the degree to which the social system is capable of organizing itself to increase its capacity for learning from past disasters for better future protection and to improve risk reduction measures (UN-ISDR 2004 qtd. in Thywissen 2010: 11). Paton and Johnston (2001: 1) describe community resilience as a community being able to "bounce back" by recovering from adversity using its own resources and potentially even experiencing growth in response; they point to a number of components of resilience that contribute to recovery, including personal and environmental characteristics, and also a sense of community, feelings of efficacy, and coping strategies. Community resilience refers to how communities are able to cope with, recover from, or adapt to hazards (Bergstand et al. 2014).

I define community resilience as the ability of community's behavior to adapt, to resist on, to cope with, and to recover from something or several disturbances that threatened their life and livelihood. Resilience is derived from human capacity and human wisdom to drive their assets in order to adjust with a problem (i.e., flooding). Resilience is the natural, human capacity to navigate life well. It is something every human being has wisdom, common sense. It means coming to know how you think, who you are spiritually, where you come from, and where you are going and the key is learning how to utilize innate resilience, which is the birthright of every human being. It involves understanding our inner spirit and finding a sense of direction (Heavy Runner and Marshall 2003 qtd. in Fleming and Ledogar 2008: 7). In this case the measurement of resilience is the magnitude of disturbance that can be absorbed before the system changes its structure by changing the variables and processes that control behavior (National Academy of Engineering 1996: 33). National Academy of Engineering 1996 emphasizes the control behavior of resilience system to explain the *ecological resilience* as supported by Walker et al. (1969). Recently, resilience has been increasingly conceived as a perspective, as a way of thinking to analyze linked social-ecological systems (Folke 2006: 22). No clear definition is suggested for the term. Rather, *resilience* can be conceived as a collection of ideas about how to interpret complex systems (Andries et al. 2006 qtd. in Fridolin and Jax 2007: 9). Resilience theory has its foundation in systems thinking, including complex systems theory, and it is essentially about understanding the characteristics of change and the interactions between human and natural systems (Gunderson and Holling 2002: 21).

2.6 Understanding Sustainable Actions as the General Framework and Specific Framework on Sustainable Livelihood Resources for Building Community Resilience for Flood Disaster

2.6.1 Understanding of Sustainable Actions as General Framework on the Economic and Policy Domain Arena

The way in which *sustainability* is socially defined has important repercussions for the "orientation and limitations" of social actions and territorial orientations (Werlen 2015). Additionally, Werlen quotes a short sentence from the Chief Executive officer (CEO) from General Electric, Jeffrey Immelt who said, "green is green" means that "green is green is very simple" (Immelt 2006 qtd. in Werlen 2015: 290). *It explains the path of sustainability which should not be thinking as easily as the majority of people perception, said that: "sustainability is adhere with ensuring the orientation of economic without considering the limitation of exploiting the natural resources." Whereas, meanwhile, it should have a balance-system between 'orientation and limitation of utilizing the natural resources for the people's benefits'. It aims at delivering the right perception in which as to not easily allowed people thought and people practical actions* (Immelt 2006 qtd. in Werlen 2015: 290). Immelt's argument formed the basis of our argumentation and allowed us to critically review the way in which sustainability is materialized through different social practices in a specific national context. In order to reach an effective path-way, people are seeking economical sustainability ideas and actions to address various kinds of disturbances, in this case, flood disasters. In this context, there has been a social practical assessment. Social impact assessment has been an arena of concern for charities and governments for some time, as more organizations aim to demonstrate the social and economic value of the programs they fund and run. In the specific case of Indonesia, for instance, *Rencana Pembangunan Jangka Panjang Nasional, RJPN* or Indonesia's National Long-Term Development for the period of 2005–2025 envisages a "green and everlasting Indonesia" *(RPJPN* 2005–2025*)*. One of the *RPJPN*'s eight national development missions is the realization of "a greener and sustainable Indonesia" (RPJPN 2013: 29). It recognizes that "the long-term sustainability of development will face the challenges of climate change and global warming which affect activities and livelihood" and requires the government of Indonesia to pursue its economic growth targets in accordance with socially balanced, resource-efficient, and environmentally friendly management. This is part of a vision to establish a country that is developed and self-reliant, justice and democratic, and peaceful and united. Economic development is aimed at achieving efficient and modern mining and agricultural sectors, a globally competitive manufacturing sector and productive service sector. In Indonesia's First Mitigation Fiscal Framework (MFF), the Indonesian Ministry of Finance estimated that the annual cost of actions in forestry and peat lands, energy and transportation sectors required to reach the 26% emission reduction target by 2020 would be between IDR100 trillion and IDR140 trillion (US$10.7 billion and US$15 billion at the time), (The United Nations Environment Programme, UNEP 2015: 11). The Ministry assumed that between one- and two-thirds of the cost of new initiatives would be financed publicly, including fiscal incentives to stimulate private investment. Mitigation cost for agriculture, industry, and wastewater were not considered in the first MFF. Indonesia's financial system is dominated by banking. The banking sector holds 78.6% of total assets of all financial institutions, which stood at IDR6 611.67 trillion (US$ 550 billion) in June 2014 (excluding venture capital firms, investment managers, and securities companies). The Figure of Banking Sectors in Indonesia can be seen in Fig. 2.1.

To develop a sustainable financial system in Indonesia, it will be important to involve all relevant United Nations Economic Programme (UNEP)-inquiry/ International Finance Corporation (IFC)/Association for Sustainable and Responsible Investment in Asia (AsRIA), sustainable financing in Indonesia stakeholders in the financial sector, in addition to overcoming real economy barriers to sustainable investments. The Indonesian financial service authority (*Otoritas Jaksa Keuangan, OJK*) is clearly in a lead role and as will be discussed below, has already taken important steps and developed a comprehensive Roadmap for Sustainable Finance.

The *OJK* has been able to build upon the work previously conducted by Bank Indonesia on sustainable finance. As macroprudential regulator, Bank Indonesia may

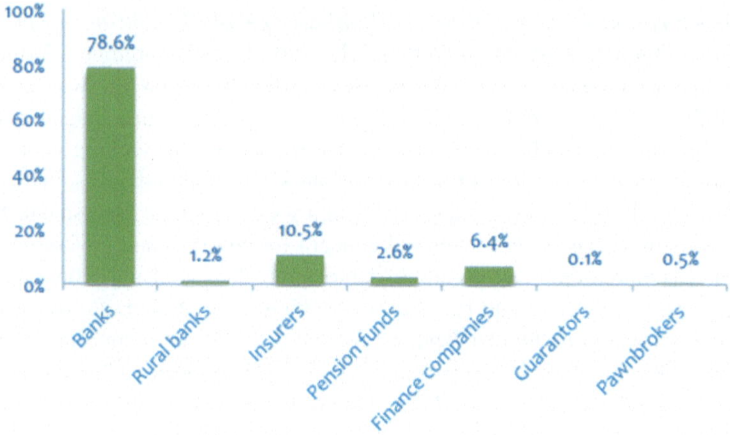

Fig. 2.1 Banking sectors in Indonesia. *Source* Bank Indonesia (2014) qtd. in RPJPN (2013)

still have an important role to play in dealing with climate and other ecological risks to the Indonesian economy (Schoemaker et al. 2014: 12; Volz 2014: 19). The Ministry of Finance can affect the lending and investment decisions of banks and NBFIs, as well as the investment decisions of individuals and of nonfinancial corporations through various tax and subsidy schemes. The stock exchange can affect corporate behavior through listing requirements. Last but not least, financial industry associations can play an important role in disseminating information on sustainable finance as well as training and capacity building activities.

A further interesting exception with respect to the analysis of environmental and climate change-related risk is general insurance companies. Over the last years, Indonesia has been exposed to several natural disasters that are commonly associated with climate change, such as flooding and landslides. Given that the latter have an immediate effect on property or motor vehicle insurers (the two most important types of nonlife insurance in Indonesia), the Association of Indonesian General Insurance Companies has started to collect data on claims related to flooding and landslides and develop risk maps accordingly. These risk maps are shared among the members, who adjust their risk premier accordingly. Given that Indonesia is highly vulnerable to the effects of climate change (Yusuf and Francisco 2009: 8; Global Facility for Disaster Risk Reduction and Recovery, GFDRR 2012: 16; Mani and Wang 2014: 22—The World Bank's Research Report), it would be reasonable for other financial institutions to also consider environmental risk. In addition to financial indicators, good corporate governance was frequently referred to as a major factor affecting investment decisions.

Adherence to environmental legislation was widely considered an important element of good corporate governance, and some interviewees even expressed the thought that corporations that consider environmental and social considerations may over the long run outperform competitors, as they are implementing practices that may sooner or later become mandatory for all firms. In this context, it is noteworthy that in January 2014 the Indonesian Stock Exchange (IDX) introduced new corporate governance requirements for listed companies based on the Indonesia Corporate Governance Manual that OJK published together with the International Finance Corporation (IFC) in the same month (*Otoritas Jaksa Keuangan, Ojk*—The Indonesian Financial Service Authority 2014: 28). In February 2014, OJK, supported by IFC, launched an Indonesian Corporate Governance Roadmap. Moving forward, IDX could play an important role in strengthening Environmental Social Governance (ESG) practices among listed companies through its listing requirements.

The institutional reform in every country nowadays still deals with economic compromises and policy responsiveness. It has been highlighted that the issue of achieving sustainable development is commonly the result of environmental pragmatics, i.e., human and natural resources. Additionally, it has also been seen that the critical issue of ecological crisis has brought constant economic and policy issues regarding ecosystem management of a dynamic society.

I add that a dynamic society characterizes a vast environment that has socio-ecological problems along the way to achieve sustainable development. For this conception, Werlen (2015) explained that we found that pressure on the environment and the reemergence of socio-ecological conflicts are the most common result of the environmental institutional reform in Latin America as an example.

It emphasizes the essence of concerning the reemergence of socio-ecological conflicts on vast environment while involving less economic issues. In the context of resilience for sustainable development, the solution-based-approach in achieving sustainable development is not merely from a single aspect (i.e., economic capital), but it is derived by multi-approaches (i.e., human, natural, physical, economic, social, cultural, religion, political, and coexisted regulation).

For this analysis, Werlen (2015: 291) has extended his idea of *sustainability* to *sustainability over-confidence*. Furthermore, he defines the *sustainability over-confidence* as the social belief that it is possible to overcome the ecological crisis by merely restricting economic development environmentally in terms of (i) the rational exploitation of natural resources and (ii) the rational management of pollution by incorporating new green technologies. He explained these two forms of criticism are raised to challenge rational management as an instrument with which to overcome the ecological crisis.

Based on Wieriks (2013: 22), applying the theory of inspired confidence to sustainability assurance a few important differences compared to financial auditing can be noted. For example, Limperg (1922–1949: 39), brings the idea that confidence of sustainability is based on management control over financial contributions by general public organizations. This social participation in financing creates a clear, tangible, or at least accountable, relationship between management of the organization and society; it reflects the obvious principal agent relationship. The principal agent relationship involves a strong collaboration among fiscal institutions, social institutions, policy and law institutions, and adaptive governance regime (institutional) and also all related institutions in Indonesia and foreigner countries in tackling the case of sustainability. The adaptive governance regime (institutional) toward management of resilience for the flood disaster in the West Malaka Subdistrict is not well established in any policy frameworks in the government of Indonesia regulation. This is one obstacle upon the sustainable path that could be implemented. To discuss this, I put forth previous research definitions on integration of political leadership in targeting a more sustainable development, and she also discusses key opportunities in an international context. Walker et al. (2004: 1), defines *transformability* as the capacity to create a fundamentally new system when ecological, economic, or social, including political, conditions make the existing system untenable. Key opportunities to set regional development upon a more sustainable path were missed because problems were repeatedly and incorrectly framed to serve other political and development agendas (Lebel et al. 2004: 21; Luangaramsi 2002: 59). An example of when key opportunity to set regional development upon a more sustainable path was missed but later corrected was in the Everglades, where leadership involved integrating the extant scientific understanding of the ecosystem, summarizing that integration, and communicating that understanding to a wider audience (Olsson et al. 2006: 3a). In the 1960s and 1970s, Arthur R. Marshall provided the same leadership at a research institute at the University of Miami and communicated his understanding with "the zeal and passion of a country preacher" (Light et al. 1995 qtd. in Olsson et al. 2006: 2). Two decades later, scientific leadership was provided by the adaptive management group consisting of Buzz Holling, Carl Walters, Lance Gunderson, Steve Davis, and Steve Light (Davis and Ogden 1994 qtd. in Olsson et al. 2006: 3b).

To generate a single action on economic sustainability is a pressure on ecosystems (i.e., human and natural dimensions and policy). The idea of this matter is pointed out by Werlen (2015) that we subsequently highlighted that many enterprises are oriented toward economic sustainability—that is, they do not make compromises about their future needs in terms of the rational usage of natural resources or pollution control—have exerted strong pressure on ecosystems in terms of the natural and the human dimension. This has drastically changed these ecosystems' properties, often leading to their destruction, and, subsequently, reinforcing historical and/or latent socio-ecological conflicts. The biggest risks we face, as collective humanity, come not from nature but from ourselves. They derive from our newfound global interdependence and the fragility of the systems that are driving it.

In order to create a specific framework while keeping in mind a general framework of sustainable development, therefore avoiding a pledge to a dominating single arena and domain analyses in the context of human and natural dimensions. I discussed the specific framework of sustainable development and community livelihood assets which have a relation with the resilience and sustainable community livelihood resources.

2.6.2 Understanding Sustainable Actions as a Specific Framework for Community Livelihood Assets on Resilience

Based on the Livelihood assessment toolkit, Food and Agriculture Organization and International Labor Organization, FAO and ILO (2009: 11) describe that assets refer to the resource base of people. The Sustainable Livelihoods Framework (SLF), it represented as a pentagon which consists of five categories; they are: natural resource (natural capital), physical reproducible goods (physical capital), monetary resources (financial/economic capital), manpower with different skills (human capital), and social networks of various kinds (social capital). The types of issues and details of these categories (FAO and ILO 2009) are covered as follows:

1. Human capital involves labor, health and nutritional status, skills, and knowledge.
2. Natural capital involves access to land, water, wildlife, flora, and florets.
3. Social capital involves those stocks of social trust, norms, and networks that people can draw upon to solve common problems. It is mediated through kin networks and group membership.
4. Physical capital involves houses, vehicles, equipment, and livestock.
5. Financial/economic capital involves savings, gold/jewelry, access to regular income, net access to credit, and insurance.

These capitals in the Sustainable Livelihood Resources Assessment are also supported by Mayunga (2007), who defines five capitals on resilience as the sustainable livelihoods approach which community resilience to disasters springs from it (qtd. in Bahadur et al. 2010: 5). She mentions the determinants of resilience as social capital, economic capital, human capital, physical capital, and natural capital. Each of these five capitals corresponds to a number of characteristics of resilient systems. In another case study in Pakistan, a result of research on community meeting with future strategy agendas by the (The United Nations Development Programme, UNDP 2013: 4–5) has taken a lot of community livelihood resource information. Information on the livelihood assets of individuals, households, and communities (sometimes collected in a separate "livelihoods assessment") can be classified according to the six capital assets introduced in Chap. 1: human, natural, physical, political, financial, and social assets. The indicative types of information (UNDP 2013: 4–5) are:

1. Human capital assets consist of skill levels and educational attainment, the extent to which local skills match the needs of existing or emerging local businesses, knowledge base and local use of science and technology, livelihood aspirations, and access to food, education, and health care.
2. Natural capital assets consist of environmental conditions in which people live and work, and natural resource management issues affecting livelihoods (e.g., land tenure and disputed natural resources). Physical Capital assets consist of access to adequate housing, access to safe water and sanitation, and access to, and use of infrastructure (roads, bridges, schools, hospitals, agriculture systems, etc.).

3. Physical capital assets consist of access to adequate housing, access to safe water and sanitation and access to and use of infrastructure (roads, bridges, schools, hospitals, agriculture systems, etc.).
4. Financial capital assets consist of access to finance, including microcredit and coping and adaptive strategies for responding to financial shock (e.g., drawing down savings, borrowing from friends, relatives, or money-lenders, and selling or trading assets).
5. Social capital assets consist of participation in social groups and networks that provide support (financial or psychosocial), e.g., community-based organizations, NGOs, faith-based organizations, women's organizations, and extended family networks and perceived levels of safety and security.
6. Political/Governance capital assets consist of access to and participation in government decision-making processes, gender relations, and power structures, influencing control over decision making within households, communities, and the wider society as well as access to judicial systems and security sector institutions.

Before discussing these capitals on resilience, I discussed the community sustainable livelihood resources assessment on resiliency systems. The community sustainable livelihood resources can be understood as the capital or asset to eagerly improve, reconstruct, and complete do self-assessment both for one's existing capitals or assets and also for plausibly creating new capitals or assets on resiliency systems to achieve longevity of sustainable livelihood resources.

Community resilience can be defined as a smart community that is self-independent, self-capable, and self-liable to overcome certain obstacles they encounter. Brown and Kulig (1997: 163) add that a further emphasis is on the community gaining the ability to rely more on internal physical resources and human capabilities rather than being dependent upon external development in meeting the community's needs. The community's needs alluded to their certain cases or obstacles and certain targets which are not succeed yet during the process and final remarks. The certain cases or obstacles and certain targets on this case study is the need of capitals or assets, as measuring and improving the community capitals or assets, their existing capitals, or assets to deal with flood disaster. These capitals or assets are expected to be fulfilled by the community if they have a resiliency system which could include being well-maintained for long-term actions, or better referred to as sustainable livelihood resources development. For this case, community resilience has a bulk of capacities to cope with for flood disaster.

Linking with this, *livelihood resources* refer to community capitals or assets. Resilience refers to a livelihood resources system that models process of each capital or asset that a community has that enables them to address certain or several disturbance events and/or catastrophic events, specifically in this case study, flood disaster. For understanding a hierarchical definition of sustainable livelihood, I put forth a self-conceived process of analyses or assessment.

This self-conceived process for resilience assessment is based upon the sustainable livelihood assessment. This major framework of resilience is a system which is based upon whole or holistic analyses started from ex-ante or before flood disaster, during flood disaster, and ex-post or aftermath flood disaster occurrences. This idea of sustainable livelihood assessment is supported by Knutsson and Ostwald (2005: 29) where they describe that by adding process trends and flexibility rates to the sustainable livelihood assessment (SLA), we are able to disregard assets that have decreasing process trends and low flexibility rates. This makes our proposed tool more selective in promoting support to assets that will generate a less vulnerable livelihood.

For instance, a strong base of social capital comes in the form of trust, norms, and networks which lead to a high degree of coordination and cooperation in a given community, as evidenced by the presence of a large number of non-profit organizations. Similarly, human capital in the form of education, health, skills, knowledge, and information will lead to, for instance, a high capacity to develop and implement an effective risk reduction strategy. Indicators of this would include high levels of educational attainment and good health.

2.7 Sustainable Community Livelihood Resources and the Building of Community Resilience

2.7.1 Improve Human Capital to Deal with a Flooding Disaster

Human capital is among the most important determinants of resilience because it can increase or decrease the efficiency of the other types of capital in resilience-building efforts by providing access to a skilled and trained workforce for economic development and capacity building (Mayunga 2007: 8).

According to Cooperative for Assistance and Relief Everywhere, CARE (2003, section iv), human capital consists of the skills, knowledge, ability to labor, and good health that are important to the pursuit of livelihood strategies. At the individual and household levels, educational attainment and health status of members shapes their ability to absorb the negative impacts of a shock and to successfully adapt to changing social, economic, and environmental conditions. At the community level, human capital reflects the collective level of access to skills, labor, knowledge, and physical and mental health; it is also key to innovation. The task of strengthening community human capital is perpetually consistent and well maintained if community is not the sole actor to adequately build or rebuild their collective level. Therefore, a community requires a unit of local entities that work together with resilience; they are namely, non-government entities, related institutions, and the government. A community seeks a platform of discussion in order to deliver their aspirations toward their resiliency system and local entities. I then formed two questions to address the problems which encountered by the community, as follows:

1. What are the community human capitals that create their own identity assets (uniqueness)?
2. How could they use and improve their human capital to empower their all-livelihood resources, i.e., the whole capitals on resilience?

These questions are presented in discussions and assessments on Focus Group discussion (FGD) and Stakeholder Meetings (SMES) in Chaps. 8 and 9.

2.7.2 Fostering Education on the Community's Natural Capital and the Availability of the Natural Capital in Order to Build a System of Resiliency

Natural capital is a community's natural resources: environmental stocks from which resources useful for livelihoods are derived (e.g., land, water, forest, rangeland, fisheries, wildlife, biodiversity, and environmental services) (Cooperative for Assistance and Relief Everywhere, CARE 2003, section iv). In addition to simply possessing natural capital, the management of natural resources and ecosystem services while maintaining a sustainable livelihood base is a key element of community resilience (Pasteur 2011; Twigg 2009 qtd. in Frankenberger et al. 2013: 12).

Natural resources are one of the key assets collective actions that try to manage natural disaster. Effective management and protection of environmental resources requires collective norms and valuation of the environment as a public good in order to avoid "tragedy of the commons." An example of a natural resource is a watershed; the manner in which one community uses watershed resources has implications for watershed health and viability, and thus for its use by other communities that rely on it. As Frankenberger et al. (2013: 13) explain their idea in the following quote that:

> Resilience at the community level may be affected by an array of factors relating to the quality of natural assets such as soil, forest cover, pasture, fishery stocks, riverine/coastal habitats, surface, and belowground water supplies. Many of the natural disasters that pose the greatest risk for vulnerable populations have an immediate, detrimental, and long-lasting impact on the natural resource base. Likewise, many man-made shocks and stresses (e.g. conflict, price increases, and disempowerment) are centered on competition for and dispute over access to scarce natural resources (Frankenberger et al. 2013: 13)

These problems or casualties due to natural impacts are considered as a versus or contradictory things with community natural capitals. As mentioned by The Association for Non-Governmental Organization, TANGO (2003), natural capital; (land, water, forest, rangeland, fisheries, wildlife, biodiversity, and environmental services) are belonging to environmental stocks. These environmental stocks would not be efficiently and sufficiently utilized by the community if there is inadequate information and a lack of educative are several lacks information and educative communication given by the non-government institutions and the government to educate and provide socialization to a community. The analysis of the quality of natural capitals and natural hazard factors that disrupt or reduce the natural capital of community has been descriptively assessed through Focus Group Discussion (FGD) and Stakeholder Meetings (SMEs) as the field approaches. These field approaches as presented in Chapter 9. The points were assessed by the community:

1. Recognizing the natural capitals of community (what are their natural capitals?)
2. Finding out the obstacles faced by the community (what are their obstacles or barriers in utilizing their natural capitals?)
3. Answering the problems by galvanizing the non-government institutions, related institutions, and government to design a solution framework for building their resiliency system on natural capital (how could their resiliency system be improved by potentially utilizing their natural capitals against or versus the abandonment of natural casualties?).

2.7.3 Provide and Discuss the Information of the Use of Physical Capital

When assessing the impact of a disaster on critical infrastructure and determining the contribution of infrastructure to community resilience, it is necessary to determine whether infrastructure supports key services (education, health, safety, economic activity), the accessibility of infrastructure to all members of the community, existing gaps in productive infrastructure, and community-based mechanisms for maintenance. Physical capital refers to the built environment, which comprises of residential housing, public buildings, business/industry, dams and levees, and shelters. It also includes lifelines such as electricity, water, telephone, and critical infrastructure such as hospitals, schools, fire and police stations, and nursing homes (Mayunga 2007: 7). Doing so requires ensuring that community infrastructure systems and other basic services can operate at a level that provides individuals and groups the means to survive and recover during natural or man-made disasters (Longstaff et al. 2010; Pasteur 2011 qtd. in Frankenberger et al. 2013: 13a). In order to build community infrastructure in the aftermath of a disaster, a community should take on more financial risk out of their own pocket. Infrastructure rehabilitation for a community requires maintainable and controlling support from the Indonesian government. This maintainable support from the Indonesian government deals with physical capitals such as public infrastructure; (roads, bridges, water, and sanitation systems), disaster shelter, media and telecommunications, health, facilities, schools, community centers, irrigation, and drainage systems.

In terms of controlling, communities are not always able to directly control some of the physical assets available to them (e.g., power systems or electrical grids); they may, however, be able to influence their use through *indirect* means. For example, a community can build in redundancies that allow alternatives when one system breaks down or by requiring maintenance/user fees. These are examples of collective action for the management of public goods. It is important to acknowledge that redundancy and maintenance of physical assets can be expensive, and communities will have to weigh the tradeoffs of allocating funds for long-and short-term reserves (Longstaff et al. 2010 qtd. in Frankenberger et al. 2013: 13b).

Based upon these problems, the community has provided answers in Chaps. 7 and 9 through Focus Group Discussion (FGD) and Stakeholder Meetings (SMEs) about the two more concerning issues:

1. Providing information for the community about maintenance and controlling of physical capital in terms of community infrastructure by themselves and the non-government, related institutions, and government as well as the beneficial assets of the use of physical capitals for the daily necessities of a community and also;
2. Supporting their social-economic activities such as drainage and irrigation.

2.7.4 Empowering Economic Capital

Empowering economic capital can be understood as a process of building community economic resources in regard to utilization of resilience' assets as well as to manage a flood disaster. Werlen (2015: 300) describes this issue into an economic thinking that: "If you are going to invest in green technologies, you are going to earn a lot of green dollars. So, if we want to make this possible, we need new economic thinking, we need smart regulation, we need smart investment, and we must understand that this is not only a challenge, but also a huge opportunity, and also for industrialized countries. And if this conference can help us in this debate, then you have done a wonderful job." In accordance with these statements, the new sort of economic idea that I propose is to relate to new economic thinking of resilience and disaster management which take a multi-subdiscipline understanding to the whole economic term.

2.7.5 Enlightening the Community—Ensuring Social Capital

The Social Capital Assessment Tool (SOCAT) is used by the World Bank to assess comparative levels and variations of social capital across countries, and as an indicator for economic development and social well-being (Krishna and Shrader 1999: 12). This tool, because of its global use, must reflect a broad definition of social capital and be applied across a wide range of contexts and cultures. *Social capital* can be described as the quantity and quality of social resources (e.g., networks, membership in groups, social relations, and access to wider institutions in society) upon which people draw in pursuit of livelihoods (Frankenberger and Garrett 1998 qtd. in Frankenberger et al. 2013: 14).

Rupasingha et al. (2006: 18) proposed a county-level model to measure social capital based on their definition of social capital as participation in associational activities. While it may encapsulate political institutions, social capital is broader than political capital because it includes informal social processes at individual, household, and community levels. Social capital has often been described as the "glue" that binds people in society together. It is based on strong perceptions of local embeddedness, self-regulating moral codes, and the norms, reciprocity and trust that exist between individuals and groups at the community level (Chaskin 2008: 9).

Close interaction between people through tight-knit communities, the ability to rely on others in times of crisis, and open communication between stakeholder groups are all generally seen as signs of well-developed social capital. Methods of measuring social capital vary. One example, Rupasingha et al. (2006: 19) used secondary data as proxies of structural indicators of social capital to describe social capital across United States (US) counties over two-time periods. Their indicators included: associational densities, presidential voter rates, census response rates, and per capita non-profit organizations. They argued that the degree of participation and density provides a good measure of social capital differences between regions.

Another, Grootaert et al. (2001: 6) discussed the challenges in measuring social capital directly and identified proxy indicators recommended for use in measuring social capital. These included: memberships in associations and networks based on the density of organizations and incidence of membership, indicators of trust and adherence to norms referring to cognitive social capital and measured through surveys, and indicators of collective action, identified as the provision of services. These indicators, collated and analyzed together, provide a good measure of the broad concept of social capital. I conclude her idea of social capital in regard to its provision service. Yet, social capital can be understood as a strong of social kinship among society in terms of their participation and engagement into several organizations. It helps them to tie up a social cohesion which could be collectively practiced by them in a disturbance event, in this case, flood disaster.

Previous research demonstrates that the extent and application of social capital is a crucial element in determining the nature of resilience at the community level (Aldrich 2012: 9). At the household level, social capital is properly viewed as one form of capital among many that have a direct bearing on household food security, nutrition, livelihood security, and resilience. However, amid the complex and dynamic interactions that take place within and between larger populations, social capital can have a predominantly strong influence on the attainment of resilience at the community level (Aldrich 2012; Cutter et al. 2008: 9). For instance, disasters may sometimes enhance social capital because they activate or give rise to neighborhood associations and collective organizations that can be used to disseminate vital information, provide community members with a voice, and afford leverage to assist in taking control of rebuilding efforts (Aldrich 2012: 11). I may agree with Aldrich's opinion that in order to do analyses on resilience, for further assessing optimization of community resilience, social capital is seen as a form which ties three distinct, they are *bonding* social capital, *bridging* social capital, and *linking* social capital.

Aldrich (2012: 12) provides a useful analytical approach for assessing the influence of social capital on community resilience by identifying three distinct but interrelated forms of social capital. The resilience of a community is dependent on social bonds and collective action based on networks of relationships, reciprocity, trust, and community norms. Social capital can contribute to community resilience by providing an informal buffer to those affected by disaster, overcoming challenges to adaptation through coordinated local processes, and enabling transformative change by strengthening the community's collective voice.

Aldrich's framework can be useful for considering the influence of social capital on community resilience because resilience at this level requires institutional reforms, behavior shifts, cultural changes, the questioning of values, challenging of assumptions, and close examination of identities, stereotypes, and fixed beliefs (Bene et al. 2012: 27; Smith and Stirling 2010: 21). Each of these forms of social capital will be discussed now. The first is bonding social capital which is seen in the bonds within and between community members.

In many rural community environments, members operate as if they were members of the same extended family. Bonding social capital assumes high levels of familiarity, involves a willingness to forego some degree of privacy, and typically entails an implicit commitment to reciprocity. Bonding social capital is perhaps most simply described as "horizontal" ties between individuals who are similar to each other and may live within close proximity to one another (Putnam 2000: 16).

An important potential downside to bonding social capital is that a strong sense of belonging to a group, tribe, or nation can create indifference or even hostility toward nonmembers (e.g., nationalism, patriotism, xenophobia). Yet, others note a potentially symbiotic relationship between bonding social capital and disaster management capacity in situations where survivors work more closely together to address the multiple problems that arise in the wake of a disaster. In such instances, the extent and quality of bonding social capital can increase because of difficult conditions (Solnit 2009: 19; Norris et al. 2008: 22). The Model of Bonding, Bridging, and Linking of Social Capital is presented in Fig. 2.2.

The second kind of social capital is bridging social capital which connects members of one community or group to other communities/groups (Aldrich 2012: 12). Bridging social capital often crosses ethnic/ racial lines, geographic boundaries, and language groups, and can link community/group members to external assets and broader social and economic identities. Bridging social capital makes a direct contribution to community resilience in that those with social ties outside their immediate community can draw on these links when local resources are insufficient or unavailable (Wetterberg 2004: 16). The other model, I arguably put on this analysis specifically on social capital, is a model on community resilience by Norris et al. (2008: 22). In the model of community resilience, social capital encompasses three elements that include (a) social support, (b) social participation, and (c) community bonds (Alvarado et al. 2020; Norris et al. 2008: 23). The first element is the development of social support that is represented by informal networks formed with family and friends. Through these networks, individuals embed themselves in social relationships that can provide a variety of types of social support. The second element represents social participation. This includes the creation of formal social networks that individuals have with groups and organizations. These networks create more structured relationships that encompass professional, social, economic, and health-related participation. Social participation also occurs across organizational levels to create, maintain, and alter functional and structural needs of individuals and

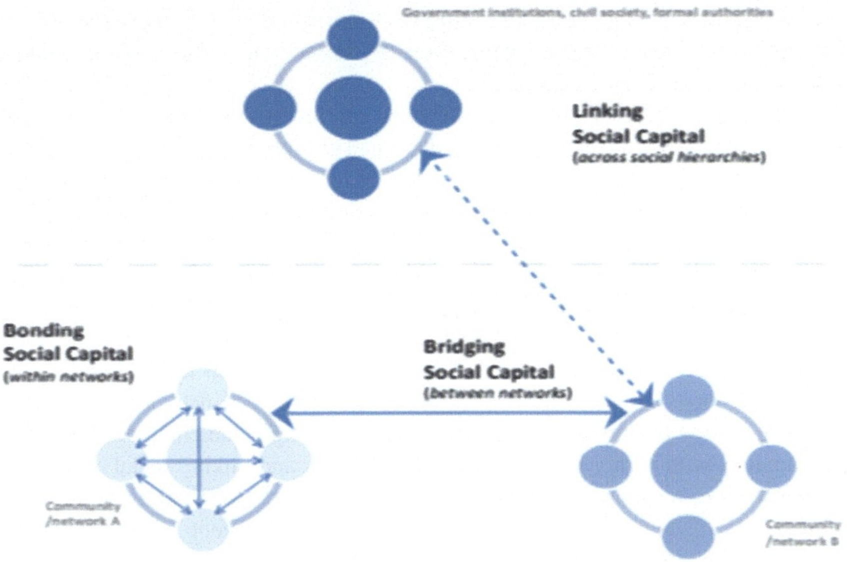

Fig. 2.2 The model of bonding, bridging, and linking of social capital. *Source* Aldrich (2012: 12)

groups. The third element is the establishment of community bonds through citizen participation in group and community activities. This represents social capital at the community level. The social capital established in this multilevel formal and informal networking can be mobilized to meet unique needs in time of stress and uncertainty. Inherent in these social relationships is an element of trust, reciprocity, shared norms and values, and the establishment of rules and obligations that define social capital as a beneficial resource for a community (Fiorini et al. 2018; Hobbs 2000: 25).

Since social capital takes its role in the formation of social interconnections among individual, household, private, and public institutions, a critical discussion, and assessment is not only on social capability but also on social adaptation. I concur with previous researchers' concepts on social capital; that is where, Norris et al. and Uphoff present the essence of social adaptivity, where *social capital* is seen as a set of adaptive capacities that can support the process of community resilience to maintain and sustain community health (Ebers 2015; Norris et al. 2008: 29).

Furthermore, Uphoff (2000: 13) defined two distinct categories of social capital: structural social capital and cognitive social capital. *Structural social capital* refers to the forms and varieties of social organizations and networks that are thought to contribute to the development of social capital and *cognitive social* capital has to do with the mental processes and perceptions resulting from norms, values, attitudes, and beliefs that foster mutually cooperative behaviors. These are the perceptions that people have about trust, cooperation, and reciprocity that facilitate behaving in that way. Additionally, because of contact with close friends and community resilience, conceptual framework and measurement relations are more regular for interaction with those outside of one's immediate network is more likely to provide new perspectives and resources. In addition to forming linkages with external assets, bridging social capital enables generation of "broader identities" among linked network members (Putnam 2000: 16).

Continuing on Aldrich's third kind of social capital, linking social capital is seen in trusted social networks between individuals and groups interacting across explicit, institutionalized, formal boundaries in society (Aldrich 2012: 13). Linked networks are particularly important for economic development and resilience because they provide resources and information that are otherwise unavailable through bonding or bridging capital (Aldrich 2012: 13). Linking social capital is often conceived of as a vertical link between a network and some form of authority or power in the social sphere. It also can involve linking across functions that would otherwise operate independently; for example, biologists working on ecosystem health may not, as matters of course, interact with engineers constructing road networks or governmental economic developmental departments promoting value chain development. Communities with higher levels of bonding, bridging, and linking social capital are inherently more resilient than those with only one type or none (Aldrich 2012: 13; Woolcock 1998: 21). It is also critical to acknowledge that none of these forms of social capital by itself is a panacea for food and livelihood insecurity.

To ensure community resilience to shocks and stresses over the long term, each of the different types of social capital must be promoted and sustained together, and communities can take actions that enhance their absorptive, adaptive, and transformative capacities simultaneously. It is also important to recognize that while rich intra-and inter-communal relationships facilitate collective action, this may have heterogeneous effects on community subgroups given the equity and power relationships inherent in the social structure, such as the often uneven distribution of power and wealth (Morduch and Sharma 2002: 13). For example, social capital may help in-group members recover from a shock, while out-group members or those with fewer social resources who live outside the mainstream experience slower or no progress toward recovery (Aldrich 2012: 15). Neighborhoods and groups with fewer social resources face negative outcomes, both inadvertent and deliberate; for example, people who are dependent on others for their well-being, such as children or the elderly, may lose their caretakers due to illness, death, or migration. The absence of any form of social capital can stall a community's recovery, compromising collective action (Buckle 1998: 11; Chamlee-Wright 2010: 20).

Therefore, it is critical to acknowledge the central role of effective leadership at multiple scales. Leadership at the community level is vital to strengthening both bonding and bridging social capitals whereas leadership at higher levels is a critical component of the enabling conditions that support linking social capital and transformative capacity. These capacities have an important role in terms of keeping a value of social relationship at the community level. Additionally, in the context of resilience and disaster, linking social capital and transformative capacity are seen also as the driving of a social mobility which prominently influences the role of community's repository actions. In the context of an examination of disasters and disaster resilience, a given example of social mobility, where place cannot easily be ignored as a primary factor in experiences of social capital. Place has most prominently figured in the research on disaster resilience as it pertains to displacement resulting from forced migrations or evacuations.

Norris et al. (2008: 139) describe the sense of community as "an attitude of bonding" that includes "mutual concerns and shared values" and that is a characteristic of resilient communities. Consistent with the research on disaster (Erikson 1976a, b; Kaniasty and Norris 2004 qtd. in Aldrich 2012: 18). Furthermore, interviewed residents' location on this continuum often reflected the nature and severity of their losses. Rural communities because of their small size, relative homogeneity, and geographic location tend to foster a strong sense of identification and belonging (Canadian Centre for Justice Statistics 2007: 5). Rural experiences of reorientation may be mediated by unique relationships to social capital, including: (1) the propensity for long term social networks (Onyx and Bullen 2000: 22), (2) extremely diverse networks due to smaller community size and close proximity to an array of community members (Erickson 2003: 8), and (3) heightened levels of social capital (Erickson 2003: 9). Disasters, because they can disturb and realign social capital, have implications for resilience and place. The analyses on social capital (social resilience) in empirical research, i.e., fieldwork for this book's project, are the three components or elements of social capital, namely bonding social capital, bridging social capital, and linking social capital, adopted by the previous researcher theory (Aldrich 2012: 16).

2.7.6 Revitalization and Enlightenment of the Community's Cultural Capital and Community's Religion Capital

Cultural and religious perspectives and or experience of community actions toward their system of resiliency have changed the way of navigating ideas into contextual problems to deal with any challenges. I self-conceived the idea to use this specific capital for the analysis in order to understand and assess the community values which are revitalized and enlightened by their self-assessment toward the space and the nature. Community has devolved their own illegally and legally decisions or constituents that they've made to the customary leaders and religious leaders' rules to resolve their essential problems, in this case, community resilience for flood disaster. The use of this specific capital in West Malaka Subdistrict of Malaka District in East Nusa Tenggara Province, Indonesia, reveals a new paradigm in the society towards resiliency system where rural-society has a strong-faith to their ancestors through the customary leaders and a well-guidance of religious leaders.

Given an example of study case in Africa where despite the challenges germane to some customary mechanisms for protection and recovery from shocks and stresses, many of these long-established institutions can be quite effective in these roles, as illustrated by the funeral societies common across Africa, community-based savings groups, and indigenous pastoral management systems (Pavanello and Levine 2011:

13). Another example is in West Malaka Subdistrict of Malaka District in East Nusa Tenggara Province, Indonesia, where community has a different system of protection and recovery from a flood disaster. These systems are dependent on the specific role of the customary leaders who have a strong leadership to guide the community. The customary leaders are coordinated and cooperated with the religious leaders (i.e., priests, sisters, and brothers) to overcome the community' problems in dealing with flood disaster.

Furthermore, in Somalia, for instance, offers strong examples of clan-based commerce and informal systems for resource transfers that have thrived in the absence of state welfare and market systems (Sexsmith 2009: 89). Shotter (2003) alluded to the possibilities inherent in such a shift in our cultural relationship to distress and disorientation (qtd. in Simao and Valsiner 2007: 15). Shotter (2003) described a move away from a problem-solving orientation to suffering and distress to a reflective approach that would open up a dialogic space in which the possibilities called forth by disorientation are explored (qtd. in Simao and Valsiner 2007: 16). Furthermore, he explained that this opportunity, painful and unwanted though it may be, is implicit in the words of one survivor of the McClure Forest Fire as he described the disorientation, he felt following "the fires you don't belong anywhere". There's this fire and all that went with the fire, and then this strange feeling, this strange, not belonging anywhere (Brian, Barriere resident). It was, he offered, a feeling of being a "like fish out of water", and, like fish out of water he and his fellow survivors were suddenly conscious of that which, until that point, had been largely out of their awareness (Shotter 2003 qtd. in Simao and Valsiner 2007: 17). Reorientation, the individual and collective negotiation of identity and belonging in the wake of disasters can be painful, stressful, and confusing, but it can also be transformative.

There is some debate as to whether formal protection schemes "crowd out" informal mechanisms whereby community members transfer resources to others in times of shocks and stresses. Yet there is some evidence indicating that formal cash transfers, for example, can support the growth or strengthening of customary social protection measures, such as by enabling people to participate in rotating savings clubs (Deveraux et al. 2008: 16). Such savings clubs in and of themselves are a form of collective action, in that the community is self-organizing itself to assist its members manage disturbances that cause loss or hardship.

Nevertheless, many customary institutions anchored in communities are in a state of decline. For example, increased migration and the resulting fragmentation of households can compromise the reliability and robustness of informal social protection mechanisms and increase the burden on those who remain behind. Seasonal or permanent migration can also remove both the incentive and the ability to care for home community natural resources on a sustained basis.

On the positive side, migration can help home communities through transfers (remittances), though the importance and impact of remittances also varies with context (Morduch and Sharma 2002: 21); this is an illustration of the value of bridging social capital. Idiosyncratic events such as illness or death, and casualties in contexts of violent conflict, can also weaken resilience at both household and community levels because of their toll on human capital assets.

The impact on resilience capacities can be quite severe when these shocks are experienced on a grand or protracted scale. Regarding the impact of conflict on social capital and collective capacities, it should be acknowledged that both "productive" and "destructive" social change scenarios are possible, which, respectively, support or hamper resilience: in Somalia, customary institutions for maintaining trust in trading relationships and for redistributing resources have been maintained, encouraging efficient economic behavior, and supporting conflict resolution.

Likewise, in cultural capital has a unique perspective where without ignoring productive and destructive of cultural changes are acknowledged into a harmony system of resiliency, can protect the values of generating several social-political and economic stocks such as partnerships constituted-policy acts and the leadership role and its influence, and monetary issues. In other opinion of informal political networks based on clan identity have played a crucial role in maintaining commercial activity after state collapse (Sexsmith 2009: 91). In contrast, in Sierra Leone, "*Informal institutions ....*" facilitated the perpetuation of a patrimonial economic system whose leaders relied on violent conflict to generate wealth, contributed to prolonged suffering, and delayed the emergence of democratic order (Sexsmith 2009: 92). I analyzed that the role of customary leaders in the West Malaka Subdistrict of Malaka District in East Nusa Tenggara Province, Indonesia, have an essential role which enables to protect their community from conflicts and catastrophic disasters. There is some evidence of increasing cooperation between customary and pastoral or religion institutions/ practices and state efforts in resilience-promoting actions. This evidence is described on the content of cultural capital and religion capital in the contextualization of community resilience dealing with disturbance events and or any catastrophic disasters in this case flood disaster. Cultural capital is a means of social norms, social trust, and beliefs that trusted by the local community that had been built the social unity (through social capital).

The cultural setting correlates the human adaptation to their environmental setting (Marfai and Hisbaron 2011: 18). Human adaptation to their environmental setting in the study area is determined by the role of community and community leaders. For instance, the role of community leaders in West Malaka Subdistrict of Malaka District in East Nusa Tenggara Province, Indonesia, can influence the local community's attitude, aptitude, and altitude of their actions and behavior in a community. The role of community leaders through a prayer to the ancestors could solve the problems of flooding in terms of reducing the level of water which can enter the customary leader's house. Additionally, customary leaders can also assist the people for their harvest in the agriculture fields. Even though it sounds not a logical one, the local people there actually knew and understand well their behavior or the way of their life. Thus, a disaster is thought not as a bane but a boon. It has been embedded in their life as local wisdom that supports resilient community.

As supposed explicitly by Debray et al. (2007), a resilient community is praised as a partnership in terms of mutual assistance to help each other and a partner of nature. The concept of a community might be considered not only how a community returns to its original state, but also how to deal with changing circumstances. This is related to social institutions are formed to the rules established by the communities themselves. So that, they could also seek to make the best use, increase public awareness that often occurs after a disaster to improve local conditions and realize the aspirations of a more proactive strategic. The similar case of partnerships in the church, Holman (2009: 13) in Tearfund International Learning Zone (TEARFUND)'s study of The Churches in Disaster Response in Karoan, India, they found that the church has its own resources to offer but its work can be enhanced by Christian Non-Government Organizations (NGOs) coming alongside in partnership. Such partnerships have enormous potential, but there are risks involved too. Insensitive approaches from the NGO or unrealistic expectations from the church can do lasting damage to the longer-term work and witness of the church.

Thus, I defined that culture is the most important element in guiding the behavior patterns of a society. The society was reflected behavior patterns of habits which is believed and lived in community's life. Custom reflects the habits which are believed and implemented in communities' life. As supported by Bria (2004: 8) which states that every society in the world has a culture, and one of the components of culture is custom. He defines custom a reference that held as behavior guidelines of each member of the indigenous people and certain indigenous groups. As supposed by Choby (2010: 1a) defines cultural capital is as forms of knowledge, both tangible and intangible, that have value in each society in relation to status and power with a more practice-oriented definition of cultural capital as the rules for engaging other types of capital (human, economic, and social). Cultural capital defines how people (humans) engage each other (social) and their resources (economic). Whether the culture of an organization is good or bad—cultural capital is created when values, traditions, beliefs, and language become the currency to leverage other types of capital (Choby 2010: 1b).

Furthermore, community cultural resilience enables us to embrace other types of resilience as it has a unique coping system to absorb a problem. Community or cultural resilience is the capacity of a distinct community or cultural system to absorb disturbance and reorganize while undergoing change to retain key elements of structure and identity that preserve its distinctness (Healey 2006: 12). Werlen (2015: 19) adds the idea of culture in the context of sustainability is that culture is regarded as the central pillar of sustainable development. He explains that culture can be considered the elephant in the room for transdisciplinary research, which addresses complex societal problems and contributes to the development of more sustainable human resources.

On this book, I mentioned and described that custom can be also viewed from two sides (i.e., in terms of spiritual of the pastoral or religion institutions; church and in terms of customary institutions; the role of customary leaders).

To understand, the definition of religion capital is a sensitive matter since several researchers and authors have been tried to invent the idea of religious spiritual and religion capital but not included all elements in the universe. Thus, on this study, I defined religion capital into two definition sides; *firstly*, a spiritual and or a faith medium of several certain religions, which is generally also known or adhered as, of individual, ordinary people/communities faith as vertically in people spiritual forms, trust and their self-defense to the Creator of the Universe that creates a set of faith-resources (intangible and tangible things) and as horizontally within people spiritual forms, trust and their self-defense could serve and or solve their own cases.

Secondly, a spiritual and or a faith medium of several certain religions that vertically people are belong and believing to the Creator of the Universe, which is particularly adhere to the servants of the Creator of the Universe (i.e., priests, clergy, leader or servant of several certain religions), as to intangible and tangible serve and or solve the surrounding humanistic cases of people in the Universe in horizontally perspective based on the rule of the religion system of ecclesiastical structure or resources (i.e., capitals) they had.

Custom in terms of spiritual/religious views is that intended of environmental spiritualism, the role of religious institutions, in this case the church in providing services to the people of the fortitude of faith and steadfast struggled with trials and the threat of live there. Additionally, other assistances in sponsorship funds and volunteer power of the Church (nuns and priests /ecclesiastical structures). For instance, flooding occurred in the study area, the nuns gave assistance as a voluntary service to indirectly or intangible serve the people who are sick (treat) due to flooding, mass ministry/worship to the community and also the church's voice is known as one of information sources in addition to preparedness disaster team and head of village in every village. While church assistance in terms of financial or directly or tangible assistance is sourced from the Catholic Relief Service (CRS) which is the central institutions in channeling funds and agencies/institutions in the country and abroad to help process of recovery after flooding. The role of the church is very helpful in the recovery process and aftermath disaster. Yet, people need basic solidarity not only from humanistic assistantship from the Indonesian government and other institutions assistantship but also strong-supports on devotional matters. The idea of this is supported by Solnit (2009: 3) who explained in her book titled "A Paradise Built in Hell: The Extraordinary Communities that Arise in Disaster" that frequently, and often, ordinary people respond to disaster with self-sacrifice, humanity, kindness, and basic solidarity.

The devotional and religion solidarity appears on religious coping systems. Chatters et al. (2009: 9) describe in other problem context that it is critical to explore how religious concerns shape major areas of one's life, including the use of specifically religious strategies to cope with significant life problems. They had examined demographic predictors of attitudes regarding religious coping (i.e., prayer during stressful times and look to God for support, strength, and guidance) within a national sample of African American, Caribbean Blacks, and non-Hispanic Whites (National Survey of American Life).

In the context of religious coping system in West Malaka Subdistrict of Malaka District in East Nusa Tenggara Province, Indonesia, community has a well-established relationship with the religion leaders in order to support the manifestation of the religious solidarity in terms of tangible and intangible services. Tangible services include the religious leaders' preventive action, alerting or alarming copings, and recovery action to help the affected community of flood disaster, while intangible services include prayers and religious motivations for the affected community of flood disaster. These two kinds of services or solidarities are seen as a faith-based partnership and brothership.

The religion solidarity in the United States is proven by the "Nonprofit organizations, including faith-based and community organizations, play a vital role in both preparing for disaster, and in ensuring an inclusive and participatory community-wide recovery from a disaster", according to Partnerships for the Common Good, a toolkit produced by the White House office (Brown 2015: 5). In his book, he mentioned that The Georgia Emergency Management Agency (GEMA) formed praise and preparedness to provide congregations with resources to help reduce the impact of a large-scale emergency when a dozen tornadoes touched down in Bartow County in 2013, several faith-based organizations jumped in and played a key role in post-disaster recovery, and GEMA officials realized these groups were a key avenue for disaster preparation that they had overlooked. Praise and preparedness strives to help churches on three levels: preparing the physical facility in case an emergency takes place during a service; preparing church leaders and volunteers to spread the word about disaster readiness among their congregation through information and readiness toolkits; and providing church leaders and volunteers the training required to help the community during disaster response and recovery.

Furthermore, The White Office of Faith-based and Neighborhood Partnership (2009: 6) on the role of faith communities in disaster and building disaster resilient communities alluded on Integral to all phases of the disaster lifecycle are the support and services provided by congregations and faith-based social service and disaster programs. For instance, the backbone of mitigation education and preparedness training for all relief and long-term recovery operations is the work of religious organizations collaborating as interfaith or faith-based networks. This action could enlighten and revitalize the values of communities' faith as to cope with disaster.

Adding by Holman (2009: 8) in Tearfund International Learning Zone (TILZ)'s Programme, the work of disaster management can best be done when local churches and their communities work together with specialist Non-Government Organizations (NGOs) and relief agencies. On the previous research in Thailand, there are eight potential roles of the local church in disasters based on The Micah Network Global Consultation (2006: 16)'s report of consultation on the community.

1. Facilitator of community action; Support assessments/mobilization and Community meetings/committees
2. Connector with the wider world; Early warning information
3. Advocate on behalf of the marginalized; Envisioning for this role and Identification/targeting of most vulnerable
4. Provider of relational care and support; Prayer, pastoral support, hospitality, acceptance, forgiveness, hope, salvation Potential role of the local church in disasters
5. Community Peacebuilder; Reparation and reconciliation; Basic training and support
6. Influencer and Shaper of values; Teaching against unjust, high-risk cultural practices

7. Provider of resources; Communications infrastructure, personnel, leadership, buildings
8. Immediate responder to sudden onset crisis; Strong Volunteer base.

The same role has been proved by the role of church in the United States. Holman (2009: 8) in TILZ's conference in California-Nevada, the annual conference disaster response ministry in 2009 which has aims to facilitate the ability of the church to be in ministry to persons in need as a result of a disaster and in furtherance of the conference, they purpose of the ten plans (Holman 2009: 8 in TILZ) are:

1. To provide pastoral care and support to disaster victims, from congregants to community members.
2. To make individual and family disaster planning and training available to the congregation and to our immediate community.
3. To identify and mitigate the risks from likely disasters.
4. To protect church property and ensure the continuation or quick resumption of worship services and ministry tasks.
5. To identify and organize resources, equipment, and facilities of the church, including its membership, that could be of benefit in the time of a disaster and recovery.
6. To plan ways to meet spiritual, emotional, and physical needs during and after a disaster, especially for those persons with special needs.
7. To offer a "ministry of presence" in the event of a disaster.
8. To create a Disaster Response Team to be used in the event of a local disaster.
9. To work cooperatively with the appropriate conference units, Sonoma County VOAD (Volunteer Organizations Active in Disasters), and local government agencies in the identification of, advocacy for, and assistance in disaster recovery.
10. To provide direct communication from the church to the district and/or conference and hence to the national offices of the United Methodist Committee on Relief (UMCOR) and United Methodist Volunteers in Mission (UMVIM) to obtain needed resources and assistance or to provide available resources and assistance to areas in need outside of the immediate community.

I thus emphasized that community resilience reflected in community pattern of empowerment and having ability to adapt the threat of flooding as to not destroy or ruin the environment to become the source impact of flooding as a human-induced. Community has to keep the nature as well as doing good for others are reflected in the system of mutual assistance "*hakawak*" which means an implication of a good behavior of community toward the Creator of the universe.

As supposed by Bria (2004: 8) which states that environmental spirituals rooted in a vision of the universe and the place occupied by humans in nature. Hereinafter described, the processing environment needs to be implemented as possible like God Himself, is to welcome the life, honor and respect the rules and rhythm, as well as helps the universe to stay lush and wonderful (Genesis 2: 15) The Lord God took the man and put it in the Garden of Eden to work it and take care of it (qtd. The Holy

Bible New International Version 2011). It has a meaning that Adam and Eve were placed in the garden to cultivate and guard not to ruin it.

The second type of custom of a combined content and contextualization of cultural capital and religion capital is that custom in terms of local wisdom of community (the role of customary leaders). People in Timor Island, Indonesia, in terms of local wisdom has customary institutions. On Chapter 4, Article 43 of Indonesian law/regulation number 22-year 1999 qtd. in Illicit Discharge Elimination Programe, IDEP (2010: 17) stated that the Indonesian institution customary has been set the position and roles of community. The community in West Malaka Subdistrict of Malaka District in East Nusa Tenggara Province, Indonesia, has the system of mutual assistance "*Manaek/hakawak*" rooted in patterns of culture values (see also Chap. 4, the role of customary institutions in research methodology).

Legally, as supposed also by the Indonesian Constitution in 1945, asserts that the Indonesian State acknowledges and respects entity of customary law on community and their traditional rights of all are still alive and in accordance with the development of society and the principle of the unitary Republic of Indonesia (qtd. in IDEP 2010: 17). According to Bria (2004: 9), customary leaders serve to empower customary institutions, preserve, and grow the customs of the society in their region. One of the cultural values that need revitalization is to enlighten society so as not to be a destroyer of nature, but a nature conservationist or otherwise maintaining the ecological balance.

In connection with the environment is the existence of prohibitions or customary law "*Badu*" in terms of deforestation, forest clearing, and logging off the system that damage the environment, which in turn have an impact on the lack of land cover, causing catastrophic flooding. With the establishment of customary law, there are penalties for those who violate it. In relation to in terms of local wisdom of community resilience lies in the spiritual power of the environment which shaped a cultural entity in terms of local wisdom of community (i.e., customary leader's traditional house in the research location to invoke the power through betel nut or *"Sirih Pinang"* in Indonesian language and *"Fuik Bua"* in Belu language (*Tetum*) for the Almighty)).

2.7.7 *Educate and Promote the Importance of the Role of Political Capital for Building Community Resilience*

When seeking to measure community levels of political capital potential areas of interest may include the effectiveness of local government in addressing the needs and priorities of the community, voter participation, involvement of women, and minorities in political leadership and decision-making, interaction between formal government and traditional authorities, and transparency and accountability among government officials (The Association of Non-Governmental Organization, TANGO 2003: 9). Political capital in both human and social actions has been an individual and society capacities which can affect policy from the Indonesian government. Both individual and society has their own aspirations to affect the policy implementation form the Indonesian government. Political capital can be understood as the capability of households and individuals in a community to get their rights of having aid and assistance after disaster occurrences. Political capital consists of power relationships, as well as access to and influence on the political system and governmental processes at local and higher levels (TANGO 2003: 11). The level of political capital at the community level determines the nature of community participation in the process of policy formulation and implementation.

2.7.8 Assess the Legal or Law Capital by Involving the Community as a Decision-Maker

Legal or law capital could be assessed by involving the community as a decision-maker. The *law* or *legal* capital is defined as the strong component systems on the government' regulations and implementations, roles and rules are created and produced, monitored, and controlled and executed and punished by the legislative, executive, judicial or judicative councils and other related institutions of the country case which also implemented the trial politics into certain policy frameworks are related to humanitarian disaster relief case.

Related to regulation and rule of institution to provide community welfare in a basis of economic sustainable development in international scale, the United Nations has passed the ways to involve themselves in the international law of disaster relief including natural disaster, i.e., flood disaster. It can be proved through their efforts through *Reaffirming its resolution 46/182 of 19 December 1991*, to which contains the guiding principles for the strengthening of the coordination of emergency humanitarian assistance of the United Nations system, as well as all its resolutions on international cooperation on humanitarian assistance in the field of natural disasters, from relief to development, and recalling the resolutions of the humanitarian segments of the substantive sessions of the Economic and Social Council (The United Nations International Strategy for Disaster Reduction, UN-ISDR 2007: 1).

Furthermore, The United Nations has also tried to forefront their efforts to be inclusive across its policy frameworks where through *Reaffirming further the Hyogo Declaration, the Hyogo Framework for Action 2005–2015: Building the Resilience of Nations and Communities to Disasters* and the common statement of the special session on the Indian Ocean disaster: risk reduction for a safer future, as adopted by the World Conference on Disaster Reduction, held in Kobe, Hyogo, Japan, from January 18 to January 22, 2005, and bearing in mind that the implementation of the Hyogo Framework for Action is coming to an end in 2015 (UN-ISDR 2007: 1). Given that effort of the United Nations (UN), an example case of legislative framework is addressed by the affected community of flood disaster in Brazil. The legislative framework most directly addressing disaster risk reduction (DRR) in Brazil is found in its laws on civil defense, and this area of the law has in recent years been frequently amended, often to accommodate changes considered necessary so as to better respond to ongoing disasters (International Federation for Red Cross and Red Crescent, IFRCRC 2013: 8).

Likewise, in Indonesian context, the law or legal capital is still not well-designed and implemented yet for the affected community' capitals on resiliency systems. The civil defense took a major step ahead in legislative framework than Indonesian government does astill a minor action. The term of law related to regulation and rule of Indonesian institution which is expected to provide community welfare in a basis of economic sustainable development. The Indonesian government regulations number 7-year 2004 and number 42-year 1998 concerning the preparation of design patterns and resource management are so far in the process of implementation, as well as design patterns based on the Indonesian government regulation number 46-year 2008 clause 16 (Baru 2011: 23).

However, according to The Watershed Management Bureau of Benanain-Noelmina (2011: 8) that states explicitly in clauses 19, 20, 21, 22, and 23 provincial regulations, East Nusa Tenggara Province number 1 Year 2001 consecutive set of protected areas, protected forest area, the area that provides protection against the subordinate.

East Nusa Tenggara Province, Indonesia, has also enacted (*Panduan Daerah Aliran Sungai Terpadu,* PDAST or Integrated Planning of Watershed Guideline) documents which contained in the regulation number 5 of 2008 which includes the planning, implementation, empowerment, guidance, and controlling of the local government in East Nusa Tenggara Province, Indonesia (The Watershed Management Bureau of Benanain-Noelmina 2011: 9).

By looking at the Benanain River watershed as part of the space that has its own characteristics which is located across the three districts, namely (*Timor Tengah Selatan, TTS,* or South Central of Timor), (*Timor Tengah Utara, TTU* or North Central of Timor), *Belu* and West Malaka and crossing one's neighbor's country, Timor Leste Republic. Benanain River watershed was recently characterized by a high number of critical areas, in terms of the lack of land cover, as well as floods and droughts that occur seasonally changing every year, it is highly recommended that cooperation initiatives and plans related with "Integrated Planning of Watershed Guideline" that involves a participative of multi-stakeholder.

The is no law monitoring of the Indonesian government that sees and controls human activities in critical area, land cover area in upstream of Benanain River watershed. The human intervention in the upstream area would lead to a negative impact to river ecosystem protection. Even though disaster risks are beyond human (people) control toward nature (Sawada and Zen 2014: 3). It is essential to have a systematic legal framework to control the river. Thus, I may suggest that the need of the law glasses implemented by the Indonesian government is important to tackle the issue of watershed management.

As added by Riwu (2005) qtd. in The Watershed Management Bureau of Benanain-Noelmina (2011: 11) who mentioned other forms of savannah, farmlands, and forests are very vulnerable to fire, especially when land clearing, as the fire became the means of production. In other side, then, on specific to the priority areas related to the watershed have not been good in the implementation of integrated management.

The need of social cooperation and coordination of the customary institutions, i.e., the role of leaders of the culture in the community area and pastoral or religion institution in the community area is important and hierarchically essential in guiding people from the upstream area, middle-stream area, and downstream area where the research location, West Malaka Subdsitrict of Malaka District in East Nusa Tenggara Province, Indonesia, is located.

To analyze this case, I invented this law capital and put into these analyses in order to assess the government legal system in tackling three cases;

(1) The case of natural and physical factors of controlling, monitoring, and protecting Benanain River Watershed and Dike System Management within discussion on executive and legislative roles and rules.
(2) The case of human actions (i.e., human activities/intervention or man-made) that has less of the Indonesian government responsibility to protect and to give penalties and sanctions in related with the existence of customary legal system and conjunction to requesting a new legal system of the land use and human actions protections of the watershed resources management in the cross-border area between West Malaka, East Nusa Tenggara Province, Indonesia, and Timor Leste Republic, since Benanain River Watershed divides these two countries; and
(3) Last but not least is the case of financial or economic assistantship in terms of micro-scale insurance for several devastated aspects due to flood disaster, which is supposed is given by the Indonesian government for the community affected by a flood disaster, not only in the phase of ex-post of aftermath flood disaster but also in the phase of ex-ante or before flooding disaster occurred and during flooding disaster occurred.

Specifically, in its ongoing application in the field, the responsibility of the Indonesian government is that implementation of watershed management does address in a question out of how a systematically integration and management of the sectors from upstream area, middle-stream area to downstream area with consideration on biophysical and socio-economic conditions that exist in a watershed in achieving common goals. It is at the core of an integrated river basin management which applies the principle that embraces "one river, one plan, and one integrated management" (Riwu 2015 qtd. in The Watershed Management Bureau of Benanain Noelmina 2011: 28). Therefore, the need for the organization and integrated river basin management requires a comprehensive plan to accommodate the various stakeholders in a Benanain watershed (Riwu 2005 qtd. in The Watershed Management Bureau of Benanain-Noelmina 2011: 29). For its implication, a discussion and assessment about three cases are dealing with community and government of Indonesia on Focus Group Discussion (FGD) and Stakeholder Meetings (SMEs) has been conducted in Malaka-Timor, Indonesia.

2.8 Discussion and Summary

The concept of *resilience* in the twenty-first century is certainly not a new concept in the assessment of the coping systems that people possess who overcome hazard or risk of natural disaster (i.e., flood disaster), as already stated by Fleming and Ledogar (2008: 8a). The concept of *resilience* was used in the last century by Fleming and Ledogar, who defined *resilience*, an English word derived from the Latin for *springing back, or jumping back up*, to take on an additional preventive meaning, in part because it helped change the focus of research from pathologies to opportunities for supportive action. The idea of resilience would be more valuable if the traditional approach through local people's understanding could be applied and analyzed in order to find significant solutions in building community resilience.

Hence, this book implements and combines the notion of resilience in a conceptual framework (resilience as a process rather than an outcome) to form the invention of resilience in the contextual framework of social-human geography, corresponding to previous researchers' conceptual ideas. The nature of resilience is absorptive and supportive, as mentioned by Fleming and Ledogar, Paton and Johnston, Bergstrand et al. and Aldrich. In the process of achieving resilience, there is effort (i.e., *springing back, or jumping back up, bounce back and proceeding collective actions*) to recover from flood disaster in the short-term and long-term. This effort is understood as the process of absorbing *a force or positive stimulus from the influence of negative stimuli* by the flood disaster.

Furthermore, it is argued that resilience moves forward as a dynamic development process with a community's efforts to cope with flooding, supporting the idea of Luthar and Chichetti (2000) who said that the term *resiliency* is derived from *ego resiliency*, which differs from *resilience* in that the former is a character trait and the latter is a dynamic developmental process (qtd. in Fleming and Ledogar 2008: 2).

However, I also combine Luthar and Chichetti's ideas with Aldrich's ideas and considers collective actions which integrate the traditional approaches of the people with their collective actions. There is no obvious single-conceptual, *ego resiliency*, with cooperative actions amongst the community members in the study area to respond to flood disaster. Synthesizing as positive adaptation instead of adversity adaptation was helpful for the actor of resilience (*i.e., affected community due to flood disaster in the study area*) in driving their own efforts to cope with flooding. In this concept of resilience, Alwang et al. (2001) mentions that *resiliency* is the ability to resist downward pressures and to recover from a shock (qtd. in Thywissen 2010: 8). Alwang states that, in the sociology literature, *resilience* is the ability to exploit opportunities and resist and recover from negative shocks.

It can be concluded in the framework concept of human geography that resilience is thus a system which integrates a set of adversity and diversity from resilience itself (*how does a community use its assets of resilience to outgrow flood disaster)* and flood disaster itself (*how does a community cope with the risk of flood disaster using its own system of resiliency*). It is also understood as human abilities that create and drive their actions as to adjust with natural environment, i.e., flood disaster. This idea is supported by UN-ISDR (2004) qtd. in Thywissen (2010: 11). The UN-ISDR puts forward a contemporary approach from resilience to building resilience and this was acquired by the local people and positively affected their ability to deal with flood disaster. Yet, resilience in a system, community, or society that has potentially been exposed to hazards has a varying capacity to adapt by resisting or changing to reach and maintain an acceptable level of functioning and structure. This is determined by the degree to which the social system can organize itself to increase its capacity for learning from past disasters for better future protection and to improve risk reduction measures (UN-ISDR 2004 qtd. in Thywissen 2010: 11).

Acknowledging the ideas of resilience and building resilience, this study provides a new synthesis between these concepts, that resilience is hypothetically not a single definition that can be used explicitly regardless of the combination of conceptual and contextual frameworks that were studied by previous researchers.

This book, I used fieldwork with communities affected by flood disaster in the study area, i.e., Malaka-Timor, Indonesia, and combines the idea of previous researchers and elaborates them in a comprehensive synthesis that resilience can be packaged in conceptual and contextual frameworks which refer to the community's actions and include a coping system, adaptation, resistance, and maintenance.

References

Aldrich DP (2012) Building resilience: social capital in post-disaster recovery. The University of Chicago Press, Chicago

Alvarado MMC, Rendon ZA, Peres VDA (2020). Integrating public participation in knowledge generation processes: evidence from citizen science initiatives in Mexico. Environ Sci Pol J 114:230–241. Elsevier. Retrieved from https://www.sciencedirect.com/science/article/abs/pii/S1462901120300575

Associated Programme on Flood Management, APFM (2008) Organizing community participation for flood management: a tool for integrated management. Flood management tool series. The Associated Programme on Flood Management

Bahadur AV, Ibrahim M, Tanner T (2010) The resilience renaissance? Unpacking resilience for tackling climate change and disasters. strengthening climate resilience discussion paper 1. First published by the Institute of Development Studies of University of Sussex, The United Kingdom

Baru P (2011) Pengelolaan Sumber Daya Air Wilayah Sungai Benenain. Direktorat Bina Penatagunaan Sumber Daya Air, Atambua, Indonesia

Béné C, Godfrey-Wood R, Newsham A, Davies M (2012) Resilience: New Utopia or New Tyranny? Reflection about the potentials and limits of the concept of resilience in relation to vulnerability reduction programmes. The Centre of Social Protection: Institute of Development Studies (IDS)'s Working Paper 405; ISSN: 2040-0209 ISBN: 978-1-78118-091-4

Benson C, Clay JE (2003) Economics and financial impacts of natural disaster: an assessment of their effects and options for mitigation: synthesis report. Overseas Development Institute, London

Bergstrand K, Mayer B, Brumback B, Zhang Y (2014) Assessing the relationship between social vulnerability and community resilience to hazards. Springerlink, The United States of America

Birkmann J (2006) Measuring vulnerabilities to natural hazards: towards disaster resilient societies. United Nations University Press and The Energy Resource Institute (TERI), Tokyo, Japan and New Delhi, India: ISBN: 81-7993-122-6

Bria F (2004) Along the way to happiness: a perspective of Belu' people about the society's structures, culture, religiosity and tourism in Belu. Caritas Publishing House, Jakarta
Brown D, Kulig J (1997) The concept of resiliency: theoretical lessons from community research. Health and Canadian society research. University of Lethbridge, Canada
Brown J (2015) Preparedness and recovery for The Georgia Emergency Management Agency (GEMA): churches play a growling role in emergency management. Emergency and Management's News, The United States of America. Retrieved from http://www.govtech.com/em/disaster/Churches-Playing-Growing-Role-Emergency-Management.html
Buckle P (1998) Re-defining community and vulnerability in the context of emergency management. In: Paper presented at disaster management: crisis and opportunity: hazard management and disaster preparedness in Australasia and the Pacific Region Con ference, James Cook University, Centre for Disease Studies, 1–4 Nov 1998, Cairns, Queensland, Australia
Canadian Centre for Justice Statistics (2007) A comparison of large-urban, small-urban, and rural-crime rates 2005. Statistics Canada, Catalogue no. 85-002-XIE, vol 27, no 3. Minister of Industry, Canada
Chamlee-Wright E (2010) Expectations of government response to disaster. Public Choice 144(1):253–274
Chaskin RJ (2008) Resilience, community, and resilient communities: conditioning contexts and collective action. National Emergency Training Center. Childc Pract 14(1):65
Chatters LM, Taylor RJ, Bullard KM, Jackson JS (2009) Race and ethnic differences in religious participation: African Americans, Caribbean Blacks, and non-Hispanic Whites. Ethnic Racial Stud 32:1143–63
Choby P (2010) The social capital: what is cultural capital. Cobalt Group
Cutter SL, Barnes, Berry M, Burton C, Evans E, Tate E, Webb J (2008) Community and regional resilience: perspective from hazards, disasters, and emergency management. Hazard and Vulnerability Research Institute, Department of Geography, University of South Carolina, Columbia, South Carolina
Debray B, Affeltranger B, Vincent G, Coutto R (2007) Risk community profile: a tool to assess a variety of hazards on a territory. In: International symposium on loss prevention and safety promotion in the Process Industry, Edinburg, United Kingdom, p 7. Retrieved from https://ineris.hal.science/ineris-00976182/file/2007-113_hal.pdf
Devereux S, Dorward A, Guenther B, Al-Hassan R, Poulton C, Sabates-Wheeler R (2008) Linking social protection and support to small farmer development. IN: A research paper commissioned by food and agriculture organization of the United Nations—FAO. Institute of Development Studies and London: School of Oriental and African Studies, University of London, Brighton
Ebers M (2015) Interorganizational relationships and networks. In: James D (ed) Wright, international encyclopedia of the social & behavioral sciences (Second Edition), Elsevier, pp 621–625, ISBN 9780080970875. Retrieved from https://doi.org/10.1016/B978-0-08-097086-8.73125-4 and (https://www.sciencedirect.com/science/article/pii/B9780080970868731254)
Erickson B (2003) Social networks: the value of variety. Contexts 2(1):25–31
Fiorini PD, Seles BMRP, Jabbour CJC, Mariano EB, Jabbour ABLD (2018) Management theory and big data literature: from a review to a research agenda. Int J Inf Manag, 43:112–129, ISSN 0268-4012. Retrieved from https://doi.org/10.1016/j.ijinfomgt.2018.07.005 and (https://www.sciencedirect.com/science/article/pii/S026840121830553X)
Fleming J, Ledogar RJ (2008) Resilience, an evolving concept: a review of literature for a relevant to Aboriginal research. J List PMC 6(2):7–23
Folke C (2006) Resilience: the emergence of a perspective for social-ecological system analyses. Glob Environ Change 16(3):253–267
Food and Agriculture Organization and International Labor Organization, FAO and ILO (2009) The livelihood assessment tool-kit. Analyzing and responding to the impact of disasters on the livelihood of people. Policy and Support Branch Communication Division. Geneva and Rome
Frankenberger TR, Nelson S (2013) Background paper for the expert consultation on resilience measurement for food security. In: A paper presented at the expert consultation on resilience

measurement related to food security. Food and Agricultural Organization and World Food Programme, Rome. Retrieved from http://www.fsnnetwork.org/sites/default/files/resilience_measurement_background_paper_2.12.2013.docx
Frankenberger T, Mueller M, Spangler T, Alexander S (2013) Community resilience: conceptual framework and measurement feed the future learning agenda. Westat, Rockville
Fridolin SB, Jax K (2007) Focusing the meaning(s) of resilience: resilience as a descriptive concept and a boundary object. Institute for Landscape Ecology, Technische Universität München, and Department of Conservation Biology, UFZ-Environmental Research Centre, Leipzig. The Resilience Alliance, Germany
Global Facility for Disaster Risk Reduction and Recovery, GFDRR (2012) Integrated flood risk management: non-structural measures in cities and flooding—a guide to integrated urban flood risk management for the 21st Century, pp 282–437
Gunderson LH, Holling CS (eds) (2002) Panarchy: understanding transformations in human and natural systems. Island Press, Washington, DC
Grootaert C, Van Bastelaer T (2001) Understanding and measuring social capital: a synthesis of finding and recommendations from the social capital initiative. Social Development Department: Social Capital Working Paper Series, The World Bank, Washington DC
Healey S (2006) Cultural resilience, identity, and restructuring of political power in Bolivia. University of Guelph, Canada
Hobbs G (2000) Social capital formation in Tanzania. DFIDs's Research Project: Department for International Development, selected studies of Civil Society in Tanzania: Policy, social capital, and networks of the vulnerable. Dares Salaal, DFID, pp 34–44
Holman SR (2009) Tearfund International Learning Zone—TILZ, Religion, Global Health and Human Rights. Published in the United States of America by Oxford University Press, New York
International Task Force on Global Public Goods (2006) Meeting global challenges: international cooperation in the national interest. A report of the international task force on global public goods, Stockholm, Sweden. ISBN: 0-9788790-0-7. Retrieved from https://ycsg.yale.edu/sites/default/files/files/meeting_global_challenges_global_public_goods.pdf
Illicit Discharge Elimination Program, IDEP (2010) Panduan Penanggulangan Bencana Berbasis Masyarakat: Buku Panduan Praktis untuk Sebelum, Saat dan Sesudah Bencana. IDEP Foundation Publication, Bali, Indonesia
International Federation for Red Cross and Red Crescent, IFRCRC, Legislative Advocacy for Disaster Risk Reduction (2013) International Federation of Red Cross and Red Crescent Societies, IFRCRCS. Legislative Advocacy Guidance Note Series No 1, Geneva
Jha MK (2010) Natural and anthropogenic disasters: an overview of vulnerability, preparedness and mitigation. Springerlink, The Netherlands. ISBN. 978-90-481-2498-5
Kasperson J, Kasperson R (2005) The social contours of risk: Volume II: risk analysis, corporations, and the globalization of risk. Earthscan, The United Kingdom and The United States of America
Knutsson P, Ostwald M (2005) A process oriented sustainable livelihoods approach—a tool for increased understanding of vulnerability, adaptation and resilience. Goteborg University, Sweden
Krishna A, Shrader E (1999) Social capital assessment tool. The World Bank. Wash ington, DC
Lebel L, Manuta J, Garden P, Huaisai D, Khrutmuang S, Totrakool D (2004) Urbanization in the Mae Nam Ping Basin: are transitions in the Chiang Mai-Lamphun Corridor contributing to regional sustainability? USER working paper WP-2004-02. Unit for Social and Environmental Research, Chiang Mai University, Chiang Mai, Thailand
Limperg T (1922–1949) The function of the accountant and the theory of inspired confidence. The Social Responsibility of the Auditor; The University of Amsterdam, Limperg Institute, Amsterdam, The Netherlands
Luangaramsri P (2002) Redefining nature: ecological knowledge and the challenge to the modern conservation paradigm. Earthworm, Chennai

Mathias SL, Jaan V (2007) Otherness in questions: labyrinths of the self. Information age. A volume in advance in cultural psychology. University of Sao Paulo and Clark University. Information of Age-IAP Inc., USA. ISBN 1-59311-232-7

Mayunga JS (2007) Understanding and applying the concept of community disaster resilience: a capital-based approach. Department of Landscape Architecture and Urban Planning. Hazard Reduction and Recovery Center, Texas

Mani MS, Wang L (2014) Climate change and health impacts: how vulnerable is Bangladesh and what needs to be done? A research report of the World Bank Climate Change Programme and Health Promotion Unit of the Ministry of Health and Family Welfare, the International Centre for Diarrheal Disease Research, Bangladesh

McAslan A (2010) Community resilience: understanding the concept and its application. Torrens Resilience Institute, Adelaide. Retrieved from http://sustainablecommunitiessa.files.wordpress.com/2011/06/community-resilience-from-torrens-institute.pdf

Marfai MA (2011) Community's adaptive capacity due to coastal flooding in Semarang Coastal City, Indonesia. Int J Seria Geografie 2:209–221

Ministry of Natural Resources (2007) Emergency management. Ministry of Natural Resources's Publication, Ontario

Morduch J, Sharma M (2002) Strengthening public safety nets from the bottom up. Dev Policy Rev 20(5):569–588

National Academy of Engineering (1996) Engineering within ecological constraints. The National Academies Press, Washington, DC. Retrieved from https://doi.org/10.17226/4919

Norris F, Stevens S, Pfefferbaum B, Wyche K, Pfefferbaum R (2008) Community resilience as a metaphor, theory, set of capabilities, and strategy for disaster readiness. Am J Commun Psychol 41(1–2):127–150

OJK, Otoritas Jaksa Keuangan (2014) The Indonesian financial service authority, roadmap for sustainable finance in Indonesia for 2015–2019. OJK's Assessment Report, Jakarta

Olsson P, Gunderson LH, Carpenter SR, Ryan P, Lebel L, Folke C, Holling CS (2006) Shooting the rapids: navigating transitions to adaptive governance of social-ecological systems. Conserv Ecol Soc Edn Pap 11(1):18

Onyx J, Bullen P (2000) Measuring social capital in five communities. J Appl Behav Sci 36(1):23–42

Organization for Economic Cooperation and Development, OECD (2009) Better aid: managing practices of DAC member countries. OECD Corrigenda Publishing

Pamlin D, Stuart A (2015) Global challenges: 12 risks that threaten human civilization: the case for a new risk category. Global Challenge Foundation Publishing. United Kingdom. Retrieved from https://api.globalchallenges.org/static/wp-content/uploads/12-Risks-with-infinite-impact.pdf

Paton D, Jahnston D (2001) Disasters and communities: vulnerability, resilience and preparedness: disaster prevention and management. Emerald Insights Publication, New Zealand

Pavanello S, Levine S (2011) Rules of the range: natural resources management in Kenya-Ethiopia border areas. Humanitarian Policy Group (HPG) Working Paper. Overseas Development Institute

Phillips BD (2009) Disaster recovery. Auerbuch Publications, USA

Putnam R (2000) Bowling alone: the collapse and revival of American Community. The production of social capital in US Counties. J Soc Econ 35:83–101. Retrieved from https://doi.org/10.1016/j.socec.2005.11.001

Rossi G, Harmancioglu N, Yevjevich N (1992) Coping with floods. Series E: applied sciences. NATO Advanced Study Institute of Coping with Floods, vol 257, Erice, Italy

RPJPN (2013) Rencana Pembangunan Jangka Panjang Nasional-Indonesia's National Long-Term Development Plan, "Green and Ever-Lasting Indonesia—Greener and Sustainable Indonesia" for the Period of 2005 to 2025. The National Document. Jakarta

Rupasingha A, Goetz SJ, Freshwater D (2006) The production of social capital in US counties. J Soc-Econ 35:83–101. Retrieved from https://doi.org/10.1016/j.socec.2005.11.001

Sawada Y, Zen F (2014) Disaster management in ASEAN. Singapore Institute of International Affairs (SIIA) and Economic Research Institute for ASEAN and East Asia (ERIA). ERIA's Publication, Singapore
Schoemaker D, Van Tilbrug R, Wijffels H (2014) If saving the world is not profitable. In: The missing ecological dimension of macroprudential supervision. A paper prepared for the UNEP inquiry-CIGI research convening on design options for a sustainable financial system, December 1–3 in Waterloo Ontario, Canada
Sexsmith K (2009) Violent conflict and social transformation: an institutionalist approach to the role of informal economic networks. Eur J Dev Res 2(1):81–94
Smith A, Stirling A (2010) The politics of social-ecological resilience and sustainable socio-technical transitions. Ecol Soc 15
Solnit R (2009) A paradise built in hell: the extraordinary communities that arise in disaster. Viking, New York
The Association of Non-Governmental Organizations, TANGO (2003) Socio-political analysis: critical information needs for addressing underlying causes of poverty and marginalization. Prepared for CARE, Draft
The Holy Bible New International Version (2011) Biblical. Retrieved from https://www.biblestudytools.com/genesis/2.html
The United Nations, UN (2009) Guidelines for reducing flood issues. National Academic and Atmospheric Administration (USA NOOAA) and The Swiss Agency for Development and Corporation (SDC) Publication, USA
The United Nations International Strategy for Disaster Reduction, UN-ISDR, Hyogo Framework for Action (HFA) (2007) 2005–2015: building the resilience of nations and communities to disasters. The HFA's research projects, in the Andean region, Central America, the Caribbean, Asia, Pacific, Africa and Europe
The United Nations Human Settlements Programme, UN-Habitat (2007) Enhancing urban safety and security. UN-Habitat series, Earthscan, London. ISBN: 978-92-113-1929-3
The United Nations Development Programme, UNDP (2013) Livelihoods and economic recovery in a crisis situation. Bureau for Crisis Prevention and Recovery Livelihoods and Economic Recovery Group, New York, ISBN: 978-921-1-126352-7
The United Nations Environment Program, UNEP (2015) Towards a sustainable financial system in Indonesia. The Association for Sustainable and Responsible Investment in Asia (ASrIA) and The International Finance Corporation, World Bank
The Micah Network Global Consultation (2006) The role of the local church in disasters and emergencies: a workshop's draft in 11–16 September, Thailand
The White House Office of Faith-Based and Neighborhood Partnership (2009) A toolkit: partnership for the good common. The Centre for Faith-Based and Neighborhood Partnership, The United States Department of Health and Human Services, Washington DC
The Watershed Management Bureau of Benanain—Noelmina (2011) Rencana Pengelolaan Daerah Aliran Sungai Terpadu di DAS Noelmina. Kementrian Kehutanan Direktorat Jenderal Bina Pengelolaan DAS dan Perhutanan Sosial, Kupang, Indonesia
The World Bank (2008) Building resilient communities: risk management and response to natural disaster through social funds and community driven development operations. The World Bank, Washington DC
Thywissen K (2010) Core terminology of disaster reduction. The United Nations University, Institute for Environment and Human Security (UNU-EHS), The United Nations
Uphoff N (2000) Understanding social capital: learning from the analysis and experience of participation. In: Dasgupta P, Serageldin I (eds) Social capital: a multifaceted perspective. World Bank, Washington, D.C.
Walker BH, Ludwig D, Holling CS, Peterman RM (1969) Stability of semi-arid savanna grazing systems. Ecology 69:473–498
Walker B, Holling CS, Carpenter SR, Kinzig A (2004) Resilience, adaptability and transformability in social-ecological systems. CSIRO sustainable ecosystem, University of Wisconsin Madison,

Arizona State University. Ecology and Society Journal—The Resilience Alliance, The United States of America
Werlen B (1992) Society, action, and space: an alternative human geography. Department of Geography, University of Zurich, Switzerland, Routledge International Publishing, Switzerland
Werlen B (2015) Global sustainability, cultural perspectives and challenges for transdisciplinary integrated research. Department of Geography, Friedrich Schiller University of Jena, Germany, Springer International Publishing, Switzerland
Wetterberg A (2004) Crisis, social ties, and household welfare: testing social capital theory with evidence from Indonesia. World Bank, Washington DC
Wieriks RHY (2013) The landscape of sustainability assurance towards further shaping the assurance approach. Eburon Academic Publishers
Woolcock M (1998) Social capital and economic development: toward a theoretical synthesis and policy framework. Theory Soc 27(2):151–208
Yusuf AA, Francisco H (2009) Climate change and vulnerability mapping for Southeast Asia. Economy and Environment Program for Southeast Asia (EEPSEA), Singapore

Chapter 3
Malaka-Timor, Indonesia

The distribution of 320 respondents was selected in this book's research project based on purposive sampling with field-observation using interviews in the research site, which is prone to flooding. The location of the research project can be seen in Fig. 3.1.

The distribution of sixteen respondents' maps in the research site depicts four geographical boundaries (i.e., East, West, South, and North) that classified each village by direction and borders. The classification of four geographical boundaries is presented below:

1. East: Motaulun, Naas, Maktihan, Lofoun, and Rabasa
2. West: Sikun, Oanmane, Motaain, and Umatoos
3. South: Besikama, Lasaen, Umaloor, Fafoe, and Raimataus
4. North: Rabasahaerain, and Rabasahain.

The description of the history of the geographic site in the West Malaka Subdistrict of East Nusa Tenggara Province, Indonesia, is an introduction to the community's socio-economic activities since floods began in 1939. It discusses the return period of flooding (*periodic flooding*) in connection with the story of the non-native people from *China-Malaka* who came to the research site for socio-economic and cultural reasons. These non-native people changed the pattern of socio-economic and cultural development through marriage with the local people, socializing with the community and the culture, occupying the agriculture area, and turning it into a settlement area, and establishing local economic trade through selling crops in the traditional market. These activities led to a new intervention by the non-native people concerning the local people's socio-economic and cultural activities in the research site and also changed the research site's spatial planning development, through a reciprocal relationship with the local people's socio-economic and cultural activities. This is an anthropological story, where the non-native people have been adjusting and influencing the local people's socio-economic and cultural values. Furthermore, the story of the community's socio-economic activities that utilize their natural asset (i.e.,

A. D. S. da Costa, *Building Resilience for Flood Disaster in Malaka-Timor, Indonesia*, Springer Natural Hazards, https://doi.org/10.1007/978-3-031-63272-3_3

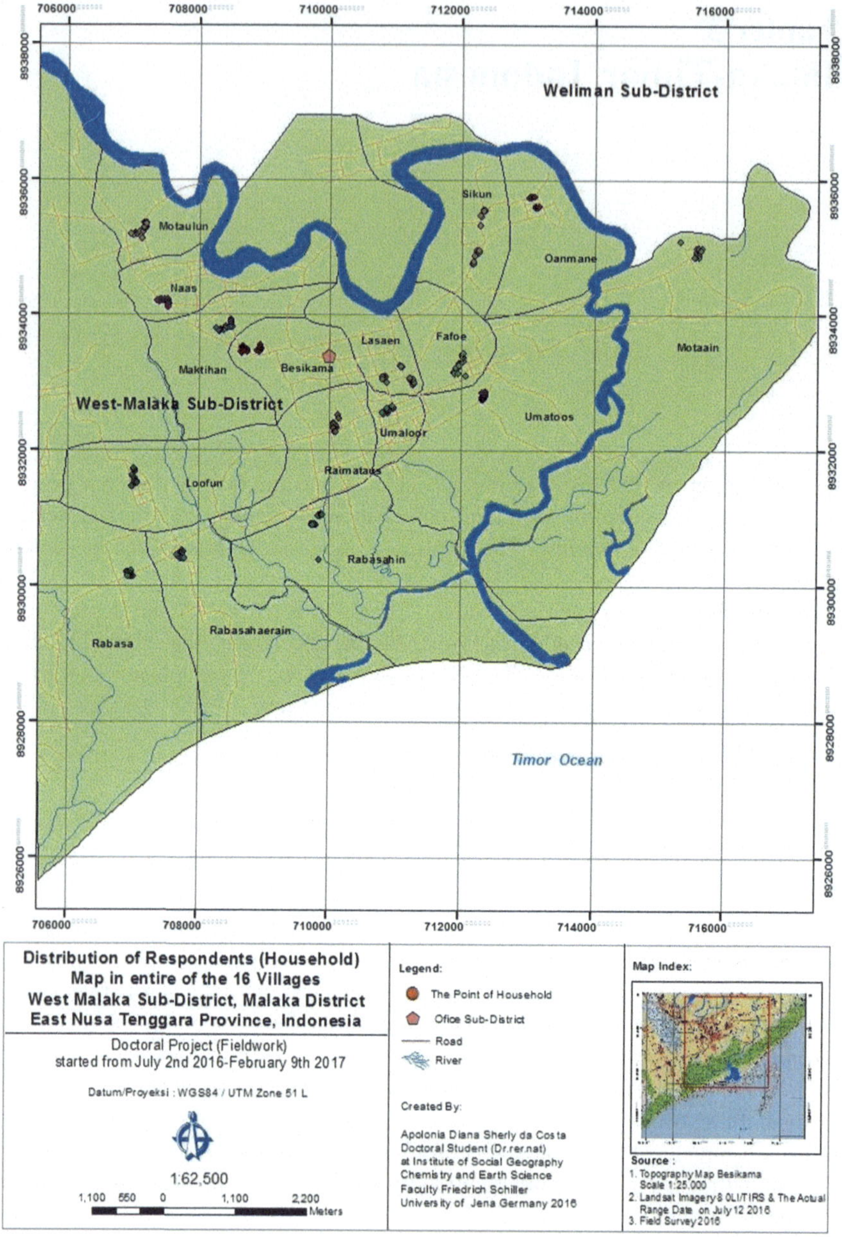

Fig. 3.1 Research location, i.e., the distribution of respondents in the 16 villages of the West Malaka Subdistrict, East Nusa Tenggara Province, Indonesia. *Sources* Author's research fieldwork, (survey and interviews), July 9, 2016 to February 9, 2017. Earth Explorer; https://earthexplorer.usgs.gov/ Landsat Imagery 8, USGS

crops) was described by an interviewee, *Roni Seran*, who said that the community's perception of flood disaster is as a "boon" not a "bane" as far as seasonal crops are concerned. This story was also told by two interviewees, *Patrisius Klau* in this chapter and *Leonardus Jefri* in Chap. 6.

In addition to the general flood story, a description of the failure of a dike in the nearby West Malaka Subdistrict in a part of the Belu District in 2011 was discussed. The key factor that influences flood mitigation through recognizing and understanding the socioeconomic characteristics was also discussed, as described by Brody et al. (2010). One of the most practical characteristics in the community's daily life, in terms of socioeconomic adaptation, is the local people's homes, which are constructed from simple materials like wood, bamboo, and a leaf rooftop. The local people's home construction, namely staged house or so-called *Rumah Panggung* or (*Uma-Fehan*) in their local language (Malaka and Belu Language: *Tetum*) enables them to adapt to the natural environment (i.e., flood disaster) and is known as one of their coping mechanisms that also allows them to store the output of cultivation and agriculture.

The discussion in this section describes an understanding of vulnerability which deals with the human ability to adapt to the natural environment and adapt the social economic environment to flood disaster. Flood disaster has been destroying the community's livelihood resources in the research site since 1939 without any resilience and flood disaster management by the Indonesian government in the site (see Chap. 7).

The physical and social characteristics of the research site are reported in population and administrative data in general and specifically from the East Nusa Tenggara Province which is divided into 24 districts (see Sect. 3.3.1). The data on Malaka District's population is discussed in Sect. 3.3.1 and details of the total population by Subdistrict in Malaka District is provided in a Table 3.1 (see Population by Subdistrict, in Sect. 3.1). The topography data and boundaries in Belu District are explained in the discussion (see Sect. 3.3.2) as well as the topography map of the Malaka District in Fig. 3.2.

In regard to the climate at the research site (see Sect. 3.3.3), the average temperature of 24–34 °C is typical of the tropics, with the rainfall in the Malaka District varying between 16 and 172 mm per month, indicating low rainfall (16–68 mm per month). Details of rainfall condition and the rainy days map in Malaka District are presented in Fig. 3.3.

The discussion of the low rainfall in the West Malaka Subdistrict and other Subdistricts in Malaka District highlights a contradiction where flood disaster happens mainly due to human intervention (i.e., socio-economic) in the nearby Benanain River watershed rather to physical factor (i.e., rainfall) (see Sect. 3.4 Physical and Social Characteristics of the Benanain River, in the next section).

The overview of Benanain River introduces the sites of the river area which is divided into two states, namely the Republic of Indonesia and Timor Leste. Details of the distribution of each area in the Republic of Indonesia and Timor are presented. The map of Territorial Division of Benanain watershed is presented in Fig. 3.4. The map of territorial division of the Benanain watershed was discussed according to provincial

Table 3.1 Population by sub districts in the *Malaka* district, 2013

No	Sub-districts	Population (people)
		2013
1	East *Kobalima*	6.285
2	*Laenmanen*	11.330
3	*Io Kufeu*	7.681
4	*Kobalima*	17.347
5	East *Malaka*	9.424
6	*Rinhat*	14.562
7	*Wewiku*	17.877
8	*Botin Leobele*	4.760
9	*Sasitamean*	8.225
10	West *Malaka*	20.030
11	*Weliman*	17.689
12	Centre *Malaka*	35.869
Total		171.079

Note Administratively, the Malaka District, which has a total area of 1.160,63 km^2, is divided into 12Districts and 127 villages and sub-villages. The research location, the West Malaka Region has 16 villageswhich is based on population, households, and population density in 2010. Detailed data is presented in Appendices.
Source Info Nusa Tenggara Timur, NTT 2011:39

data in the East Nusa Tenggara Province, Indonesia (see *Info Nusa Tenggara Timur, NTT* 2011: 40). The three districts administrations that cross the Benanain River watershed within each geographic coordinates are: South Central of Timor (Timor Tengah Selatan or TTS; it is known as the upstream area), North Central of Timor (Timor Tengah Utara or TTU; it is known as the Middle-stream area), and Belu district (it is known as the downstream area), which are located at the coordinates 124° 11′ 45.64″ east longitude, 125° 07′ 31.22″ west longitude and 8° 56′ 33, 21° south latitude and 9° 58′ 34.60″ north latitude. This section also introduces discussion of the Benanain River's condition and its tributaries, topographic and boundary, and land activities of socioeconomic development and sustainable communities in the whole ecosystem in TTS, TTU, and Belu District (i.e., West Malaka Subdistrict).

The overview of the physical characteristics in the upstream, middle-stream, and downstream area of the Benanain River presents the length of the river which is approximately ± 128 km. It is therefore known as the longest river in East Nusa Tenggara Province, Indonesia. Benanain River has nine sub-rivers, namely *Baen river, Mau Besi river, Bun river, Bikoni river, Muti river, Buna river, Fatu river, Noni river, and Besi river*.

The next discussion is on the social overview of the upstream, middle-stream, and downstream area of Benanain River and the border area between Malaka District, East Nusa Tenggara Province, Indonesia, and Timor Leste Republic. The social overview

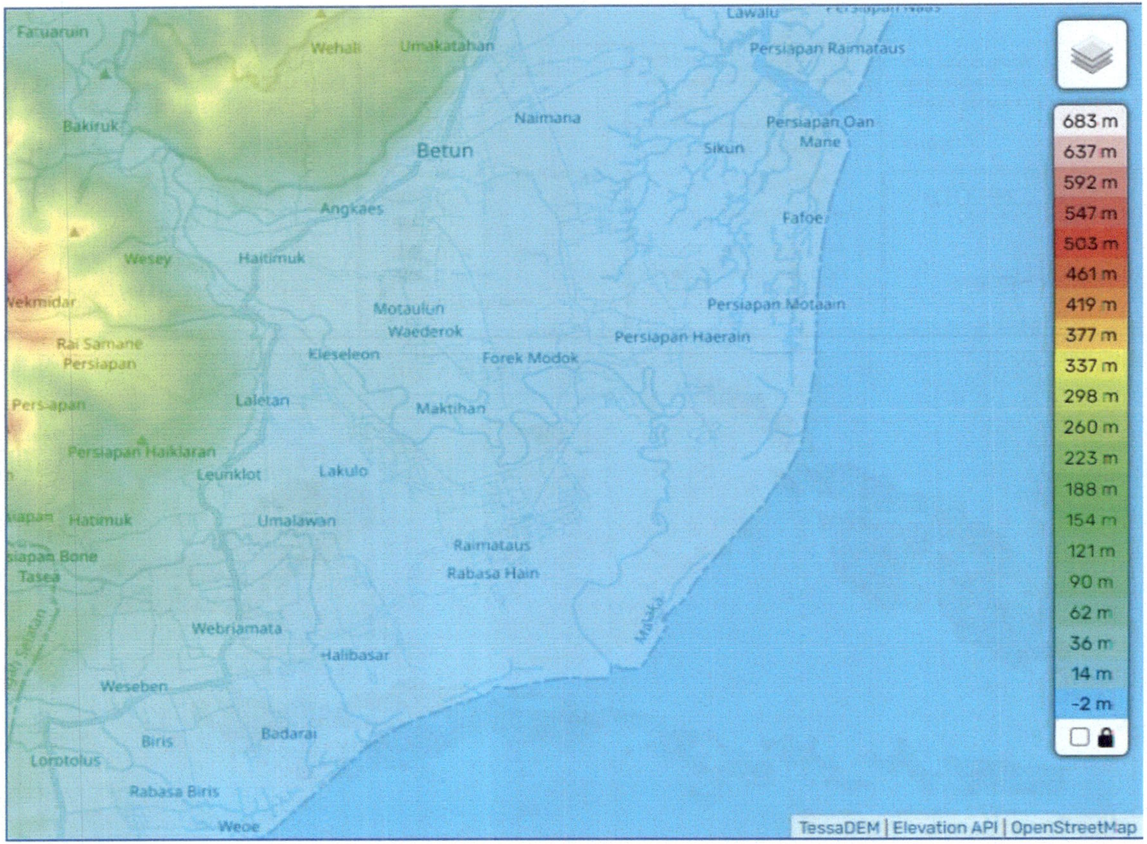

Fig. 3.2 Topography map of Malaka District, Indonesia. *Source:* Malaka Topography Map. Result for Understanding the Elevation and Terrain of the Region, Indonesia (2023)

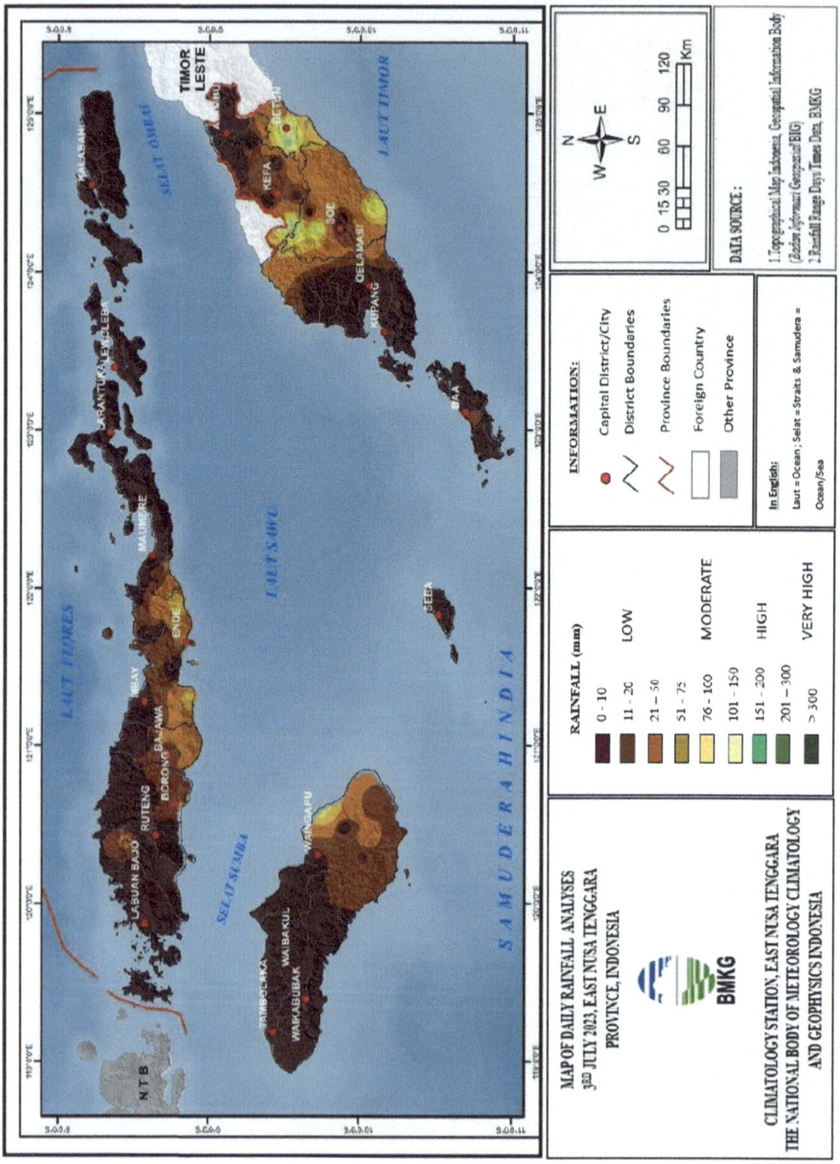

Fig. 3.3 Rainy days map of Malaka district, East Nusa Tenggara Province, Indonesia. *Sources* National Body of Meteorology, Climatology, and Geophysical, Indonesia; Map of Daily Rainfall Analyses July 3, 2023, East Nusa Tenggara Province, Indonesia, Climatology Station

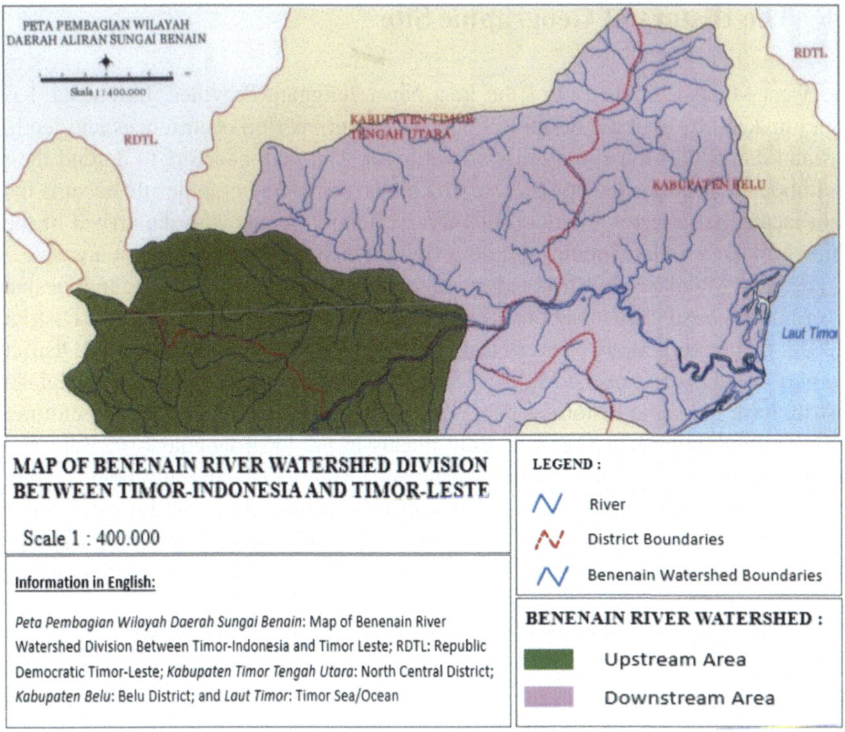

Fig. 3.4 Map of territorial division of Benanain Watershed. *Source* Badan Nasional Penanggulangan Bencana, BNPB, Indonesian National Disaster Management Body 2011:8

discusses environmental degradation and river floods caused by the intervention of people who are living in the area surrounding the Benanain River watershed, particularly in the upstream area. Such environmental degradation is described by Damayanti et al. (2010: 13), through Decentralisation as Contribution to Good Governance (DeCGG), which mentions that the Benanain River brings frequent flash floods. Such environmental degradation in the Benanain River watershed has been negatively impacted by other types of natural disasters due to potential climate change effects, such as landslides and drought. It leads to environmental resilience problems for the local people. The last discussion of this chapter presents the research site in overview and explains the suitability of plants (i.e., technical treatment), the forest, local people awareness, and land rehabilitation efforts carried out by people who live near to the Benanain River watershed.

3.1 The History of Geographic Site

The West Malaka Subdistrict of the East Nusa Tenggara Province, Indonesia, has been impacted by extreme flooding for twenty return period occurrences and led to loss and damage of the community's livelihood. The major sectors to defend their livelihood are farming and cultivation. The history of the geographic site became the main factor that affected the flood history. Its history begins with the arrival of the China Malaka who live around southern Belu, which is known as the flat areas.

The West Malaka Subdistrict belonged to the Belu District before the Indonesian government's legal decision about area extensions of the Belu District into a District in 2013. One of the extension areas is the Malaka District. The West Malaka District is a part or subdistrict of the Malaka District. Therefore, Belu District and Malaka District have the same historical geographic site that ties the local people, culture, and social circumstances together. Belu means in the local language of, Belu and Malaka, *Tetum*: "*Belun*" or translated to English, "friend." The migrants believed that this area could result in windfalls for their efforts, besides they could inherit. Since 1939, 1959, and 1979 floods that occurred in a periodic routine in the area indeed have no caused deaths and the economic slump (*Author's Fieldwork on September 22, 2016; Interviewee, Roni Seran*).

Local residents regard the floods as not *a bane but a boon.* In the years of 1974–1975 and 1999–2000, the largest death roll in the history of flooding occurred in the Belu district. This was during to the arrival of the East Timor refugees to the Belu District who altered patterns of behavior, especially when such a disaster strikes. The refugees from the Ainaro District (East Timor) were an indigenous community that did not want to flee to a safe and secure place during the flood disaster due to a belief (local wisdom). Some of people have ever left their area and they have withstood the rigors of life and death in such circumstances. (*Author's Fieldwork on September 22, 2016; Interviewee, Patrisius Klau*).

A large-scale flood occurred in the prone area to flood disaster due to a dike failure in the nearby West-Malaka Subdistrict in a part of the Belu District in 2011. In February 2011, the dramatic year of big floods began after the dike in this subdistrict broke. However, the community did a simulation of a flood and designed a practical assessment of resilience by building a local-levee barrier. The flood disaster could have been prevented with the traditional action of the community. Community perception has knowledge about the cycles/periods of flooding that occurred (20 years), precisely in 1939, 1959, 1975, 1979 and most recently in 1999–2000 along with the arrival from East Timor Refugees to the Belu District.

The period of 1999–2000, was the largest flood event that occurred in the Belu District, especially in the western Malaka Subdistrict which has killed over a 100 people.

In terms of socio-economic conditions, Marfai et al. (2007) define the socio-economic characteristics of the local community, as one of the key factors influencing the mitigation action regarding coastal hazard which can be seen by the way that they construct their homes. Likewise, in Belu generally and the West-Malaka

Subdistrict in particular, most of the local people's homes are constructed from simple materials like wood, bamboo, and a leaf rooftop. The Construction of their house is a staged house or so-called *Rumah Panggung* in Indonesian language or (*Uma-Fehan*) in *Tetum*. From the community point of view, this pattern of home construction typically distinguishes indigenous communities; it also allows them to adapt to the environment as a coping mechanism that helps to protect the crops of rice, maize, and tubers. The house structure of the *Rumah Panggung* can help them avoid being affected by floods. Considering this, I thus have attempted to formulate a comprehensive interpretation on vulnerability that combines several aspects of disaster, development, and environment, paving the way for understanding the role.

3.2 Physical and Social Characteristics

This research project of this book also found substantial information about the connection between physical and social characteristics in the research site that was provided by the local and national governments in Indonesia (*Info Nusa Tenggara Timur, NTT* 2011) & (Ministry of Public Work 2015). This information on physical and social characteristics is limited in the archived data provided by the Indonesian government. This empirical research uses a descriptive qualitative method, through survey and interviews and Focus Group Discussion (FGD) and Stakeholder Meetings (SMEs), where the community provides information about physical, socio-economic, and cultural characteristics in the study area. This information was used to synthesize an explanation and concept about the connection between resilience and floods, building on previous research and using the present study. Furthermore, this information is also important and relevant for understanding and assessing community social-economic and cultural activities in respect of the physical characteristics of the research site (see also Chap. 5). Thus, this research provides more detail in describing social-economic and cultural characteristics as discussed in Chaps. 5 and 6. This section also encompasses the collected population data and administrative data of the inhabitants in Malaka District (see Sect. 3.3.1).

A comparison of the highest and lowest populated areas of the Malaka District is provided in Table 3.1. Also, topography and climate data are presented in Sects. 3.3.2 and 3.3.3, showing the geographical position of the research site (i.e., coordinates: *longitude and latitude*), the boundaries of the river (i.e., Benanain River) and the mountainous characteristics with a flat area in the southern part of Malaka District; the flat area is the research site (i.e., West Malaka Subdistrict). The topography map of Malaka District is provided in Fig. 3.2. In addition to topography data, this research provides climate data, including temperature and rainfall in Malaka District (see Fig. 3.3, rainy days map). In regard to this physical data (topography and climate), the social characteristic provides substantial information about the nature of the people who live in the flat or downstream area (i.e., West Malaka Subdistrict). Their behavior is both to protect the ecosystem near the Benanain River and educate and protect themselves to adjust to flood disaster. Communities educate and protect themselves through valuing their cultural assets; practicing, obeying, and preserving their local cultural rules taught by their leaders (i.e., customary leaders).

These cultural values are unfortunately not considered sufficiently important by the people who live in the upstream area to lead them to be more aware about educating themselves and protecting the surrounding ecosystems in forests and Benanain River. They lack awareness due to the problems they face in water scarcity, which becomes a substantial constraint on socio-economic development in an area that is in between semi-rural and urban. The result is that the people in the upstream area are in a phase of transition to living in a modern society and they have no figurehead (customary leaders) who should be obeyed regarding the rules for protecting watershed resource management. Even though, there is an Indonesian government regulation or *peraturan pemerintah* (PP) (number 37 point a, b, and c from 2012) concerning *river watershed resource management and protection and environmental protection* that was provided by the Indonesian government, key local people (customary leaders) are also important in guiding and giving a cultural sanction to those who disobey the cultural rules (see also Sect. 4.2.9 in Chaps. 4 and 7). Moreover, the people in upstream area have no authority actors, such as customary leaders, who have full control of the people's intervention, unlike the people and their customary leaders in the downstream area (told by an *interviewee, Roni Seran*). Human intervention can be undesirable when there is yet no control, monitoring, or management by the related river watershed management institutions in Indonesia. The people in the downstream area rely on their awareness of the area's hazard and consistently practice collective actions (i.e., assets of resilience) (see Chaps. 4, 5, 6, 7, and 9 for more details about finding and discussion). The discussion on the physical and social characteristics of Benanain River is presented in Sect. 3.4.

3.2.1 Population and Administrative Data

The East Nusa Tenggara Province with an area of 2445.57 km^2 or 5:16% has an overall form of land which is divided into 24 districts. Based on data from the district, in 2013 the population of Malaka was 171.079 inhabitants. The subdistrict with the largest population was Centre of *Malaka* that had as many as 35,869 people, while the subdistrict with the least population is *Botin Loebele* that registered as many as 4,760 people. Details of total population by subdistrict in *Malaka* can be seen in Table 3.1.

3.2.2 Topography

Belu is located at coordinates 124° 35′–126° 12′ East longitude and 8° 57′–9°49′ South latitude. It has a strategic position because it is located at the junction of Timor

Leste with another state: the East Nusa Tenggara Province; the East Flores and the North Central of Timor also intersect in Belu. The boundaries of Belu are as follows:

Northern: Ombai Strait Southern: Timor Ocean.

Eastern: State of Timor Leste (*Republica democratio de* Timor Leste).

Western: North Central Timor and East South Timor.

The topography of Belu is indicated by flat area to rolling hills, mountain to the rivers, which flow to the north and south and follow the direction of the slopes. Rivers that flow from the south empty into the Ombai Strait and the Timor Ocean. Of the fourteen rivers which empty into the north, the Baukama River, Malibaka, and Talau are widely used for agriculture (*Info Nusa Tenggara Timur, NTT* 2011). Furthermore, the flat area is in the south and extends to the southeast coast of the Timor Ocean with a slope of less of 2%, while the flat area waves, undulating from 3 to 40% and almost evenly distributed throughout region, reaching 55.86% of the area.

The mountainous region is greater than forty percent of the total area, and it is contained in the central region, spread into the east with an area of approximately 17:40%. The topography map of Malaka District, Indonesia, can be seen in Fig. 3.2.

3.2.3 Climate

The area of the Malaka District with an average temperature of 24–34° C tropics generally changes in every half-year and altered from the dry season and the rainy season where the dry season is more prevalent. The Rainfall in this regency is varied between 16 and 172 mm per day. The low rainfall (16–68 mm per month) dominates the eastern part of the region of this regency, namely East Kovalima, Kovalima, Botin Loebele, East Malaka, Central Malaka, West Malaka, Weliman, and Wewiku with the width area 72.788,56 Ha. While Moderate rainfall (69–119 mm per month) is located in the western part of the region of this Regency, namely Rinhat, Io Kufeu, and Sasitamean with the width area which is 33.389,17 Ha. The high rainfall (120–172 mm per month) is located mostly in Laenmanen with the width area which is 5428.25 Ha. Details of rainfall condition in Malaka District can be seen in Fig. 3.3.

3.3 Physical and Social Characteristic of the Benanain River (An Overview)

3.3.1 The Sites of Benanain River

Benanain River area is in two states, namely the Republic of Indonesia and Timor Leste.

The distribution of each area in the Republic of Indonesia and Timor Leste is as follows:

1. Benanain River area in the Republic of Indonesia is 6460.12 km^2
2. Benanain River area in the State of Timor Leste is 3158.16 km^2
3. The total area of Benanain River is 9618.67 km^2.

The map of Territorial Division of Benanain River watershed can be seen in Fig. 3.4.

Geographically, Benanain River that is located in Timor Island includes three districts administration; South Central of Timor (*Timor Tengah Selatan or TTS*), North Central of Timor (*Timor Tengah Utara or TTU*), and *Belu* District, which are located at the coordinates 124° 11′ 45.64″ of east longitude, 125° 07′ 31.22″ of west longitude and 8° 56′ 33, 21° of south latitude and 9° 58′ 34.60″ of north latitude (*Info Nusa Tenggara Timur, NTT* 2011: 40).

Watershed is an area of land that is an integral part of the river and its tributaries, which serves to accommodate, store, and drain water from rainfall to the lake or the sea naturally, the limit on land is a topographic and boundary separation in the sea until the water area is still affected by land activities. Otherwise, watershed is the unity of the whole ecosystem from upstream to downstream of the main elements of the soil, vegetation, water, or air and has a major role in the economic development of sustainable communities.

East Nusa Tenggara Province *(ENTP)* with land area of 473.499 km^2 of the province island is made up of the 566 islands comprising 307 watershed areas with a diversity of biophysical properties and characteristics, the unit of land and natural resources.

The entire watershed area in *ENTP* is about 4.735 million ha, which is divided into Six Regional Management Unit (SWP) of Watersheds, namely Timor Island with the total of width is 1.512.079 ha, in Rote Island with the total of width is 128.011 ha, Alor Island with the total of width is 286.470 ha, Lembata Island with the total of width is 126.638 ha, Sumba Island with the total of width is 1.576.562 ha and the last is Flores and Sumba Basin with the total of width is 1.10524 million ha (*Info NTT* 2011: 40).

3.3.2 The Physical Overview of the Upstream, Middle-Stream, and Downstream Area of the Benanain River

The upstream area of Benanain River is in the South Central of Timor, while the middle-stream area is located in North Central of Timor and the downstream area is located in Belu and Malaka which empties into the south of Timor Sea. It has a length of approximately ± 128 km. Benanain River is one of the major and the longest rivers in East Nusa Tenggara Province and has relatively high flood peaks characteristic of the relatively short time. Some of the tributaries that flow into the Benanain River include Baen river, Mau Besi river, Bun river, Bikoni river, Muti river, Buna river, Fatu river, Noni river, and Besi river (*Info NTT* 2011: 41).

The width of the basin is 37.8 km. The slope of the stream is 5%. Then the watershed was elongated shape, dendrites stream pattern and density of stagnant flow. Based on data from Watershed Management Institute about Benanain and Noelmina Rivers in South Central of Timor flow along the entire land area of 126.136 ha watershed Benanain. The area that is known as marginal lands included in the protected forest area of 9.558 ha. Besides, the land area of 54. 213 ha is rather critical. While outside the forest area included in the agriculture area of 2.518 ha is critical lands and as much as 53. 308 ha and 6.539 ha are very critical (*Info NTT* 2011: 42). North Central Timor, which is known as the middle area between upstream and downstream has the overall plot of land along the basin of Benanain River, is 150.080 ha. From the area, the critical potential of land included in the protected forest area of 1.446 ha, land is rather critical is 18.743 ha and critical land is 43.081 ha. Meanwhile, outside the forest areas into agricultural cultivation in the region, the land area of 28.819 ha is rather critical, critical lands are 53.818 ha and 4.173 ha (*Info NTT* 2011: 42).

In Belu, the entire plot of land along the river basin area of 108.115 ha. The area has the critical potential of land included in the protected forest area of 2.651 ha. The land is rather critical is 17.542 ha, and the critical land is 5.491 ha. Meanwhile, outside the forest areas into the agricultural cultivation areas in the region, recorded a bit of critical land area of 18.228 ha, 53.260 ha, and 10.944 ha. This sad condition causes flooding every year in Benanain watershed (*Info NTT* 2011: 43).

3.3.3 The Social Overview of the Upstream, Middle-Stream, and Downstream Area of the Benanain River and the Border Area Between Malaka District, East Nusa Tenggara Province, Indonesia, and Timor Leste

The issue of Benanain watershed is a case for the local community who lived in its surroundings. According to Damayanti et al. (2010: 13) through Decentralisation as Contribution to Good Governance (DeCGG), together with the precious water and fertile soil, Benanain River brings frequent flash floods.

The environmental degradation in Benanain watershed has also intensified other disasters like landslides and droughts in three districts. With potential climate change effects, hydro-meteorological disasters, tropical diseases, and weakening environmental resilience are likely to exacerbate and become the leading obstacle to economic and social development in the area in the future. Further, on a broader scale, the issue is no longer limited to costs from natural disaster but extends to the continuation of environmental support capacity and services provision by the Benenain watershed. These functions have strong links to socioeconomic interests like water, food, and natural resources availability. Upstream activities can affect downstream areas positively or negatively, creating interdependency between districts. This flow of social benefits/costs across space and time is known as externalities.

The biggest damage to the watershed is due to anthropogenic factor of land conversion to farms. This is caused by the limited livelihoods available for the people besides agriculture and the high poverty there. The trend of increasing converging population near water sources also exerts rising pressure to the environment capacity. It becomes apparent that most issues center in which around population management, spatial and land use management, forest and soil rehabilitation, and communities' wellbeing.

Furthermore, biophysical aspects, as discussed by Damayanti et al. (2010: 14), the rate of degradation of land and water is very high. Watershed management issues of the biophysical aspects, among others: the extreme where the rainy season is much shorter than in the dry season which is supported by a high level of criticality of land, lack of forest, and land rehabilitation activities in the upper and middle basins have been made the downstream always get floods. The forest and land rehabilitation efforts are sometimes not noticed or considered the suitability of plants. It is proved by planting location also technical treatment of the watershed. Socio-economic aspects, that population which is inversely proportional to the area of land, economic demands and low public awareness of the environment led to pressure on the land. Riparian communities' dependence on natural resources is not comparable to public awareness of forest and land rehabilitation.

The length of the borderline between Malaka District, Indonesia, and Timor Leste is 21 kilometers (km). Topographically, along the border area is a hilly area and some areas are valleys with large seasonal rivers. Upmost about 80% area is a barren land with the shrub coverage *Malakakar* or dry land, while in the southern part of the area there is about 20% of arable land. Also, along the border area there is one cross-border post which is planned as a result of bilateral agreements and economic-trade, guarded by the officers of the Armed Forces, Police, Immigration, Customs and Quarantine, namely Cross-Border *Motamasin,* the Post boundaries in East *Kovalima* (*Info NTT* 2011: 45).

3.4 Discussion and Summary

The local people's socio-economic characteristics have been practiced and maintained by them since flood disaster began with a twenty-year cycle in 1939, 1959, 1979, and 1999, until an annual flood began from 2000, 2011, 2012, 2013, 2014, 2015, to the current year of conducting the fieldwork in 2016. The flood history was explained by two interviewees, named *Rony Seran and Patrisius Klau.*

In each period when a flood disaster strikes the research site, local people have independently applied local, practical mitigations from their own local practices for reconstructing a dike failure through a technical-traditional steps and ritual with the female leaders. Their local practice describes their adaptation to cope with flood disasters. In this section, similarly, the local people's adaptation includes their socioeconomic adaptation (i.e., staged house or so-called *rumah panggung* in Indonesian language or *uma fehan* in Belu and Malaka Language (*Tetum*). The staged house was built by local people with their own technical-traditional steps and was constructed from simple materials like wood, bamboo, and a leaf rooftop.

The local people in the downstream area of the research site have a good level of flood awareness and practical ways of understanding the context of vulnerability, but that is not the case for the people who live near the Benanain River watershed in the upstream area. Furthermore, vulnerability data is not available. Considering that the physical data and social-economic and cultural data about the community who live near the Benanain River in the downstream area (i.e., West Malaka Subdistrict) and in the upstream area (i.e., Belu District) are lacking, this chapter provides a report and discussion.

Looking at the community efforts and practices based on local adaptations, this chapter explains that little socio-economic and physical data is available from the Indonesian government about the socio-economic characteristics of the local people correlated with physical data in the upstream area, middle-stream area, and downstream area of the Benanain River watershed. The reliance on old data may mean that local issues are not getting the attention from the Indonesian government and humanitarian institutions in improving local people's resilience system for flood disaster.

References

BNPB, Indonesian National Disaster Management Body (2000) The topography map of the Besikama village of the West Malaka subdistrict, scale 1:25.000. Geo-spatial of the National Body of Disaster Management. Jakarta

Brody SD, Kang JE Bernhardt S (2010) Identifying factors influencing flood mitigation at the local level in Texas and Florida: the role of organizational capacity. Environmental Planning & Sustainability Research Unit, Hazard Reduction and Recovery Center, Department of Landscape Architecture and Urban Planning, Texas A&M University, TAMU. Original paper. Nat Hazards - Springer (2010) 52:167–184. https://doi.org/10.1007/s11069-009-9364-5

Damayanti F, Rusgiarto S, Riveira M, Riwu M, Hertz T (2010) Inter local government coorporation based integrated watershed management: Benenain River, Nusa Tenggara timur (NTT): decentralization as contribution to good governance (DecGG) in coorporated with German technical coorporation (GTC) and Ministry of Home Affairs. Non-Government's Publication. Kupang, East Nusa Tenggara Province, Indonesia

Indonesian National Disaster Management Body (2011) The Map of Territorial Division of Benanain Watershed, Timor Indonesia

Info Nusa Tenggara Timur-NTT (2011) Belu dalam Angka. East Nusa Tenggara Province, Indonesia

Malaka Topography Map (2023) Understanding the Elevation and Terrain of the Malaka Region, Indonesia. Retrieved from: https://en-gb.topographic-map.com/map-39sq9m/Malaka/?center=-9.62857%2C124.85915&zoom=12

The National Body of Meteorology, Climatology, and Geophysical (2023) Map of Daily Rainfall Analyses 3rd July 2023, East Nusa Tenggara Province, Indonesia. Kupang, Indonesia

The United States Geological Survey, USGS, Earth Explorer: Landsat Data Access. (2016) Retrieved from https://earthexplorer.usgs.gov/ https://landsat.usgs.gov/landsat-data-access

The Watershed Management Bureau of Benain Noelmina (2011) Rencana Pengelolaan Daerah Aliran Sungai Terpadu di DAS Noelmina. Kementrian Kehutanan Direktorat Jenderal Bina Pengelolaan DAS dan Perhutanan Sosial. Kupang, Indonesia

Chapter 4
Research Methodology

4.1 Discussion of the Research Methodology

This chapter involves research methodology in frameworks which I neatly organized. *Methodology in frameworks* of this research on this book encompasses research outline, data collection, fieldwork equipment, and research method, the research outline, data requirement and data source, fieldwork equipment, technique of sampling, research method of collecting the data, survey and interviews and research questionnaire, Focus Group Discussion (FGD), Stakeholder Meetings (SMEs), and Reflection and Assessment on FGD and SMEs. This chapter also discusses the behavior and instrument or sequence of methods of the empirical research as a whole (see Sect. 4.2).

The first introduction section of research methodology in frameworks is the research outline. It discusses three phase activities, namely pre-fieldwork (input process), ongoing fieldwork (amidst process), and post-fieldwork (output process), (see Fig. 4.5). The content of the research outline was completed in this research with the help of primary data in the fieldwork from the Indonesian institutions (see Table 4.2).

The collected data was derived from the Malaka District and the West Malaka Subdistrict, such as Non-Government Organization (NGO) and Government Organization (GO). This research uses the tools of collecting the primary data. They are Global Positioning System (GPS), a tape measure, a digital camera, and Landsat Imagery 8 (see Table 4.3). This section also presents the technique of sampling that this research used in the fieldwork and its reason, that is, purposive sampling. For discussion of the research method of collecting the data, the research project of this book provides flood and resilience data from respondents using survey and interview, research questionnaire, and FGD and SMEs (see Tables 4.1 and 4.2).

It also discusses the way the questions were delivered to respondents in the fieldwork using a questionnaire (i.e., a preplanned set of questions designed to yield specific information to meet particular needs of the research information) as

A. D. S. da Costa, *Building Resilience for Flood Disaster in Malaka-Timor, Indonesia*, Springer Natural Hazards, https://doi.org/10.1007/978-3-031-63272-3_4

Table 4.1 Data requirement and data source

No	Data requirement	Data source	Data Available	
			Yes (√)	No (x)
1	Resiliency data and community resiliency data	National, i.e., Indonesia and international books and journals	√	–
2	Flood report data	Local disaster agency, community-based disaster management, statistics body, Indonesia and River Region Centre, Nusa Tenggara II	√	–
3	Landsat imagery 8	USGS 2016 (based on the actual of geographical condition by starting the research fieldwork from July 2016 to February 2017)	√	–
4	GPS (Global positioning system)	Private	√	–
5	Administrative map	Indonesian National Disaster Management Body	√	–
6	Demography data and statistics of population of West Malaka regency in figure	Indonesian Local Disaster Management Body, Community-Based Disaster Management and Statistics Body, Indonesia	√	–
7	Topography map	Ministry of Public Works and Natural Resource Management in Benanain River, East Nusa Tenggara Province in Indonesia	√	–
8	Hydrology map	Ministry of Public Works and Natural Resource Management in Benanain River, East Nusa Tenggara Province in Indonesia	√	–

supposed by (O'haire et al. 2011: 52). This yield of specific information refers to local socio-economic and culture characteristics in which these are enabling local community in entrusting their self-conceptions based upon the dominant role of the local community's leader (i.e., female leader) [see Sect. 1.3. number 1.3.17, about disaster management law; clause 26 (1:e) no 24 year 2007, and Sect. 1.3. number 1.3.10 point c about revitalization of community cultural capital and religious capital which discusses devolving decisions to the customary leaders and also Indonesian Constitution or *Undang-Undang Dasar*, *UUD*, 1945 clause 1 (18:b), regarding customary law and *UUD* number 6 from the year 2014 concerning the community's customary law qtd. in Mulyono (2014: 69c) and the synoptic policy phases theory explained by qtd. in Mulyono 2014: 68a)].

Hence, in delivering the questions for the interviewees, these were well-planned, set and guided by the chosen people (i.e., gatekeeper whose still has a family relationship) assigned directly by the female leader. Since the research site is perceived by the local people and their female leader as the cultural site (sacred place), the role of the

Table 4.2 Fieldwork equipment

No	Tool's name	Function	Specification	Picture	Source
1	Global Positioning System (GPS)	The use of GPS in my research is to know the position of household buildings (Collecting the geographic information's site from the flooded-villages)	Garmin GPS 60i		Private
2	Tape Measurer	The use of this measurement tool in my research is to measure the signed level of water that is inside the house of respondents in the dry season and for further assessment on respondent's human, physical, and economic resilience (i.e., house building or settlements). While, if in the rainy season during the fieldwork, the use of the tool is to directly measure the level of water that is inside the respondent's house	*Tajima*-Engineer *Tan*-Measurement Tool	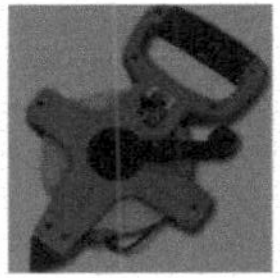	Private
3	Digital Camera-Video	The use of the digital camera in this research firstly is to capture several important aspects of geographic sites to derive old and new information from the wall print of flood mark/signs in respondent's homes, settlements, buildings, and public, prone areas. Secondly, it is used to create a short movie documentation for the FGD process, MEs process and a dissertation research movie about community disaster resilience in the prone area	Fujifilm X-A2 Brown + 16–50 OIS II		Private
4	Landsat Imagery 8, 2016	The use of Landsat Imagery 8 2016 in this research project is to map the households of the sixteen districts in flood hazardous areas as a result of visual data for analyses on community resilience	Resolution 0.5 m	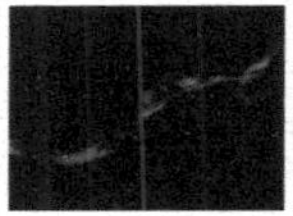	United States Geological Survey (USGS) United States of America (USA)

female leader in guiding the local people is therefore substantial (see Chap. 5 Discussion Section, Gender). The research project of this book also applied two practical methods in the research, namely FGD and SMEs in order to build a platform of the policy-planning process. It aimed at gathering policies information, supports and cooperation, as supposed by (The Policy-Oriented Marine Environmental Research in the Southern European Seas, PERSEUS 2012: 1) that mentioned the role of the stakeholder in community decision making processes and final outcomes in regard to building resilience for flood disaster.

This book introduces such discussion where the requirement of outcome of a stakeholder analysis was also important, as mentioned by (O'hara 2009: 5) and The Western Regional Air Partnership, WRAP (2013: 1b), where such participatory strategy is important and reliable in establishing group discussion so as to gain more useful results based on the people perception. This book uses and puts a flowchart into decision-making processes with the community and stakeholders in the group discussion as provided in Fig. 4.2, Reflection and Assessment on FGD and SMEs.

4.2 Research Methodology of the Book

In this book, I applied research methodology. It deals with a sequence of methods applied in fieldwork. The sequence of methods is the behavior and instruments that this research project of this book used in selecting, constructing research technique, and performing research operations such as making observations, technique of processing data, and the like. In this section, I present and discuss the behavior of the study as a whole. Methodologies in frameworks contain research outline, data collection, fieldwork equipment and research method (e.g., technique of sampling, survey and interviews, questionnaire and FGD and SMEs).

4.2.1 The Research Outline of the Book

The research outline of this book draws three phase activities of conducting the research. It is designed into a flowchart framework that involves three major activities of pre-fieldwork, ongoing fieldwork, and post-fieldwork. The research outline is shown briefly in Fig. 4.1.

4.2.2 Data Requirement and Data Source

For the data requirement and their resources, I was contacting and meeting directly the Indonesian institutions, and accessing the secondary data during the fieldwork

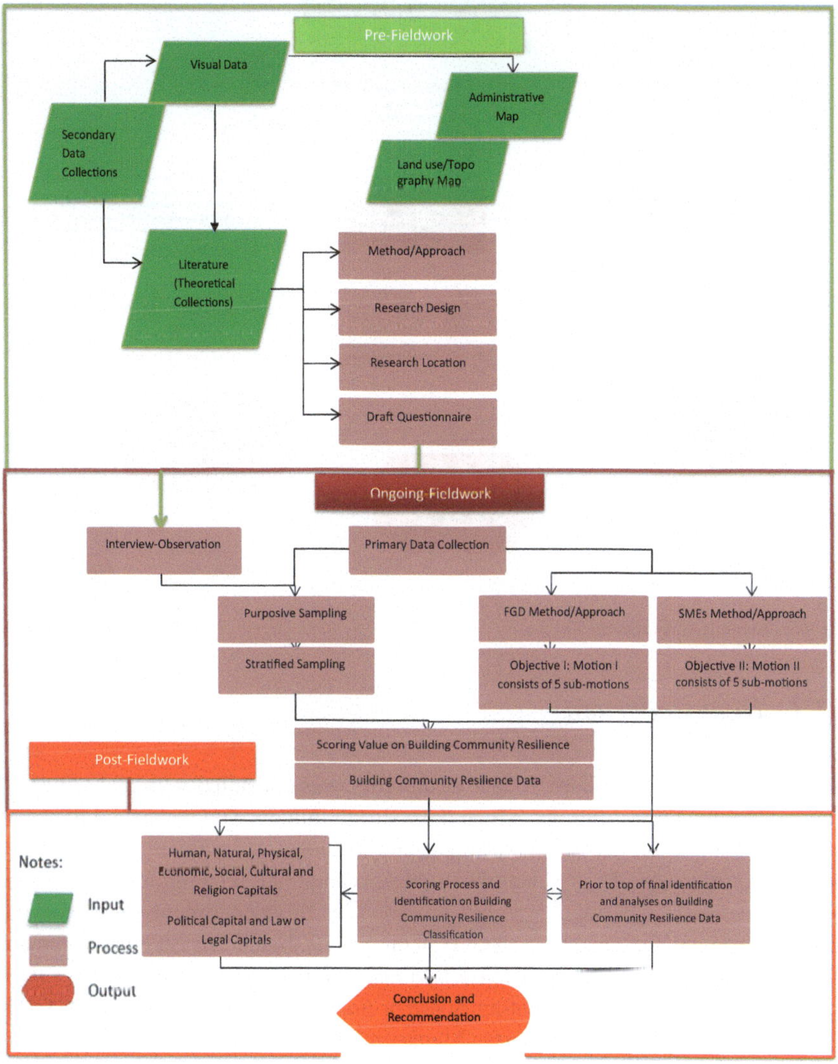

Fig. 4.1 Research outline

in Malaka-Timor, Indonesia, and also adopting the secondary data from the foreign institutions (Table 4.1).

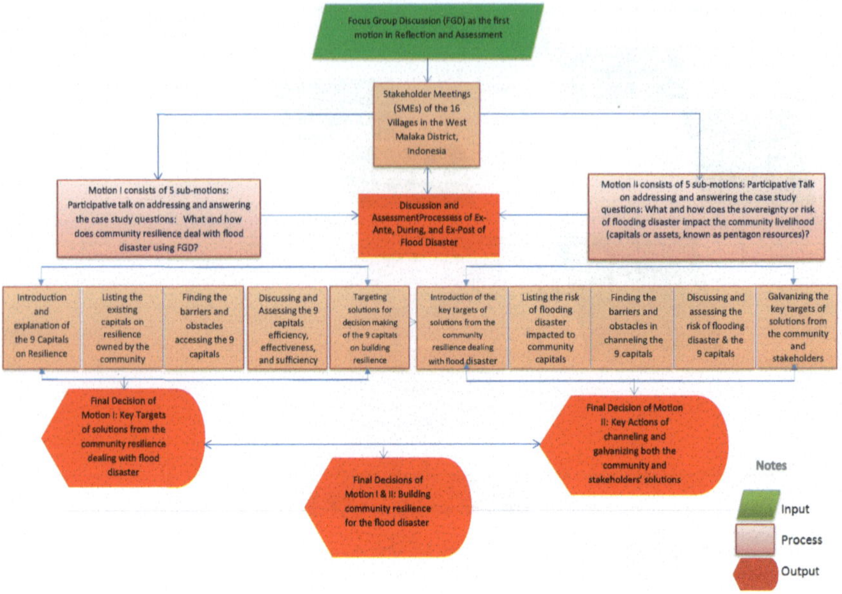

Fig. 4.2 Scheme of FGD and SME's

4.2.3 *Data Collection*

In the West Malaka Subdistrict, data availability was consulted with and collected from governmental institutions, which have strong influence on watershed management, including the Forestry Department and the local government. The Forestry Department manages the area actions of conservation and rehabilitation. However, the data of flood assessment was not provided by this department. However, the Non-Government Organization (NGO), Community-based Disaster Management (CADM) that was created by Illicit Discharge Elimination Program (IDEP) with supports from National Coordinating Board or *Badan Koordinasi Nasional* (BAKORNAS), Community-Based Disaster Management Guideline or *Panduan Penanggulangan Bencana Berbasis Masyarakat* (PBPM), Indonesian Society for Disaster Management or *Masyarakat Penanggulangan Bencana Indonesia* (MPBI), has several management goals related to disaster management dealing with preparation and prevention, disaster management, and disaster recovery for the affected community of flood disaster in the West Malaka Subdistrict of East Nusa Tenggara Province, Indonesia. The major data is sourced from the local community and the minor data is sourced from Indonesian institutions (i.e., NGO, CADM). The community, which is known as the source information for the building system of resilience and flood disaster, has also acquired the role of executive actions for their own cases. The information was not yet complete, since the executive actors (i.e., local authorities in the district level, subdistrict, and the village) have not yet taken their important role in coordinating and managing watershed management in the local level and regional level. Furthermore, some of data, such as the base map and the topography map are available from the Result for Understanding the Elevation and Terrain of the Region, East Nusa Tenggara Province, Indonesia.

4.2.4 Fieldwork Equipment

In this book, the tools that I utilized to collect the primary data are a Global Positioning System (GPS), a tape measure, a digital camera, and Landsat Imagery 8. The table of fieldwork equipment can be seen in Table 4.2.

4.2.5 Technique of Sampling

The kind of sampling that has been used in this research project of this book is purposive sampling. The reasoning for choosing this kind of sampling relies upon the judgment in selecting people in study area based on the target people of interest (See also my discussion I presented regarding why flood disaster affected each of the sixteen villages, and why the affected-villages/prone to flood disaster in Malaka-Timor were selected for my empirical research, and some discussions regarding the calculated radius of the Benanain River's distance from the settlements, as well as how adaptive the settlers living near flood hazard sites (Benanain River) are in implementing their resilience assets and coping mechanisms, in Chap. 7. "Community-Based Disaster Risk Management," Fig. 7.8 "River buffer class map," and Chap. 10 "Final Discussion, Conclusion of the Book, and Recommendation for Further Authors," Sect. 10.1.2 Finding of the Book's Part 2, Number 1 and Number 2). In addition to the reasons for choosing the location and the number of participants, this kind of sampling fits with the objective of this onsite-empirical research which can be observed in the research questions on this book. In my opinion, for this empirical research, purposive sampling focuses on particular characteristics of a population that are of interest, as in the characteristic of population referring to the people/community who are living there and have experienced/adjusted to flood hazards. The reality of the truth of experiencing flood hazards/disasters from local residents who are the sample of this empirical research is subjective from the point of view of residents and those of us who do not live in the flood hazard areas and experience the day-to-day application of resilience assets and mitigation strategies like them, but in their reality, they find their objective truth in the results of building resilience assets and mitigation strategies against flood hazards/disasters that we cannot deny. Thus, through my research, their risk perception is recognized, accepted, and acknowledged for local and global knowledge on building resilience and flood disaster management. Meanwhile, the area of interest here is the areas prone to flood hazard. Stratified sampling was chosen by me in order to sample each subpopulation (stratum) independently; this is the process of dividing members of the population into homogeneous subgroups before sampling. For the research of this book, the homogeneous subgroups of the population in the West Malaka Subdistrict of the East Nusa Tenggara Province, Indonesia, include all 16 flooded-villages (i.e., *Rabasa, Rabasahin, Umatoos, Fafoe, Sikun, Lasaen, Besikama, Umalor, Naas, Maktihan, Motaulun, Rabasahaerain, Motaain, Loofoun, Raimataus, and Oan Mane*) where within each of these villages, there were 320 respondents (per household) taken from the Subdistricts. The sample of subdistricts in each regency was taken purposively. The strata sampling seeks to be mutually exclusive and collectively exhaustive where every element in the population must be assigned to only one stratum and population elements are included.

4.2.6 Research Method of Collecting the Data

In collecting the flood and resilience data from respondents, I applied survey and interviews and research questionnaire, Focus Group Discussion (FGD), and Stakeholder Meetings (SMEs). The research method can be seen in Table 4.3.

4.2.7 Survey and Interviews and Research Questionnaire

Survey and interviews are conducted with an open framework which allow for focused, conversational, two-way communication. Structured questions are designed and phrased ahead of time and are typically asked of during all interviews. Questions are also created during the interview, allowing both the interviewer and the person being interviewed the flexibility to probe for details or discuss issues (O'Haire et al. 2011: 52). O'Haire et al. defined a questionnaire is usually concise with a preplanned set of questions designed to yield specific information to meet a particular need for research information.

In collecting the flood and resilience data from respondents, survey and interviews and research questionnaire were applied. I already wrote down and prepared the draft questionnaire in Indonesian Language or *Bahasa Indonesia.*

The community' information on social economy characteristics, livelihood, resilience experience, each capital or asset on resilience owned by the community, the community' risk knowledge, i.e., flood impact, flood consequence, flood experiences, flood perception, process of recovery and recovery system in the whole resilience assessments, types of mitigation actions that have been implemented whether by the Indonesian government and society were asked by applying in-depth-interview to each of the total 320 households in the 16 villages of West Malaka Subdistrict in West Timor, East Nusa Tenggara Province, Indonesia.

Table 4.3 Research method

No	Research aim	Data source	Method	Data analysis
1	To analyze building community resilience for the flood disaster *(Human activities on each capital of resilience deal with risk of flooding and build their system of resilience)*	Community Leaders, Key Informants and Households, stakeholders, The Government Institutions such as regional executive and legislative councils, The representative of the United Nations (UN) Local, The Non-Profit Organizations, The religious institutions, The representative key persons from the Local Bank, Businessman/woman	Focus group discussion (FGD) with community leaders, and governmental officers, The representative of the United Nation Local Officers, Civil Servant Questionnaire, survey of Household	Descriptive Analysis; quantitative (i.e., percentage of respondents' assets), spatial distribution analysis; and Scoring
2	To analyze the sovereign risk of flooding disaster impacted to community livelihood (capitals or assets, known as pentagon resources) (the risk of natural dimensions, i.e., flooding either could change or improve human perspective and experience on each capital of resilience to deal with)	Community Leaders, Key Informants and Households, stakeholders, The Government Institutions such as regional executive and legislative councils, The representative and key persons from the UN Local, The Non-Profit Organizations, The religion Institutions, The representative key persons from the Local Bank, Businessman/woman	Focus group discussion (FGD) and Stakeholders Meetings (SMEs) with community leaders and key informants Questionnaire survey of Household	Descriptive Analysis: quantitative (i.e., percentage of the flood risk impacted on respondents' assets); spatial distribution analysis; and Scoring

4.2.8 Focus Group Discussion (FGD)

Focus Group Discussion is useful to define local people's perception toward flood events. Community participation and application of local knowledge have the advantage of positively addressing the local socio-economic concerns in each village. It can also empower members with knowledge and skills, which can further strengthen their capacity to contribute to development initiatives. The need of participatory mapping is important to enable local people to share enhance and analyze their knowledge of life and conditions, to plan and to act.

On this research, FGD in the 16 villages (i.e., *Rabasa, Rabasahin, Umatoos, Fafoe, Sikun, Lasaen, Besikama, Umalor, Naas, Maktihan, Motaulun, Rabasa Haerain, Motaain, Loofoun, Raimataus and Oan Mane*) of West Malaka Subdistrict was attended by different key informants. One Sub-village was taken from the head of the Sub-village and The Head of the Sub-village so-called as *RT* and *RW* while others were taken from the head of the Sub-villages so-called as *Dusun* and the leaders of the customary institutions for the community so-called as *Pamong adat,* the leader or representative of religion in the study area, the people of the Non-Government Organizations (NGOs) and Government Institutions (GOs) and The Local and Regional of the Political and Law Councils or their Representatives such as the Leader's Representative of the Executive Council [i.e., *Majelis Permusyawaratan Rakyat (MPR)*] so-called as *Bupati,* the Leader's Representative of the Legislative Council (i.e., Local and Regional House Representative) so-called as *Dewan Perwakilan Rakyat (DPR),* the actors of fiscal Institutions such as Businessman/woman, Representative Staff of Local Bank, Investors, Sellers and Buyers of the local market from these 16 flooded-villages.

4.2.9 Stakeholder Meetings (SMEs)

Stakeholders can be called together at the beginning of a policy-planning process to explain the purpose of the policy and elicit support and cooperation (The Policy-Oriented Marine Environmental Research in The Southern European Seas, PERSEUS 2012: 1). Stakeholder engagement broadly refers to a framework of policies, principles, and techniques which ensure that citizens and communities, individuals, groups, and organizations could be engaged in a meaningful way in the process of decision-making that will affect them, or in which they have an interest. The outcome of a stakeholder analysis helps to determine which participatory strategy is suitable for which stakeholders. For some stakeholders, information and consultation may be adequate and efficient (O'Hara 2009: 5). In addition, the purpose of SMEs is to elicit greater involvement of stakeholders in the policy-making process to clarify the purpose and objectives of the process, to present findings or outcomes to the selected stakeholders from completed relevant steps in the policy cycle to inform, elicit comments and potentially identify necessary corrections. I formed the SMEs agenda based on the Focus Group Discussion' Motion of Assessment which has been established a meeting on SMEs. The FGD' actors of discussion are the same persons and representative of the related Institutions which is with and without interviewed community.

However, this is a SMEs, I focused on the decision-making process on stakeholders to have a final solution based on FGD' motions that concludes the final remark of research' assessment. The SMEs performs are mandatory for invited stakeholders who meet and share ideas of making decision and processing out the final solution. The Western Regional Air Partnership, WRAP (2013: 1b) added that the council officer attends a meeting of an established group as an invited guest where the meeting will be held in the usual meeting place for the group with recycling forming part of the agenda. An established group helps to make some useful results based on the people's perception. A focus group or workshop is usually based on a relatively small number of people who have been selected to discuss a single topic (WRAP 2013: 1a). In this study, FGD and SMEs' motions discussed the community's problem (i.e., building resilience for flood disaster). I play a role as a guide and facilitator to assist the people in delivering their ideas and opinions in FGD and SMEs. Facilitators assist the discussions using a topic guide and may record the session to ensure that information on all opinions expressed is captured. This method of consultation is useful at the start of the process to help develop and test ideas or to obtain feedback on the performance of a service at key points throughout its delivery (WRAP 2013: 1a).

In Indonesia, participatory strategy in obtaining information related to community's social-economic and cultural development was undertaken by the role of stakeholders (i.e., customary leaders and/or *pamong adat*). I also involved in the role of customary leaders in the decision-making process. In Chapter 4, article 43 of Indonesian law/regulation number 22-year 1999 stated that the Indonesian institution customary has been set the position and roles of community (qtd. in Illicit Discharge Elimination Program, IDEP 2010: 17). Communities in Malaka-Timor, Indonesia have a local customary institution that enables them to give authority regarding community's social-economic and cultural development to customary leaders. (see also, Chap. 2). It is evidenced by an establishment of the Indonesian regulation. Indonesian Constitution or *Undang-Undang Dasar*, UUD, 1945 clause 1 (18: b) normatively recognizes and respects the unities of society with customary law with its traditional rights (qtd. in Mulyono 2014: 68a). In regard to community's customary law which seems to self-governing community, the synoptic policy phases theory is introduced by Mulyono (2014: 69c); *"the synoptic policy phases gives direction that in the formulation of policies must implement methods of systemic thinking and where the aspects that can be handled and resolved by local institutions should be submitted to local creativity (local authority); while things that should be handled by the central government, the things should be pursued in the government center"*. Communities in the West Malaka Subdistrict upholds some traditional systems about obedience to the rule of customary leaders to protect the natural environment, social environment, and cultural environment which help the community to achieve a sustainable livelihood resource (Result of FGD and SMEs, see Appendix 6). The idea is again supposed by Mulyono (2014: 69a) who mentioned that "*Furthermore, Indonesia recognizes that the unity of indigenous or tribal peoples is a combination of genealogical and territorial in which the term of 'gemeinschaft' in Deutsch language is 'community' in English or 'paguyuban' in Indonesian language with emotionally tied characteristics, has a wider tradition, exists before the State, and its nature is bottom up*". This is different from the term of *"gesselschaft"* in Deutsch language or *"society"* in English *or "patembayan"* in Indonesian language and *"reinu bodik"* in Belu and Malaka language, (*Tetum*) *in terms of mutual assistance amongst society (hakawak*) with rational bound nature, autonomy, limited, after the State, and top down (Indonesia Constitution, 1945 number 6 from year 2014) qtd. in (Mulyono

2014: 69b). Mulyono adds his argument that: "*Therefore, the village in Indonesia is a community then ideally its approach is also a self-governing community; not dominated by the State or Local as local-government and local self-government approaches are insisted*". The idea of customary law with its traditional rights from the Indonesian Constitution, 1945 number 6 from year 2014 and the synoptic policy phases theory from Mulyono, reinforces that the need of an integrated and systematic policy framework through public participation and customary leaders is a substantial mode in the making decision process. Therefore, in this empirical study which involves key informants, local executive, representative of local legislative, representative of private and business sectors, representative of the local United Nations, Non-Government Organization (NGOs) and Government Organization (GOs), representative religious institutions, educators or teachers, public participation customary leaders in decision-making process in the FGD and SMEs are reliable sources to create solutions in framing a policy platform. Besides, given the role of customary leaders are significant and community obeys and values its role in the practicing of resilience actions to cope with flood disaster.

4.2.10 Reflection and Assessment on Focus Group Discussion (FGD) and Stakeholder Meetings (SMEs)

On discussion and assessment agenda are divided into two motions: firstly on FGD and secondly on SMEs. The first motion and the second motion are presented in a systematic schema on FGD and SMEs. The scheme of FGD and SMEs is presented in Fig. 4.2.

4.3 Summary

I concluded that the local people in the research site, i.e., Malaka-Timor, Indonesia, have a unique and strong cultural wisdom and described its evidence from using Focus Group Discussion (FGD) and Stakeholder Meetings (SMEs) that were found to be suitable in this study. This study provides data based on reliable information sources of secondary data and primary data.

For this book, the primary data is based on my field participation with the stakeholders and community through FGD and SMEs, survey, and interviews with 320 respondents (see findings and discussion based on survey and interviews in Chaps. 5, 6, 7, and FGD and SMEs in Chap. 9). Secondary data is from the Non-Government Organization (NGO) and Government Organization (GO) and the Indonesian government (see Chap. 3 Malaka-Timor, Indonesia). This chapter also explains that community risk perception regarding their resilience assets for coping with flood disaster produces some significant solutions. In addition, the use of FGD and SMEs helps the community to be more aware of their flood mitigation system and to create new resilience assets. Furthermore, the use of FGD and SMEs in this research project of the book can help eliminate a negative stigma stemming from ignorance by the Indonesian government and the humanitarian institutions and encourage them to be concerned about this case (i.e., community resilience for flood disaster).

References

Illicit Discharge Elimination Program, IDEP (2010) Panduan Penanggulangan Bencana Berbasis Masyarakat: Buku Panduan Praktis untuk Sebelum, Saat dan Sesudah Bencana. IDEP Foundation Publication. Bali, Indonesia

Mulyono PS (2014) Kebijakan Sinoptik Penerapan Hukum Adat Penyelenggaraan Pemerintahan Desa. Faculty of Law, The State University of Semarang. Paper: Yustisia vol 3 no 2. Indonesia

O'Haire C, McPheeters M, Nakamoto EK, LaBrant L, Most C, Lee K, Graham E, Cottrell E, Guise J-M (2011) Methods for engaging stakeholders to identify and prioritize future research needs. Methods future research needs report No. 4. (Prepared by the oregon evidence-based practice center and the Vanderbilt evidence-based practice center under contract No. 290-2007-10057-I.) AHRQ Publication No. 11-EHC044-EF. Rockville, MD: Agency for Healthcare Research and Quality

O'Hara P (2009) Enhancing stakeholder participation in national forest programmes: tools for practitioners. National Forest Programme Facility Food and Agriculture Organization of The United Nations

PERSEUS (2012) The adaptive marine policy (AMP) toolbox: "Oceans of Tomorrow". Greece

The Western Regional Air Partnership, WRAP (2013) Methods of stakeholder consultation. Information Sheet. WRAP. The United Kingdom (2013)

Chapter 5
The Social and Economic Data of the Households in Malaka-Timor, Indonesia

This chapter presents and describes the introduction of each component of the social economic data using survey and interviews with 320 households in the research location, Malaka-Timor, Indonesia. The components are gender, age, education, job, and religion, and they form the baseline of references for this project and for introducing the social and economic data of the households. Building resilience for flood disaster in Malaka-Timor, Indonesia, can be found through gathering this data (i.e., survey and interviews with 320 households). There is no supporting data on the research study of building resilience of affected communities facing for flood disaster in Malaka-Timor, Indonesia, provided by previous authors and researchers using survey and interview, either in Indonesia or other countries. Being a human geographer researcher/scientist for the first time in conducting this flood disaster studies-based empirical research with various interdisciplinary analyses and discussions in southern coast of Central Timor, i.e., Malaka-Timor, Indonesia, in addition to the reason of my origin from the land of East Timor (one ancestor with the Land of West Timor; *Wewiku-Wehali*), and also the high sense of humanity and moral responsibility of my profession that motivates me to introduce to the scientific world and the wider society wherever it is about this empirical research on building resilience for flood disaster. I pioneered the use of social and economic data as a baseline study of building resilience for flood disaster in Malaka. This research project of this book on building resilience for flood disaster is the first book in the national academic context (i.e., Indonesia) that considers social-economic characteristics as its baseline data. The social and economic characteristics in the context of the local geography need to be explored and acquired directly from the internal actors of resilience who possessed flood hazard, namely 320 households or representative communities in the research location, Indonesia. The study aims to introduce the social economic characteristics from the community in the research location to global academics who are concerned with the interdisciplinary book's project of building resilience for flood disaster or other types of natural disaster in other locations.

A. D. S. da Costa, *Building Resilience for Flood Disaster in Malaka-Timor, Indonesia*, Springer Natural Hazards, https://doi.org/10.1007/978-3-031-63272-3_5

Table 5.1 Gender of respondents

No.	Gender (female and male)	Frequency	Percentage (%)
1	Female	49	15
2	Male	271	85
Total		320	100

Source Author's Research Fieldwork, July 9, 2016 to February 9, 2017

The introduction of the socio-economic conditions of the community represents the actual conditions of local communities in the research location that applied their knowledge and experience in adjusting to flood disaster using their own system of built resiliency. The female role as a female leader and the role of Catholic religious institutions in responding to flood disaster and the community's own system of built resiliency are discussed in this chapter.

The representation of the actual of social and economic characteristics shows the condition of the local community in the research location, where its strength of purpose was part of the local character and developed in response to flood disaster with a humanist and socialist approach. This approach can be related to global ideas and perspectives from both the northern and southern hemispheres and to the characteristics of local communities as evidence in those communities of a resilience actor which aligns with the values of humanism. In this chapter, each component of the social-economic conditions is described.

Firstly, the introduction to the social-economic conditions of the community in Malaka-Timor, Indonesia, can be placed in a global context, showing the unique aspects of the local system of disaster management through building community resilience where the gender role is noteworthy (see Sect. 5.1.1). The role of gender in the context of building community resilience is considered as one of the determinants in decision-making processes that guides the community to remain on the moral corridor or avoid the "lack of moral hazard" that interacts with the social, cultural, and natural environments. In Table 5.1 and Fig. 5.1, data on the gender of respondents shows that the percentage of women is lower than that of men, indicating that the role of women is more dominant in taking care of households while I was gathering the survey and interviews with 320 households in the research location. The purpose of discussing the dominant role of women is not to demonstrate the hierarchy and status of woman as the leader of household or any discord of authority, but rather to show a cultural identity reflected in the ability of women to play their role in putting the community's knowledge into practice. The context of understanding gender is interpreted as a "feminist political ecology" that integrates the role of women's advocacy and capacity in addressing the natural environment and natural disasters as well as the duties and responsibilities of women in the society. The explanation of women's advocacy is supported by the opinions of previous researchers as described by Enarson and Meyreles (2004: 13), Ajibade et al. (2013: 5), Neonbasu (2013: 320).

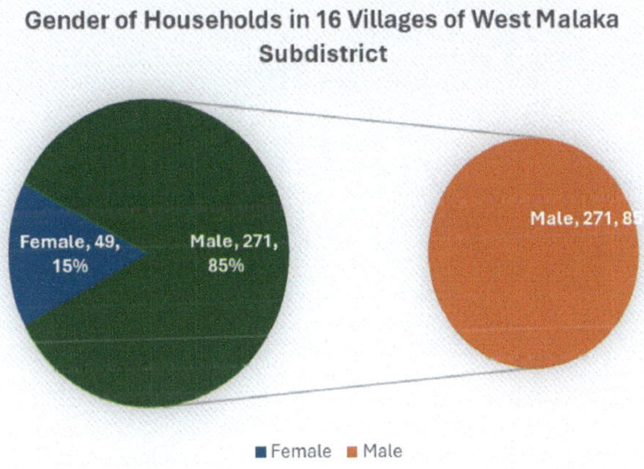

Fig. 5.1 Gender distribution overall households in 16 villages of the West Malaka Subdistrict. *Source* Author's Research Fieldwork, July 9 2016 to February 9 2017

Table 5.2 Age of respondents

No	Age (years)	Frequency	Percentage (%)
1	Young (< 35)	42	13
2	Middle (35–50)	166	52
3	Old (> 50)	112	35
Total		320	100

Source Author's Research Fieldwork, July 9 2016 to February 9 2017

Secondly, it is emphasized in Sect. 5.1.2 that the prevailing indicator of age in understanding the context of flood disaster and resilience capability is also a substantial part of assessment in this book. The assessment from the point of view of age uses three age categories:

young, middle, and old (see Table 5.2: Age of Respondents and Fig. 5.2: Diagrams of Respondents). The three age categories represent the vision and self-perception of the community in its development of its daily activities in dealing with flood disaster. However, differences in the views of each age group in the community's assessment (i.e., resilience ability to cope with floods) are not significant enough to affect the community's harmony in building a local system (i.e., local wisdom). This section also explains the flow of thoughts or perceptions in the community that support its location and realization of its own resources for integrated resilience in flood disaster management.

Thirdly, this section, I described the third social-economic component; education, that also became important to this study around building resilience for flood disaster. Educational levels are assessed from knowledge and skills on how the community responds to flood disasters with the strength of resilience it possesses. Education

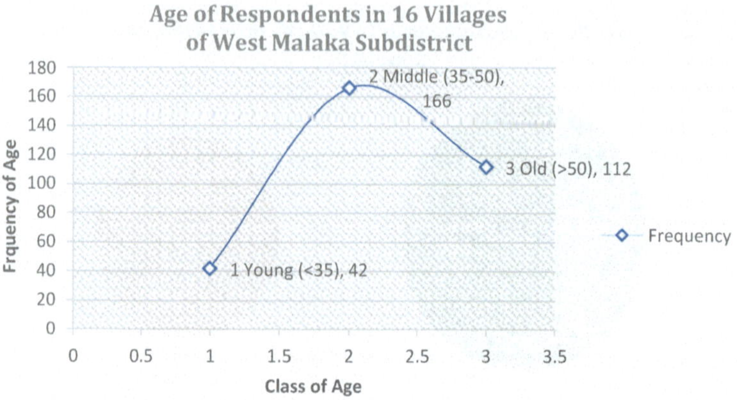

Fig. 5.2 Graphic of age of respondents. *Source* Author's research fieldwork, July 9, 2016 to February 9, 2017

can be obtained through the learning process and adaptation to flood disaster that generates knowledge and skills (see Table 5.3 and Fig. 5.3, Educational Level of Respondents in the West Malaka Subdistrict). The knowledge and skills possessed by the community in the research location are the foundation of its resiliency and strength in flood mitigation and resilience. Both community knowledge and skills can support each other in aid-responsive and flood mitigation efforts, as well as in resilience measures to cope with flooding. Discussions within the community with people who have different levels of education are presented in Table 5.3 (Educational level using survey and interviews with 320 households). The level of education is from non-schools, primary schools, junior high schools, and high schools to university graduates and university postgraduates. Different educational levels in the community determine different disaster mitigation and resilience responses. In addition, three factors in this section are also discussed: the economic, social, and cultural factors that accompany the community's forming of knowledge and skills as a platform for contributing to the community's build-resiliency system to respond to flood disaster.

Fourthly, job (source of income) is one of the key indicators in disaster management assessment and has a reciprocal relationship to community resilience in encountering floods. Sect. 5.1.4 describes each type of income that the community can access in overcoming flood disaster.

There are twelve types of income, namely civil servant, household, non-public worker, fisherman, fisherwoman, farmer, entrepreneur, honorary employer, house builder (workman), stonesman, car driver, and motorcycle service (see Table 5.4 Job, Source of Income of Respondents in the West Malaka Subdistrict). The introduction of these twelve types of community-owned incomes can help to understand the context of local people's lives in the research location where they were displaying their own characteristic pathways within a system that conforms to their natural environment. The 12 types of community-owned income show their efforts in seeking

Table 5.3 Educational level in the West Malaka Subdistrict

No.	Educational level	Frequency	Percentage (%)
1	No School	25	8
2	Elementary School	169	53
3	Junior High School	35	11
4	Senior High School	76	24
5	University Graduate	15	4
6	Above University Graduate	0	0
Total		320	100

Source Author's Research Fieldwork, July 9 2016 to February 9 2017

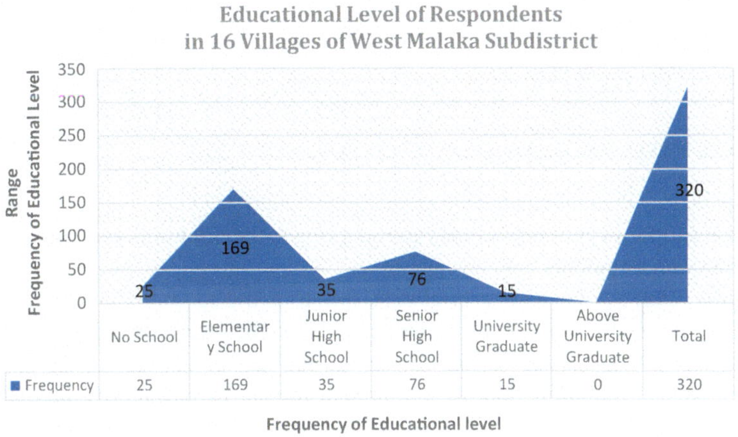

Fig. 5.3 Graphic of frequency of educational level of respondents in 16 villages of West Malaka Subdistrict. *Source* Author's Research Fieldwork, July 9, 2016 to February 9, 2017

a local resource-livelihood system for improving resilience for flood disaster. Income is one of the social-economic components in this study and presents the situation of rural communities in the West Malaka that have evolved in order to interact with their natural environment, while taking account of their homes' location close to the Benanain River watershed. I analyzed that this prominent feature of the natural environment is understood by the community as having a positive impact on their livelihoods and reflects the community's position in living with flood disaster while deriving a source of income from nature. Building a self-sustaining community that fulfills its daily needs and provides a resiliency system in the face of flood disaster is therefore one of the substantial issues in community self-reflection. The reflection of the community and these twelve types of income might be different compared with the lives of people in rural and urban areas that are affected by natural disasters, especially flood disasters. Such self-reflection also reflects the community's independent efforts

Table 5.4 Job (source of income) in the West Malaka Subdistrict

No	Job (source of income)	Frequency	Percentage (%)
1	Teacher	3	0.9
2	Civil Servant	13	4.2
3	Household	156	49
4	Non-Public Worker	1	0.3
5	Fisherman/woman	5	1.6
6	Farmer	123	38
7	Entrepreneur	13	4.2
8	Honorary Employer	1	0.3
9	House Builder (Workman)	2	0.6
10	Stonesman	1	0.3
11	Car Driver	1	0.3
12	Motorcycle Service	1	0.3
Total		320	100

Source Author's Research Fieldwork, July 9, 2016 to February 9, 2017

and social actions in earning incomes and adapting to their cultural, social-economic, and natural environments.

Finally, Sect. 5.5 presents a brief description of the religious institution's role (i.e., Christian Catholic) in which ecumenical services were provided by priests, nuns, and brothers in order to accelerate the process of alleviating the community's problems in handling flood disaster management. As the majority are Christian Catholic, the ecumenical aid system has been influential in improving disaster relief and resilience assets. Such assistance involves tangible and intangible aspects, which were concentrated with the role of the communities themselves, where the actors of resilience; priests, nuns, and brothers enabled minimization of the shortcomings of the community's resiliency assets while enhancing the community's resilience assets to floods. Thus, this section introduces the religious institution as a salient part of the community's build-resilience in response to flood disaster.

5.1 The Components of Social-Economic Conditions in Malaka-Timor, Indonesia

5.1.1 Gender

Gender is one of the benchmark assessments of community resilience in the West Malaka Subdistrict. The benchmark is based on the representation of local culture in which women have a special place as a leader in terms of their leadership of the households, employment (livelihood), and position at every level of society. They also possess the important role that can guide public morality in accordance with the local culture, which is matrilineal. The West Malaka Subdistrict is perceived by the community as a female land or *Ra'i Feto* in Belu Language (*Tetum*) which stands for *Wehali. Wehali* is located on a broad alluvial plain on the south coast of central Timor and situated on a plain *(fehan)* (Neonbasu, 2013: 320a). *Wehali* refers to *Tetum* live in the West Timor which is a matrilineal area while *Wewiku* refers to *Tetum* live in the East Timor which is patrilineal area or *male land or Ra'i Mane (an interviewee, Roni Seran). Wehali* stands opposed to the mountains (*foho*) where most *Tetum* live, and it is opposed to male land or *Ra'i Mane* (Neonbasu 2013: 320b). Neonbasu adds that the idea that all ultimate authority can be defined as female was equally difficult, if not wholly incomprehensible, Timorese idea for most European administrators (Neonbasu 2013: 320c). This cultural feature in the West Malaka Subdistrict in the West Timor Island is a unique picture of gender roles that determines the order of the cultural system which is applied from the household to the environment at large. In this area, women play a crucial role in guiding the community to seek a solution in case of a flood disaster. It might be termed as a community with a "feminist political ecology". *Feminist political ecology* integrates ideas about gender relations in specific environmental contexts with an emphasis on women's practical environmental knowledge and issues of gender inequalities, environmental degradation, and disaster vulnerability (Enarson and Meyreles 2004: 13). Gender has the rights to property, resources, space, and decision making and women's roles and responsibilities in the household and community (Ajibade et al. 2013: 5).

Advocacy of the woman's role in the family is also helpful for their husbands and children in the delivery of opinion and decision-making process and decision results. The gender of respondents is presented in Table 5.1.

As shown in Table 5.1 "Gender of Respondents" above, of the gender data that was captured, there are significantly fewer female participants. Through interviews with 320 respondents in the 16 villages it became apparent that women have limited time to engage in the interview; this is due to the role of women as a worker who takes responsibility in fulfilling household necessities. The frequency of interview data has shown that there were 49 females' respondents in every 49 homes that could be reached to provide information about their resilience system and disaster risk management. Meanwhile, 271 men were involved in this research study. Figure 5.1 shows the gender distribution of overall households in 16 villages of the West Malaka Subdistrict.

The observation that most respondents being men (85%) does not prove that the role of men is greater than women. Rather, it represents a specific role internally, in household terms (i.e., household task). Additionally, the low representation of female participation (15%) as shown in the diagram above does not indicate thoroughly that the role of women is relatively inferior, but instead, they are very strong in providing new ideas and informative things because of their local knowledge related to resiliency systems and flood disaster management. A household wife named *Martha Hoar Funan* (41-years-old) residing in Lasaen Village has candidly expressed her opinion about gender roles in the subdistrict level and specifically in her own village. She said:

> People in Lasaen village actually have similar gender roles with the 15 other villages in the West Malaka Subdistrict where women also have a strong role besides men in the internal management (i.e., as the advocates or fulfilling the needs of household necessities) and external management (i.e., as the stakeholders in the community, position holders and certain important positions in the public society), and also as the policy makers in taking one or several decisions related to community strategies to cope with flood disaster case by increasing the capacity or resilience and integrated disaster management solutions. So, no wonder that the predominance of women has become a benchmark of the strength and the role. Obviously, it has two positive effects both for the benefit of internal or private interest and external or public interests.

5.1.2 Age

In order to know the assessment indicator based on community perception, the understanding of a community's resilience system and flood disaster management could be described by the age of respondents. The age of respondents in the research location has been classified into three categories including young (> 35), middle (35–50), and old (> 50). The data is presented in Table 5.2.

Based on the table above, the middle-aged category (age 35–50) in this study is prominent with a total frequency of 166 respondents. It is followed by the old-aged category (> 50) with the total frequency of 112 respondents. While, lastly, the young-age category (< 35) was the less prominent by the total frequency of only 42 respondents. This indicates that the community is mostly populated by the middle-aged category and old-aged category. Meanwhile, the young-age category was not settled in the research location. The younger population is mostly living in the capital city of East Nusa Tenggara Province (i.e., Kupang City) and the other big cities in Java Islands, Western Indonesia. This is because the young age category desires to pursue their education; they explained their reasons for pursuing their education: In order to develop their education and get an advanced experience on education. From a different perspective, it is obviously described that the age of respondents has also represented the actual condition of social characteristic and education movements.

The social characteristic of the community in the research location is inseparable from their ways of living and attitudes about the social environment in the face of a challenging natural environment (i.e., flood). It is inadvertently triggering movement patterns of society's mindset; that is, the more rapidly a place is developing and advancing in technology the more malleable the younger-aged mindset is in thinking and pursuing a better education. Figure 5.2. "Diagram of Age of Respondents" is presented.

According to the interviews with 320 households, young people in this community tendentiously think more rationally than middle-aged and old-aged people. However, young people that were interviewed in the research location agreed that they will not ignore the source and characteristics of their culture that have prevailed and so far, been implemented by them. The movement pattern in the way of the young people and their inferred point of views can be seen and proven from the above data that there were only 42 (13%) young-aged respondents. The above data has shown a diversity of age of respondents in the way of thinking and implementing the ideas. The young people thought to develop their careers in the advanced and big cities in Indonesia without ignoring their distinctive culture. They have a vision to build and raise their culture and the place they settle although their place is a flood area. Whereas middle-aged and old-aged categories faithfully preferred to remain settled in the flood-prone area and did not want to move to other places due to their sincerity toward their ancestors. They have a self-perception that what they have been practicing is a manifestation of keeping culture. The revealed age data of the community is based on the factual research data retrieved through the interviews with the 320 respondents in the research location, i.e., Malaka-Timor, Indonesia, as presented from the 16 villages.

5.1.3 Educational Level of Respondents

Education is one of the things that is important for the assessment of community capacity in the face of floods in the West Malaka Subdistrict. This parameter enables one to measure the level of socio-economic vulnerability of the community. Education becomes a benchmark in the process of community resilience that can be used as a symbol of community strength. Both proficient knowledge and skills that are developed based on experiences bring good and helpful information. Specifically, within the research area, skills possessed by people who are natives of the West Malaka Subdistrict bring mitigation information and information assets of resilience. Knowledge can support the skills possessed by the people. The skills can support the knowledge possessed by people. Essential knowledge and skills that are possessed by the people are responsive energy and flood mitigation efforts and resilience actions of the community to deal with a flood. Social unity has functioned as power and the driver to social resilience of the community. The community in the West Malaka Subdistrict has different levels of completed education ranging from no school, elementary school, junior high school, senior high school to university graduate and above university graduate. Different levels of education possessed by people in the West Malaka Subdistrict can be seen in Table 5.3.

Based on Table 5.3 above, there are six levels of education possessed by the people in the West Malaka Subdistrict. People who are able to achieve the level of a university graduate are only fifteen respondents, or 4% of the total respondents that were interviewed. This indication of a small percentage is based on three indicating factors: economic indication, an indication of age, and cultural indications. The indication factor of economic communities that have a relatively low income to achieve university education is a logical reason why people are only able to achieve a level of education of senior high school; in this study there are 76 respondents or 24% of the total respondents interviewed who achieved this level of education. The indicator of the age factor is an obstacle for people who even already have the skills to adapt and process natural assets because they are reluctant to attend school at a higher level. Another underlying reason is that people above 20 years old believe that it will be their successors, not them, who will continue to a better education. Economic, age, and culture indicating factors or components are supported by an integrative culture which brings noble of cultural values of society in this subdistrict. In addition, the economic and age indications trigger people to not continue their education to a higher level is an indication that culture functions as a key trigger for people to preserve their culture without fleeing into another territory. Social culture is related to livelihood and social environment, blending natural environment and cultural environment. Natural environment in these terms refers to effective adaptation of the community to natural resources while cultural environment in these terms refers to obeying the local rules which co-exist in the research location. This community has created a sense of culture as the most important element of their capacity, building into the realm of social environmental and natural environmental aspects. Cultures maintained by the community were passed on to their children or their successors to achieve a better life through education in big cities or abroad (outside the area of culture-site, West Malaka Subdistrict). Thus, the three indicators and three aspects of the above can describe people who portrayed each percentage with different levels of education. The different educational levels also represent the level of community knowledge about the risk of flooding and proper understanding of the assets of resilience that they have managed against the floods in the West Malaka Subdistrict. As can be seen in Fig. 5.3.

According to Fig. 5.3, 25 respondents from this study were not in school. On the other end of the spectrum, the highest frequency in this study was 169 where the participants had reached the education level of elementary school. At the junior high school level, there were fifteen respondents who had undergone training at this level. Further, for a higher education university graduate or above, no respondents had ever reached and were educated at this level. Levels of education above show that the higher the education level, the higher the level of knowledge possessed by the community in terms of the process and flood mitigation measures and utilization of assets of resilience that is implemented for flood disaster relief. However, as discussed earlier in Table 5.3, in addition to the knowledge, skills support the knowledge possessed by the people. Skills are essential for flood disaster management and the utilization of assets from resilience. Thus, the knowledge that comes from education and skills which come from experience make a social-cultural force in support of the public good for steps in the process of adaptation flood disaster management and asset utilization process of resilience.

5.1.4 Job (Source of Income)

Income is the primary earnings possessed by the people to fulfill their daily needs, especially for flood disaster management and optimization of assets of resilience.

Communities in the West Malaka Subdistrict have a source of income and it can be seen in Table 5.4 Job (Source of Income) in the West Malaka subdsitrict.

Based on Table 5.4, it can be observed that the employment sectors of household (49%) and farmer (38%) are the major occupations that help to generate the main source of income for the locals. "Household" has the highest percentage of employment, opening more opportunities in this society for farming; in this society, there is the opportunity for farming capital in every area of 4.5 × 4.5 m in each home, and residents particularly farm corn but it is not put on the market to be sold. The 38% of total respondents in West Malaka Subdistrict are farmers. "Farmer" refers to those who farm corn and vegetable with a land area > 5 × 5 m. The land held by households and the results of cultivation and agriculture can be sold to traditional markets to access additional revenues for people's daily lives and to save for the purposes of prevention and mitigation and to improve the quality of utilization of the assets of their resilience.

The community also recognizes that a high level of skills in managing natural assets to economic assets is needed to support the process of flood disaster management and utilization of the assets of their resilience. There 4.2% of respondents had a job as either an entrepreneur or as a civil servant. This indicates that there were 13 respondents who had a job as an entrepreneur in the business of woven belt making where they market their capital for income. Meanwhile, 13 other respondents are employed as civil servants who work as village officials. The importance of these jobs in community resilience is to improve community human capital which assists them to fulfill the other capitals of resilience (i.e., economic resilience).

Furthermore, the next largest percentage of employment type is fishers. There are five fishers (three fisherman and two fisherwomen) with 1.6% of the total percentage of respondents in the West Malaka Subdistrict. These people are an example of those who are familiar with their natural environment by utilizing their natural assets in a conscious manner for the continuation of their lives; that is fishers utilize the asset of fish and shrimp ponds which bring an economic benefit for them. Furthermore, there are three respondents who are teachers or 0.9% of total respondents in the West Malaka Subdistrict. The scarcity of educators in the District of Malaka is longstanding and not a new issue although it is general knowledge that education is able to improve if it is supported with adequate teaching staff.

Finally, the people who hold jobs as house-builder are representatively small with 0.6%, followed by five types of jobs: non-public worker, honorary employers, stone man, car drivers, and motorcycle service. These last five sectors of employment are only 0.3% of the public interviewed in this research study. These types of employment support the building system of their resiliency to cope with flood disaster. Community has been seeking for the solutions of their resiliency and flood prevention through earning several incomes which represented into their types of employment. The more the community seeks income the less economic recovery should be fulfilled.

5.1.5 *Religion*

Religion has become a moral's mirror of people's attitudes in the implementation of their life in seeing, understanding, perceiving, experiencing, and behaving/treating the natural environment, social environment, cultural environment, and economic environment. I emphasize that the definition of religion in the context of the studies of flood disasters/hazards and people resilience in my book is that religion is not just an institution founded by the human for the purpose of teaching a set of practices and beliefs and adhering to them, but religion has also been a public force in reducing the adverse impacts on natural environment (i.e., a natural disaster like the flood) (See also Chap. 6 "Building Community Resilience Through Survey and Interviews in Malaka-Timor, Indonesia," Sect. 6.2.2 Religion Assets).

According to in-depth interviews, 320 respondents from Malaka-Timor, Indonesia, are Catholic. The majority people in this research location belong to Catholic religion due to a dominance of the Catholic in Eastern part of Indonesia. Geographically, the West Malaka Subdistrict which is in Timor Island is part of eastern part of Indonesia. A dominance of the catholic for the people in the research location has led to a positive impact where there are tangible and intangible assistantships were provided by Catholic in terms of pastoral charity. Tangible things are the direct assistance from the Priest, Nun, and Brothers of Catholic to support the alerting system before flood disaster (ex-ante), facilitate transportation in order to flee to a disaster shelter during flood disaster (during) and provide several services of recovery systems such as medical supports, foods, and drinks, after flood disaster (ex-post). These kinds of assistantships are known as the ecumenical services which were provided by the Priest, Nun, and Brothers of Catholic to accelerate community resiliency process in handling flood disaster management. Whereas intangible things are the indirect assistance from the Priest, Nun, and Brothers of Catholic to create a harmony soul and thought through prayers, worships to the Lord. It aims at helping the community to get more strength of spirits in order to remain utilizing their resilience capitals to deal with flood disaster. These two kinds of supporting things (i.e., tangible and intangible) are the manifestation of both the community and Priest, Nun, and Brothers of Catholic to minimize the lack circumstances of community capitals to endeavor for improving their resiliency for floods. Thus, this book denotes that religion is important for community resiliency process to deal with flood disaster management.

5.2 Discussion and Summary

This chapter discusses results for the five types of socio-economic components owned by the community: gender, age, education, job, and religion. An overall view of the social economic components shows that all the 320 households are Christian Catholic. There is no gap that stands out between a dominance of the Christian

Catholic and gender role while it is creating mutual cooperation. Where women become the female leaders for the community, including the phase of building resilience for flood disaster, the relationship of mutual cooperation in building resilience for flood disaster between Catholic religious institutions and gender roles is affected. The explanation is given by a household, namely *Martha Hoar Funan (41-years-old)* (see Sect. 5.1.1). She represents her role as the female leader in the household and her age in the middle age (range 35–50 years), with her job as a household wife who completed her study only to junior high school level. She is a Christian Catholic.

The content of the age's components, education, and job in this chapter shows a significant variation within each component's type. It can be observed that there is a significant total frequency of 166 households for middle age that dominate jobs as households, farmers, and entrepreneurs and who occupy a high frequency in the employment component. The educational component is dominated by households who graduate at elementary school and junior high school levels and none of them study above university graduate level.

This chapter concludes that Malaka-Timor, Indonesia, has a strong cultural environment as represented by the role of women in the community's daily life and the process of increasing resilience for flood disaster. Flood disaster is a part of community life that is manifested in the community's daily activities in carrying out resilience efforts that can be used to overcome flood disaster (see Chaps. 6 and 7). However, age, education, and job become issues that require the attention of the Indonesian government to assist the community's efforts in implementing their resilience assets for flood disaster. The factors of age, education, and job represent a reflection of the lives of an underdeveloped community in accessing jobs with higher incomes (communities are still friendly to nature and earn income through the fulfillment of daily needs sourced from the natural environment). In addition, it is also considered that the age of the community in the flood disaster area is dominated not by young people but by middle-aged and older-aged people. Young people are a less prominent part of the resilience toward disaster than older people who value their location as a traditional site that must be preserved in the process of attaining resilience for flood disaster.

Of all the components represented by the community, this study provides new academic evidence of the context of geography in Indonesia as part of the underdeveloped community's access to jobs and education. It depicts different characteristics in the living resources to northern hemisphere countries. In southern hemisphere countries such as Indonesia, the rural community upholds its cultural values and names these values as their cultural characteristics that are closely linked with the system of interaction with the social, economy, and natural environments. Every environmental interaction from the people becomes a reciprocal relationship that can be represented to northern hemisphere countries as their characteristic life that underlies their adaptation in Indonesia and increases their resilience efforts toward the flood disaster. The social geography context in this chapter becomes the most *substantial representative* or *baseline part* where the sample of research project of this book that was presented by the community contributes to a significant difference of the social economic components, a unique manner of living resources and a moral corridor of the social, cultural, economic, and natural environment.

References

Ajibade I, McBean G, Bezner-Kerr R (2013) Urban flooding in Lagos, Nigeria: patterns of vulnerability and resilience among women. Glob Environ Chang 23:1714–1725

Enarson E, Meyreles L (2004) International perspectives on gender, and disaster: differences and possibilities. Int J Sociol Soc Policy 24(10):49–93

Neonbasu G (2013) Kebudayaan: Sebuah Agenda dalam Bingkai Timor dan Sekitarnya. PT Gramedia Pustaka Utama. Jakarta, Indonesia. ISBN 978-977-22-9343-2

Chapter 6
Building Community Resilience Through Survey and Interviews in Malaka-Timor, Indonesia

In this section, I implement the main nation and concepts of the sustainable livelihood resources from the Flood and Agriculture Organization (FAO) and the International Labor Organization (ILO), which are also supported by Mayunga. The community sustainable livelihood resources are assessed and discussed in this chapter and Chap. 9. An understanding of sustainable livelihood resources was introduced by the FAO and ILO, The livelihood assessment toolkit regarding analyzing and responding to the impact of disasters on the livelihood of people, Policy and Support Branch Communication Division, Geneva and Rome, 2009, in their Sustainable Livelihoods Framework (SLF). This SLF mentions *pentagon* resources (five resources): natural resource (natural capital), physical reproducible goods (physical capital), monetary resources (financial/economic capital), manpower with different skills (human capital), and social networks of various kinds (social capital). Communities in the West Malaka Subdistrict of Malaka District in Timor, Indonesia, have implemented their five assets of resilience or so-called pentagon resources and nine assets of resilience or so-called *nine livelihood resources* instead of their *five livelihood resources or pentagon resources* as mentioned by Mayunga. This book discovers the community's *nine livelihood resources* which are one component of the community's *pentagon resources* (see Introduction and Understanding of each of the Assets in Building Community Resilience in Sect. 6.1.1 point one to point five). The community's *nine livelihood resources* are human assets, natural assets, physical assets, social assets, economic assets, cultural assets, religious assets, law, or legal assets, and political or policy assets. The four additional applicable assets of community resilience were owned and implemented by the community (see The Additional Applicable Assets of Community Resilience in Sect. 6.2 point one to four). Each of the nine assets of resilience is assessed and discussed based on previous authors' and researchers' concepts in connection to this book's findings through surveys and interviews with 320 households in Malaka.

A. D. S. da Costa, *Building Resilience for Flood Disaster in Malaka-Timor, Indonesia*, Springer Natural Hazards, https://doi.org/10.1007/978-3-031-63272-3_6

6.1 Five Assets of Community Resilience (*Pentagon Resources*)

An asset or capital is the result of efforts in the form of human wealth resources that aid in dealing with the risks of flood. Assets refer to the resource base of the people (Food and the Agriculture Organization and The International Labor Organization, FAO, and ILO 2009: 11). Communities in the West Malaka Subdistrict have created their assets or capitals and deployed them to deal with flood risks. There are five existing assets or capitals known as the *pentagon resources*: namely, human asset or human capital, natural asset or natural capital, physical asset or physical capital, economic capital or financial capital, and social asset or social capital. These existing assets or capitals were applied by the community before or ex-ante period, during and after or ex-post period of a flood disaster occurring in the West Malaka Subdistrict.

The Sustainable Livelihoods Framework (SLF) as defined by FAO and ILO (2009: 11) is represented as a *pentagon* that consists of five categories, namely natural resource (natural capital), physical reproducible goods (physical capital), monetary resources (financial/economic capital), manpower with different skills (human capital), and social networks of various kinds (social capital). The types of issues and details of these categories are covered as follows:

1. Human capital involves labor, health and nutritional status, skills, and knowledge.
2. Natural capital involves access to land, water, wildlife, flora, and forest.
3. Social capital involves those stocks of social trust, norms, and networks that people can draw upon to solve common problems. It is mediated through kin networks and group membership.
4. Physical capital involves houses, vehicles, equipment, and livestock.
5. Financial/economic capital involves savings, gold/jewelry, access to regular income, net access to credit, and insurance.

These capitals in sustainable livelihood resources assessment are also supported by Mayunga (2007) who defines five capitals for resilience through the sustainable livelihoods approach from which community resilience to disasters springs (qtd. in Bahadur et al. 2010: 5). Mayunga mentioned the determinants of resilience as social capital, economic capital, human capital, physical capital, and natural capital. Each of these five capitals corresponds to a number of characteristics of resilient systems.

A resilient community embraces integrity resources. To achieve sustainability of livelihood resources, a resilient community not only has to make good use of their resources but also to sustain them for long-term actions. In order to achieve a long-term action of community, it is important that the existing resources are introduced and discussed as they facilitate the process of assessment in a sustainable community and help to apply the resiliency system in the study area. Furthermore, based on the interview data collected in 16 villages of the West Malaka Subdistrict, it is revealed that the five assets or capitals (*pentagon resources,* i.e., human assets, natural assets, physical assets, economic assets, and social assets) had been implemented by the community. These pentagon resources are discussed further in detail below.

6.1.1 *Introduction and Understanding of the Human Assets in Building Community Resilience*

1. **Human Assets**

Human assets refer to human resources of individual, household, group community, and community to address a variety of problems such as natural disaster, socio-economic, and cultural conflicts, and upcoming negative events. Human assets are also called human capital. I define human capital as all human resources or efforts (i.e., skills, education, experience, and other capabilities) in order to protect from disruption, to recover from natural disasters, and to improve the resources for next sudden events. Human capital is the capacity and capability of human beings to contribute to socio-economic growth (Sustainable Development Policy Institute, SDPI and Pathways to Resilience in Semi Arids Economies, PRISE 2015: 16). They point to the need for resilience-building and investing in people to strengthen human capital. In this study, I defined human assets are known as human resources of individual, household, group-community, community, and local and national governments, in this case, Indonesia to address the building system of resilience for the natural disaster issue (i.e., flood disaster) in the West Malaka Subdistrict. Additionally, human assets and flood disaster in the West Malaka Subdistrict have a connection to each other through the concept of consequence of flood risks for their implemented human assets or resources, specifically as a coping mechanism in dealing with flood risks.

2. **Building Community Resilience Through the Sustainable Use of Human Assets**

Building a community's human assets for resilience can be categorized into sustainable actions if the implemented human resources contribute to the development of the human resiliency system. According to Okuwa et al. (2016: 1), the dynamic nature of the environment makes it imperative for organizations to embrace proper human capital development to achieve resilience.

According to Cohen (1999), the human capitals include knowledge, skills, and experiences that an employee has which helps pave the way for developing an organization (qtd. in Okuwa et al. 2016: 2). More often, the human capital is referred to education and includes knowledge and skills that are accumulated through forms of education attainment, training, and experience (Mayunga 2007: 5).

Human capital resiliency can be defined as an organization's ability to respond and adapt rapidly to threats posed to its workforce (International Business Machines, IBM 2009: 1). The need for human adaptation is not new and has been happening since individuals had to deal with the environments they inhabit (Smithers and Smith 1997: 27). An individual may present high resilience and at the same time be considered vulnerable (Miller et al. 2010: 32). Furthermore, exploring human vulnerability, resilience and adaptation is key in shaping individual responses to reduce or mitigate health risks or impacts and may help develop and implement policies and actions to reduce vulnerability, increase resilience, and improve adaptation (Nunes 2016: 13).

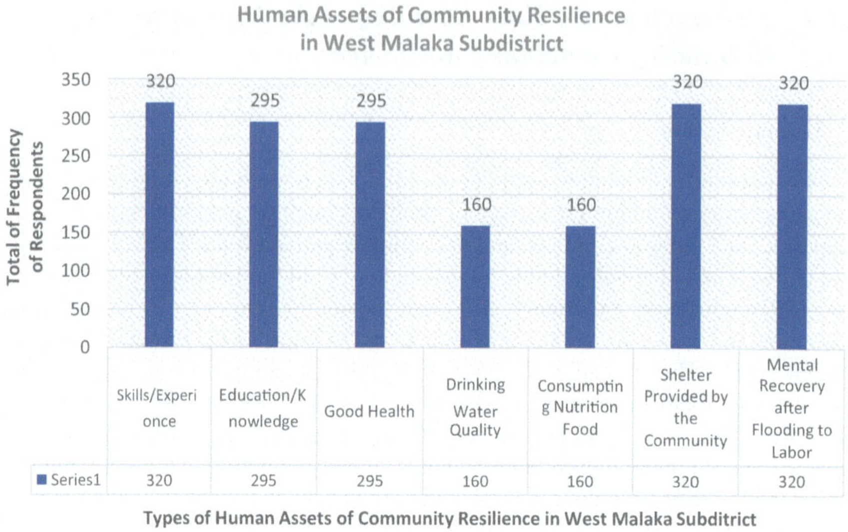

Fig. 6.1 Human assets of community resilience in West Malaka Subdistrict. *Source* Author's Research Fieldwork, July 9, 2016, to February 9, 2017

Human capital contains human competency to deal with natural and social challenges. For more detail on this discussion on human assets, the data results of human assets from this study are presented in Fig. 6.1.

Based on Fig. 6.1, there are seven subhuman assets that the community was in possession of and that were applied by the community in the West Malaka Subdistrict. The seven subhuman assets are discussed in further detail in the next sections.

(a) **Skills (Experience)**

All 320 respondents (households) in the 16 villages in the West Malaka Subdistrict have the skills (experience) to deal with flood disasters in every flood season in their area. The more times a community deals with floods, the more experience they obtain to protect themselves, both nature and their social environment. Communities which have been settled for more than twenty years (native people) in the West Malaka Subdistrict commonly have more experience in facing flood disasters than migrant people (non-native people) who have less experience in dealing with a flood disaster. Community skills and experience can create new strategies for handling a flood disaster. It is also important to note that advanced skill formation is linked to basic education and foundation skills (Jeniffer and Palmer 2009: 13). Therefore, skill (experience) has become one of the factors on community resilience which can drive their assets and apply their preventive and mitigation strategies in handling flood disaster.

(b) **Education (Knowledge)**

Education (knowledge) is also known as one of the indicators of community resilience. Based on the interview data, there were 295 respondents (households) who have passed formal education and have the knowledge of coping with the risks of flood disaster in the living area. It means that there were five respondents who did not pass formal schooling. However, the five people are trained due to the frequent flood disasters and thus are capable of managing the flood risks.

Knowledgeable people have the capability to understand and adapt to challenging situations or natural catastrophes in this case, flood disaster. Schultz (1979: 5) postulates that human capital entails investing in individuals through education and training. Human capital development entails adding value to people by educating, mentoring, supporting, training, internships, and organizational development.

(c) **Good Health**

One of the most important factors that enables people to adjust themselves to face flood disaster is having good health. There were 295 respondents who have good health standing while the other five respondents have problems with their health, such as malaria. However, this kind of disease has not so far interrupted their capacity to adjust to flood disasters. A person's health status is generally seen as an essential thing in their life, allowing them to pursue certain strategies and coping mechanisms to deal with flood disasters. Morrone (2003: 11) suggests that human capital can also be understood in terms of competencies and knowledge embodied in an individual, including their health status as well as non-cognitive skills and personality traits such as self-confidence, perseverance, adaptability, or dependability.

(d) **Drinking Water Quality**

Every individual requires water for living. The communities in the West Malaka Subdistrict are no different; they need water that is drinkable. However, in the research area there are only 160 respondents out of the 320 respondents which have efficient and portable drinking water. Each of the 160 respondents has a well and a water pipe while the other half of the 160 does not have these structures. One-hundred-and-sixty respondents are not able to get access to water from a well or a water pipe in their home and must take the water from a "nearby" river. As can be seen in the picture below, there are some kids in the Oanmane Village which take water from over three kilometers by foot from their house to the river site (Fig. 6.2).

(e) **Consuming Nutritious Food**

In order to increase the capacity and capability of humans to be more productive, consuming nutritional food is also essential. Unfortunately, not all people are able to consume nutritional foods as their income is less than the price of nutritional foods. Consuming nutritional foods is a demand for everyone including the people in the West Malaka Subdistrict who seasonally face flood disaster. It becomes a problem for the communities in the West Malaka Subdistrict as only half (160 respondents) of the 320 respondents have capability to consume nutritional foods. The low income of these communities is seen as a contributor to the lack of ability to access nutritional foods. The 160 respondents who had access to consume nutritional foods are those

Fig. 6.2 Children in Oanmane Village of the West Malaka Subdistrict were walking on foot from their home to the Benanain River, which is approximately 3.5 km away, to fetch water with their "Jerry-Can." *Source* Author's Research Fieldwork, July 9, 2016, to February 9, 2017

who have both medium and high incomes every month where the medium income ranges from 500,000–1,000,000 million rupiah (*Indonesian Rp*) which in US Dollars (American $) is $50–$100 and the high income is ranging more than $100 (> $100). Whereas the other 160 respondents who are not able to consume nutritional foods have the lowest incomes every month ranging from $15–$49 (low income) and $5–$14 (lowest income). Thus, the higher the income that is made, the more accessible it is for a family to financially be able to consume nutritional foods. However, nutritional foods can be accessed through each household's independent effort to cultivate and farm in yield-agriculture; they can improve their stock of foods, store food, and sell food in the traditional markets to receive additional incomes. This is an inherent way to consume nutritional foods and it is applied by the 160 respondents who do not have access to consume the nutritional foods.

(f) **Shelter Provided by the Community**

Shelter provided by the community in West Malaka Subdistrict functions as a secure place to live temporarily due to flood disaster. The community in the study area has a strong group of disaster communities to create a mitigation system. One of the mitigation system phases is an evacuation phase which means picking up all the people in the ex-ante of flood disaster and taking them to the secure places (i.e.,

Center of Malaka, Malaka District). Disaster relief (DR) shelters play a vital role in large-scale disasters and are an important part of disaster response and recovery (Bashawry et al. 2014: 19). Disaster shelters not only provide immediate and short-term shelter for the victims of a disaster, but also help them to recover from the trauma of a disaster as well as provide a base to start the process of rehabilitation (Bashawry et al. 2014: 20). The disaster shelter in the Center of Malaka is a permanent building and it could secure more than 300 households in the midst of a flood occurrence in the study area. The strong group of disaster community that community has created is also encouraged to cooperate with other institutions (i.e., Church) to allocate and provide support such as food, drinks, and water. These actions-based-community-approaches are not so far in the financial means of the community nor provided by the Indonesian government to help in the aftermath of a flood disaster in order to aid the affected people in the ex-ante phase. It can be assumed on the one hand that was successful in allocating and providing supports and on the other hand it was not well sustained yet since the sole resource is still from the community side but not from the Indonesian government. It has been shown in Africa and provided by IFRC that successful shelter repair and construction has to go hand-in-hand with finding solutions for a number of other issues such as water and sanitation, fuel for cooking and heating, waste management, and settlement planning (International Federation for Red Cross and Red Crescent, IFRCS 2011: 8). Following a disaster, shelter activities immediately focus on saving lives. This is particularly urgent where the affected people are exposed to harsh climatic conditions such as extreme cold or heat. Rapid shelter solutions include tents and shelter kits, or materials to build or repair homes. Alternatively, temporary accommodation in public buildings or with host families all provides support. Beyond survival, the key considerations for disaster shelters are providing protection from the climate, ensuring privacy and dignity, and providing personal safety and security (IFRCS 2011: 15). A participatory approach to safe shelter awareness (PASSA) aims to raise the awareness of the "everyday vulnerable" of the "everyday risks" related to their built environment and foster locally appropriate safe shelters and settlement practices.

Felix et al. (2013: 12) explained that post-disaster temporary housing solutions seem to have sustainability problems in two ways. Firstly, they are unsustainable in terms of costs. Secondly, they are unsustainable in terms of environmental issues. Most of the time, the solutions provided are not produced in the region where the disaster has occurred. Therefore, importing and transporting the units or materials required to re-house all the victims may involve extremely high costs. This way, temporary housing becomes a very expensive kind of solution in relation to its lifespan because it implies huge investments in units that will only be used during a short period of time (Johnson 2008: 22). As a consequence, much discussion has been centered on the question of whether temporary housing is really needed, due to longer reconstruction processes, or rather an overspending that represents an important consumption of resources that are crucial for permanent reconstruction (Steinberg 2007: 25). The community in West Malaka Subdistrict has a typical shelter which represents their cultural settlement. The typical house is a local community house that the community uses as a temporary housing for saving their food stocks

Fig. 6.3 Typical shelter of the community in the West Malaka Subdistrict. *Source* Author's Research Fieldwork, July 9, 2016, to February 9, 2017

such as maize or corn, cassava, tubers, and the cultivation facilities. Additionally, the typical of local temporary housing that community uses for relocating themselves in the city center of Malaka (i.e., Betun City) has a similar appearance building but it is larger than a temporary housing for saving the community's food stocks (Fig. 6.3).

(g) **Labor Helping Mental Recovery After Flooding**

Mental recovery after flooding can be understood as human mental recovery in post-traumatic events. For example, after a flood, the question is whether the affected person still embeds the flood event in their memory and is rethinking it for long terms (i.e., recovery illness) or rejected it and rethinking it to be more strategic in coping with disasters. Coping with disaster through optimizing the capacity is the best way to neglect the traumatic experience. Increasing the traumatized person's labor efforts helps to optimize their individual capacity, both indicatively and actively.

The more productive the labor is, the less past traumatic experiences would be mirrored in the memory. It is a path of developing mental health capacity in terms of psychological impact in the ex-post of flooding. The data in Fig. 6.1 entitled, "Human Assets on Resilience that were Impacted by Flood Risks in the Study Area," has shown that all 320 respondents in the study area agreed that their mental recovery after flooding was only disrupted in a short term.

This case study in Malaka-Timor, Indonesia, is different with a case study in Haiti. The example case study of Haiti showed that when developing a sustainable mental health care system for children in Haiti a bio-psycho-social approach that capitalizes on the strengths of Haitian children, their families, and communities is demanded (Legha and Solages 2003: 16). As added by Pierre et al. (2010: 25–351) for the Haitian community their idea of mental recovery includes the community view that floods are a deeply traumatic experience for those affected. Fortunately, from the findings in Malaka-Timor, Indonesia, a flood is not a deeply traumatic experience even for those who were affected by the flood. However, several case studies about mental health issues and post-traumatic experience after flooding highlight the higher occurrences of mental health issues (such as anxiety, depression, and post-traumatic stress disorder) in populations that have experienced flooding (Ahern and Galea 2006; Caroll et al. 2007; Alderman et al. 2012 qtd. in Jones and Garde-Hansen 2012; Fernandez et al. 2015). All the 320 respondents in the study area agreed that mental recovery after flooding to labor is short-term disruption rather than long term.

6.1.2 Natural Assets

1. **Introduction and Understanding of the Natural Assets in Building Community Resilience**

In this context of Malaka, I define the natural assets as the individual, household, group community, and community to drive the human assets' strategies (i.e., human skills, experience, education, and capabilities) in utilizing the natural assets for fulfilling their daily necessities and also specifically for resiliency management for floods. Natural assets require effective management as well as protection from the people in order to prevent an overwhelming exploitation upon the benefits of natural assets. Since 2005, the protection and management of the natural assets have taken place under a framework established in the Natural Resources Management act (University of Adelaide 2004: 8). Groundswell International (2010: 3) in their work for Natural Resources Management and Resilience to Climate Change has been exploring one of the natural assets (i.e., agriculture) for the communities to farm more sustainably and productively and to become more resilient.

2. **Building Community Resilience Through the Sustainable Use of Natural Assets**

Natural Asset includes access to land, water, and wildlife, from which households engage in agricultural pursuits and/or resource collection for both sustenance and income generation (Carney 1999: 19). Lee and Neves (2011: 15) argue that natural resources vary in the degree to which they are "naturally" available versus being altered by human actions. It means that there should be a balanced or mutual corporation from the people to utilize the natural resources. The degree to violate should be less than the degree to protect the natural resources. Resource quality and quantity are heavily influenced by human behavior and the sustainability—or lack of sustainability—of management practices. To achieve a sustainable resource of natural aspect of the community, the solution-based approach must be implemented and supported concerning agro-diversity and biodiversity. Advisory and support services for the people focus on the sustainable and equitable use of natural resources to improve livelihoods and conserve agro-diversity and biodiversity (Gitec-IGIP GmbH 2016: 1).

Rural population in the remote area, the West Malaka Subdistrict, Indonesia, has played a vital role in their livelihood system through utilizing the natural resources, so-called as natural assets. A livelihood system is not perceived only by the rural population which has utilized natural resources but also, urban populations use natural resources (land, water, agricultural produce, and wild resources) to support their livelihoods (Gordon et al. 2000: 7).

Although natural resources play a central role in rural livelihoods across the globe, little research has explored the relationship between migration and natural capital use, particularly in combination with other livelihood capitals (i.e., human, social, financial, and physical) (Nawrotzki et al. 2014: 12). They explained that there are three key findings emerging in this field of study. First, higher levels of proximate natural resources are associated with greater financial, human, and social capital for both migrants and non-migrants. Second, migrants have, on average, greater financial, physical, human, and social capital than non-migrants, and urban-to-rural migrants do exceptionally well on all capital asset categories. Third, migrants residing in areas with higher levels of natural capital tend to have significantly higher levels of human capital (education). As supposed by Lee and Neves (2011: 16), natural resources vary widely in the rules that govern access to them. Access to some resources is primarily held by individuals, while access to other resources may be shared across larger groups, including the state, and some resources are not effectively held by anyone.

Sustainable intensification is achieved through sound management of natural assets including land, water, forests, and fisheries which provide multiple benefits to food production, environmental health, and nutrition. Properly managed watersheds, rangelands, agricultural lands, forests, and fisheries enhance ecosystem functions that boost agricultural productivity, replenish aquifers, retain soil nutrients, mitigate damage from storms and floods, and reduce environmental vulnerability to the shocks and stresses associated with climate change (Frankenberger et al. 2013: 15).

For a study case in Queensland Australia, the Queensland Government, Department of Environment, and Heritage Protection has embarked on a range of projects to improve policy and raise awareness of innovative approaches to natural hazard management. This review forms a component of the "Natural assets for flood and cyclone resilience" project. The project provides an overarching policy framework with recommendations on how to better use natural assets in mitigating flood and cyclone impacts (Queensland Government; Department of Environment and Heritage Protection 2012: 7). New approaches to the mitigation of natural disasters involve developing a more holistic, multi-hazard, adaptable approach to planning strategy based on local needs and catchment characteristics (World Meteorological Organization, WMO and Associated Programme on Flood Management, APFM 2009: 9; Hawkesbury-Nepean Floodplain Management Steering Committee 2006: 23).

Local people do have a well awareness of building their assets of resilience for facing floods. Community has an adaptable way to implement their assets of resilience. The implementation of their assets of resilience was measured by an integrated assessment they have conducted to deal with flood disaster management. Yet, community's natural asset enables to absorb the impacts of flood disaster. Ecosystem integrity and the natural potential of catchments to absorb the impacts of disasters are crucial to the success of integrated hazard management plans (Becker-Goss 2007: 5).

Another case study from Fragility to Resilience in Africa, several major global declarations and reports over the past 15 years recognize the potential of Natural resource management to strengthen resilience in fragile situations (African Development Bank Group 2016: 1).

Brown et al. (2016) assert that an increase in the area (and geographical distribution) of horticultural crops with commensurate increases in yield would represent a significant opportunity for the agricultural industry.

Any expansion in total crop-able land would need to be in association with sustained investment in elevation of yields, as well as geographical diversification of crop production and crop diversification within rotations. An example of this can be seen in The Scottish Climate Change Adaptation Programme (2016: 9) which includes an objective to support a healthy and diverse natural environment with capacity to adapt. Scottish Natural Heritage promotes a set of adaptation principles to help nature adapt, ranging from reducing other pressures, and making space for natural processes to planning for change. In conclusion, Australian Wool Innovation and Meat and Livestock Australia, AWI and MLA (2016: 13) explained that healthy and resilient natural assets including soil, water, and vegetation underpin livestock production, increase the capital value and profitability of the enterprises, and add immeasurably to the comfort, satisfaction, and temperament of all who live or work there.

(a) **The Ownership of the Field Agriculture (Crops)**

Land serves as a natural asset that brings economic value for the fulfillment of community needs through the utilization of agriculture such as food. The benefits of land can be effective if there is no land degradation. Land quality is affected by degradation, or enhancement, as a function of prior use and current management patterns (Lee and Neves 2011: 17). Major gains in food production over the past half-century have been achieved through widespread adoption of technology packages and policies associated with the green revolution. These include the introduction of semi-dwarf high-yielding varieties of wheat and rice, associated with irrigation and higher levels of inputs such as inorganic fertilizers and pesticides.

Governments put forward supportive policies and investments that provided small farmers with a secure, remunerative, and low-risk environment. They invested in infrastructure and ensured that farm credit got to farmers, and they subsidized and, in some cases, distributed inputs (i.e., fertilizer and water). They also invested substantially in agricultural research, provided farmer extension services, and intervened in markets to stabilize farm-gate prices International Fund for Agriculture Development, IFAD 2011: 12). Rural households often pursue diverse livelihood strategies including farming, herding, off-farm employment, and the exploitation of natural resources through hunting, fishing, and gathering (Nawrotzki et al. 2014: 13). Brown et al. (2016: 17) mentioned that using the developing concept of "precision farming," there are opportunities for innovative approaches to optimized regional land (and water) use. In the case study area of the West Malaka Subdistrict, after flooding the soil becomes more fertile than before flooding. This can be seen by the number of households production of yield crops (i.e., Corn or Maize) in the third planting and harvest period system which lasted through three months of January–March every year.

A Household, *Leonardus Jefri* in Fafoe Village in the West Malaka Subdistrict has shared his experience about his successful yield crops (i.e., maize during the third planting system). He explained that:

> In the rainy season that starts from November to April each year, people are affected by the flood that paralyzes their natural assets of corn as their main food consumption. Leonardus was really aware of the impact of floods in each rainy season, so the coping mechanism applied is a daily pattern based on adaptive knowledge of the frequency of floods that occurred twice a year, according to him. The daily pattern in question is the pattern of cultivation of corn and other crops such as tubers and cassava during the third harvest. The third harvest season or so-called in their local language of Malaka as Batar Aheuklan which was carried out in the dry season starting from May to October. In January to the month of March as well as in July to the month of September is the third harvest where I and so the other people here were starting to plant crops in muddy conditions after the floods occurred in the dry season during May to June. The harvest from our plants after the rainy season and coinciding in the third harvest period was quite satisfactory to us. Production of corn and tubers and cassava can help our daily supply of food and food supplies to cope with flooding. The daily pattern to adapt to the natural assets that are vulnerable to flood risk has been done since the beginning of the flood disaster was occurred in 1939 until now. So, this daily pattern has become our local wisdom. We know that floods are very threatening to all the assets we have, especially the food. However, we continue to practice our local wisdom as a form or symbol of sincerity in our life. As much as we need is the assistance of the Indonesian Government and other Agricultural Department and related Institutions in Indonesia or Foreign Countries in support of our adaptive flood-wise local wisdom. Leonardus has added that: I think this is a new science or approach in agriculture study where its source is based on our local wisdom as a local community living in the watershed and affected area by floods. My hope is to improve the resiliency system from our recovering power, we need resources from the Indonesian Government such as paddy-yield tractors so it can help the natural engine that we use, i.e., cow and buffalo. One tractor for one village is sufficient or per farmer group in each village in West Malaka Subdistrict, given the relatively high price of one tractor, we could not be able to purchase it. Because the supporting power for food supplies is the availability of good facilities and infrastructure in agriculture. Sadly, we do not have it and we are just still relying on natural machines. So, we rarely enjoy harvesting paddy on land rather than harvesting corn on land that does not require both natural engine and tractor. May our voices be heard by the Indonesian government.

(b) The Ownership of Orchards

The community in the West Malaka Subdistrict has utilized one of their natural resources (i.e., soil) for more than a 100 years for planting orchards next to agriculture. They have their own-land resources for planting orchards. Based on the survey and interview data of the 320 households, there is no Indonesian government intervention toward their land ownership. Thus, a coping strategy through orchard planting to deal with flood disaster is implemented. A secure accessibility of land in the West Malaka Subdistrict allows the community to easily have their own natural assets. As opposed by Ellis and Allison (2004: 15), landless people in rural areas are particularly vulnerable, because without secure access to land and other natural resources, they encounter more difficulty in obtaining food, accumulating other assets, and recovering after environmental and economic shocks or misfortunes. The ownership of land and its use to plant orchards enables the people in this study area to utilize it as a natural resource to cope with flooding.

(c) The Ownership of the Fisheries Equipment (Resources) and the Utilization of the Benanain River

The implementation of the community ownership of the fisheries equipment (resources), and the utilization of the Benanain River have been carried out by the community since a flood disaster began (i.e., first periodic flooding in 1939). The community in the West Malaka Subdistrict sought their own solutions by utilizing the natural resources for their food stocks (i.e., food security). The community makes use of their fisheries equipment (resources) to facilitate and to accelerate their efforts to produce foods and income (i.e., fish production and its marketable in the traditional market). Like, in developing countries account for some eighty percent of global fish production and about half of the global fish trade, thus making fish essential to food security as a source of both food and income (International Fund for Agriculture Development, IFAD 2012: 7).

Additionally, water is a critical input for human consumption as well as for crop production but grossly neglected by the community. The major sources of water supply are rainfall, lakes, rivers, snowy mountains, and underground storage. Except for wells and small tanks, the community owns other sources of water collectively. However, the powerful lobbies and vested interests have been taking advantage of these water resources for their own benefits, while the poor have no means of utilizing their share. This has been accelerating the economic imbalance between the small and large landholders (Sundaram and Jha 2004: 13). Forest and fishery resources have been the object of human activity for thousands of years (Lee and Neves 2011: 22). Furthermore, they described that water availability is highly influenced by irrigation infrastructure and management in many regions, while water quality is affected by human actions which may lead to soil erosion and sedimentation, and pollution by agricultural, industrial, and human waste.

(d) The Ownership of Forests and Wildlife (Cultural Rights)

The community in the West Malaka Subdistrict has their own cultural rights (i.e., forests and wildlife). They enhanced their natural superiority through an agreement that has been agreed upon by them with the customary leaders to protect their nature (i.e., by not cutting down protected forests; trees). It is also agreed by the community and the customary leaders in a custom agreement where there are sanctions or customary punishment in the form of financial penalties, and social punishment if one member of the communities violates the agreement, they have mutually agreed upon it. Although the community knows that there are natural advantages that can be utilized in their resilience management process to floods, the community has so far remained within the corridor of custom law. An example is that the forests have been providing many direct and indirect benefits to rural communities (Sundaram and Jha 2004: 14). Long term (potential change in economic/environmental situations) is implemented by improving the rangeland health and wildlife habitat through collaborative efforts between ranchers, land management agencies and environmental organizations that use planned grazing as a tool to create a future landscape (Don et al. 2015: 8). In contrast to these two ideas, in the context of this study, according to the interviews with 320 respondents, they consider a crucial thing in the protection of nature, the power of custom law "*badu*" is still strong in society and thus it has so far not involved other institutions to tighten the protection of nature. It aims to keep the people obedient to the mutual agreement they have agreed upon it and for the people to keep the various efforts to protect the types of natural resources that provided by God to them. The community in Malaka has four types of natural assets (i.e., owner of field agriculture (crops), owner of kind of orchards, owner of fisheries equipment (resources), and owner of forest and wildlife (cultural rights). As can be seen in Fig. 6.4.

Based on the survey and interviews with 320 respondents, there are four types of natural assets, namely the owner of kind of orchards, the owner of fisheries equipment (resources), and the owner of florets and wildlife (cultural rights). People have an

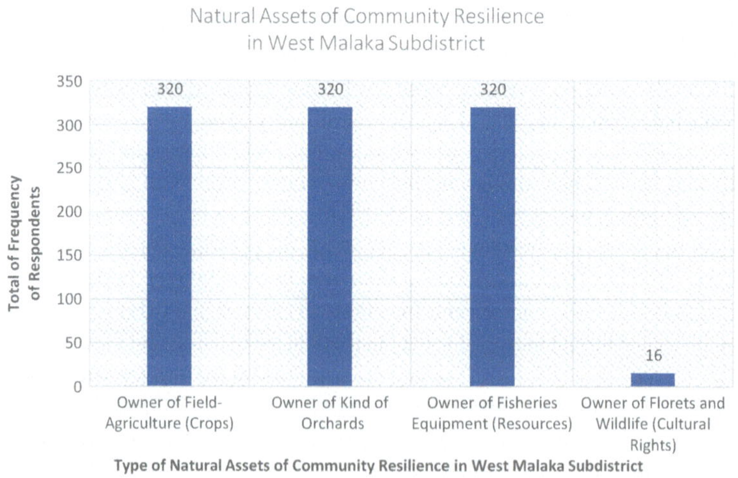

Fig. 6.4 Natural assets of community resilience in West Malaka Subdistrict. *Source* Author's Research Fieldwork, July 9, 2016, to February 9, 2017

understanding that having their own land for agriculture is a priority for their survival. With the land, they can grow crops, farm, and produce food production that can be consumed and used for the management of resilience against flood disasters. All 320 respondents agreed that the assets that assist them in carrying out the food recovery are the natural assets (i.e., owner of field agriculture (crops)). Additionally, people also consider that they have a hobby of planting and farming various types of orchards in the front and back of their yield-house such as rose flowers, pare fruit, pumpkin fruit, cabbage, cucumber, and tomato. This is very positive to stimulate the mind that remains positive in carrying out activities every day, especially the management of resilience to flood disaster. These types of orchards that the community planted also supported the community's food stock even though these types of orchards could not be able to help them in the process of their resiliency management for floods as the flood's mud swept all their orchards. However, all 320 respondents voted that they have a kind of orchard (e.g., rose flowers, pumpkin fruits, cabbages, cucumbers, and tomatoes).

Communities seek to regulate their resilience process for flood disaster through the utilization of natural assets (i.e., Benanain River) as a source of economic livelihoods. All 320 respondents said that this type of natural asset as a fisherman/woman is essential in improving their coping mechanism of resilience for floods (i.e., food stocks and incomes preparation before a flood disaster occurs or ex-ante period, during a flood disaster occurs, the phase of relocation in disaster shelters, and after a flood disaster occurs or ex-post, for food recovery).

Meanwhile, other 16 respondents argued that they are the owners of florets and wildlife (cultural rights). The 16 respondents are the head of the village and subvillage, and customary leaders. Obviously, not all communities have ownership authority over florets and wildlife (cultural rights) because of the indigenous people's factor (i.e., custom, or traditional leadership) that positively influences traditional leaders to guide the community in improving their resilience in their daily life and in facing life difficulties such as difficulties in leveraging resilience assets to tackle flood disaster management.

I assumed that the community has an independent way (i.e., resiliency understanding) of utilizing their natural assets despite their poor natural environment condition (i.e., mud of floods which are routinely impacted to their natural environment). Although the community in the West Malaka Subdistrict has a low income in terms of financial power, and thus they categorized as the poor ones, they are indeed independently seeking their own solutions through utilizing the natural resources become economic resources. It is now more generally accepted that the conditions under which the poor can manage natural resources are contingent on internal and wider institutional structures, as well as on the specific character of the natural resources themselves (Baumann 2002: 13). Furthermore, he explains that access by the poor to natural resources (land, forests, water, fisheries, pastures, etc.), is essential for sustainable poverty reduction and the livelihoods of rural people without access, or with very limited access to natural resources are vulnerable because they have difficulty in obtaining food, accumulating other assets, and recuperating after natural or market shocks or misfortunes. In connection with this study, however, a poor and rural community could be able to access their natural resources without facing any difficulties in obtaining foods and accumulating other assets if there is an intervention of the Indonesian government to create several solutions in terms of accelerating the fulfillment of community needs (i.e., the sustainable use of the natural assets).

6.1.3 Physical Assets

1. **Introduction and Understanding of the Physical Assets in Building Community Resilience**

In this context of this study, I define the physical assets as the resources that created by the individual, household, group community, and community to facilitate the regularity of life and a resilient processing in the overcoming of the negative impacts such as floods. Physical capital includes basic infrastructure (e.g., transportation, shelter, energy, communications, and water systems, health facilities, and markets); production equipment; and other material means that enable people to maintain safety and enhance their relative level of well-being (Mayunga 2007: 5). Doing so requires ensuring that community infrastructure systems and other basic services can operate at a level that provides individuals and groups the means to survive and recover during natural or man-made disasters (Longstaff et al. 2010; Pasteur 2011 qtd. in Frankenberger et al. 2013: 13a). Physical assets have an engineering function created by humans to repair the damage of the surrounding physical environment caused by natural disaster. When assessing the impact of a disaster on critical infrastructure and determining the contribution of infrastructure to community resilience, it is necessary to determine whether infrastructure supports key services (i.e., education, health, safety, economic activity), the accessibility of infrastructure to all members of the community, existing gaps in productive infrastructure, and community-based mechanisms for maintenance (Frankenberger et al. 2013: 13).

2. **Building Community Resilience Through the Sustainable Use of the Physical Assets**

In order to improve the sustainability of physical assets, a community needs an effective channel and networking among the members of the community. It aims to accelerate the channeling of the community's initiative and creative ideas in which the community itself as an independent and a resilient community, has already conducted a disaster management action through voluntary actions to overcome the problem of flood disaster (i.e., building a community-based-disaster-management, rebuilding a local drainage and a canal, community's media and telecommunication, and community center for health facilities).

The community in the West Malaka Subdistrict has given their opinion about the implementation of their own physical assets and the Indonesian government's physical assets. As can be seen in Fig. 6.5.

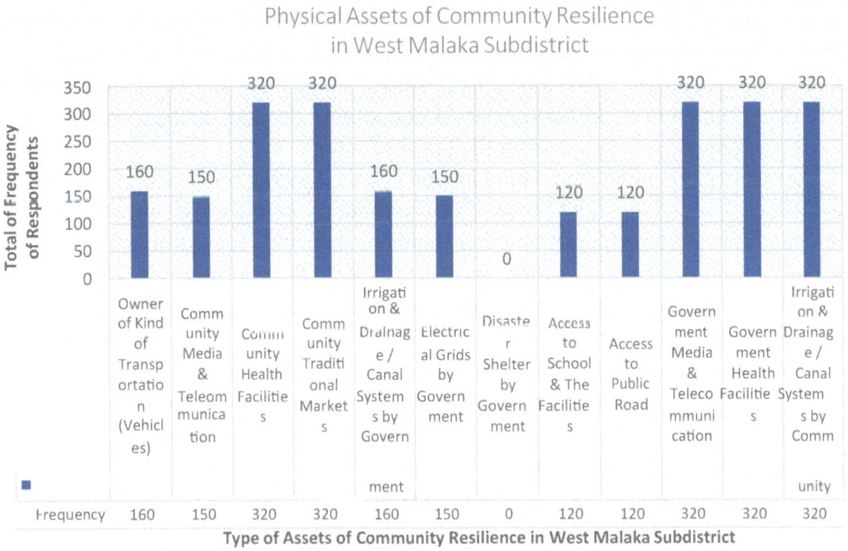

Fig. 6.5 Physical assets of community resilience in West Malaka Subdistrict. *Source* Author's Research Fieldwork, July 9, 2016, to February 9, 2017

According to Fig. 6.5, the community explains each section of the twelve subnatural assets in detail and its benefits in their daily life as well as the management of resilience for floods. The discussion is provided below.

(a) **The Ownership of the Means of Transportation**

Motorcycles, as well as cars and trucks, allow access to distant resources that can then be sold (Quiroz-Carranza and Orellana 2010 qtd. in Nawrotzki et al. 2014: 671). Based on the interview with 320 respondents in the West Malaka Subdistrict, there were 160 respondents that had their own vehicles (i.e., motorcycle) to facilitate the mobility of everyday activities. The use of the motorcycle supports their daily mobility in the ex-ante period or before a flood has occurred and after ex-post or after the flood occurred in the case of early mitigation phase and the preparation phase to save themselves from the flood hazard. The transportation usefulness has short-term benefits because of the risk of flooding. It can be assumed by the greater flood in the rainy season in which on the one hand that was not always appropriate in the phase of mitigation measures and on the other hand that was not succeed on early implementation phase or inappropriate targets. Thus, it can be said the use of other modes of transport (i.e., motorcycles) does not have a long-term positive impact in terms of early phase mitigation and preparation phase of securing the people themselves in a safe place from flooding. The community in the West Malaka Subdistrict has a "cultural transportation," horses. Ten respondents in the community own horses but also have motorcycles. This type of asset owned by the community helps them to get to a safe place rather than taking place in the early mitigation phase and securing themselves to a safe place where the horse as a means of cultural transportation. Horses are also treated as a cultural transportation asset that brings them benefits in terms of livelihood. Specifically, in the phase of ex-ante and ex-post of flooding where in these two phases, a community designs their preparation and recovery measures to recover from the flood disaster.

The community in the West Malaka Subdistrict is looking for food out of the cattle business (i.e., pigs, cows, and horses). The community makes use of horses as one of the main means of transportation other than a motorcycle. They utilize the horse to pick up trimmed green grass which can be used to feed their horses, pigs, and cows. In maintaining three types of livestock, the community seeks to sell them to raise money and to re-add the economic assets as a source of community revenue. This kind of transportation is thus useful in the long run and has to be maintained. As a means of cultural transportation for the community, livestock is paramount importance to the West Malaka community in the phase of ex-ante or before flooding occurs and in the phase of ex-post or after flooding has occurred.

Therefore, the process of community resilience in the use of cultural transportation is a reflected physical asset through the community's daily routines that generate for their livelihoods. Transportation assets preserve the uniqueness of cultural transportation of physical assets through three types of livestock (i.e., pigs, cows, and horses) in the application of a form of long-term resilience.

(b) **Community Versus Indonesian Government in Providing Media and Telecommunication**

In general, the community in the West Malaka Subdistrict is connected to the public media and telecommunications facilities which are applied in their daily living such as public internet in Umatoos Village which can be accessed by the community in other villages as well. Besides, there is a public telephone as a telecommunications facility of the community. The usefulness of these types of physical assets is temporary or can be useful when ex-ante or before flooding occurs, during flooding and ex-post or after flooding occurs and these cannot be sustained in the long run. The benefit of these types of assets is for connecting with relatives who settle down in the upstream area of the Benanain River, located in the South Central of Timor (*Timor Tengah Selatan, TTS*).

Every person in the community has a mobile phone (handphone) where the price for a cell phone is below 750,000 rupiah ($75) and over 100,000 rupiah ($10). The mobile phone is very useful for information exchange about the high rainfall and flooding estimation with relatives who settle in the South Central of Timor; therefore, it is helpful in early precautions and mitigation measures in the ex-ante process or before flooding occurs. When flooding occurs, communication media was led by a group of volunteers in the West Malaka Subdistrict. Community becomes a strong foundation in terms of information and communication for the preparedness, mitigation measures, and evacuation toward the relatively safe area in the Central Malaka Region, namely Betun City. Betun City is the capital city of the Malaka District and is known as the nearest shelter for the affected community if a flood disaster in the West Malaka Subdistrict occurs. As a result of the strong connections, information and communications guided by a Disaster Group Volunteers in the West Malaka Subdistrict have been assisting the villagers throughout the resilience process. Strong communication and information exchange is needed for early preparedness, mitigation, and evacuation when ex-ante or before flooding occurs, as well as during flooding. Information including the duration of the flood, how long the refugees should continue staying in a shelter in Betun City and also how long it takes for refugees to return to their homes in the West Malaka Subdistrict after flooding, should be filtered carefully. A similar path of process and review are applied by the community in the field in the ex-post period or after flooding in the subsequent phase of recovery actions. Thus, the community media (i.e., Disaster Group Volunteers) which led by the young leaders of 16 villages in the West Malaka Subdistrict has been a benchmark for the success of implementing the system of disaster risk management and resilience starting from early preparedness, mitigation, and evacuation in the ex-ante or before flooding occurs, during flooding and in the ex-post or after flooding occurs (i.e., recovery process).

The recovery process is problematic due to coexistence of this kind of physical asset that is implemented by the 150 respondents, but which could not be implemented in long-term actions. The community which experiences flood hazard requires real action taken by the Indonesian government. The Indonesian government can provide media equipment and technology information for disasters that can further help the community carry out the alerting system and preparedness, early mitigation, and evacuation in the ex-ante period or before flooding, during, and the recovery actions in the ex-post or after flooding.

The community explained that they did not find significant assistance by the Indonesian government in terms of media and telecommunication. All 320 respondents pointed out the importance of media and telecommunication which is believed to be necessarily provided by the Indonesian government since it brings positive impacts to the community. Although public media has a complex and significant role in the process of community resilience it should not be the single solution in the long run since the single solution is only driven by the community.

(c) **Community (Public) Versus Indonesian Government's Health and Facilities**

Public health facilities have become a risky thing to be considered carefully in the process of disaster management and community resilience. Based on the interviews with the 320 respondents, the data revealed that all of the respondents have claimed that the availability of public health facilities that were provided by the Indonesian government (i.e., Integrated Service Post-Centre (*Pos Pelayanan Terpadu or POSYANDU)* and Community Health Center (*Pusat Kesehatan Masyarakat or PUSKESMAS*)) are helping the sustainable development of health care for the maternal toddlers, elders, and common age issues such as infants, toddlers, pregnant woman, and breastfeeding mothers, the disabled people, the suffering people of a serious illness, public health for the old age (elders), and health for the people at any age. The availability of public health is also one of the types of physical assets where medical personnel in collaboration with a group of volunteers for community disaster help people in the ex-ante period or before flooding occurs, during flooding, and ex-post or after flooding has occurred.

In the ex-ante period, the phase of preparedness, early mitigation, and evacuation for the vulnerable groups such as infants, toddlers, pregnant woman and breastfeeding mothers, the disabled people, people suffering a serious illness and elders is needed. In the phase of flooding, vulnerable groups become a major priority for keeping them safe from the danger of the flood. Whereas, in the phase of ex-post or after flooding, fulfilling the nutritional needs of breastfeeding mothers for infants, nursing mothers and pregnant women becomes a further necessary step in the recovery phase. However, availability of the *POSYANDU* and Community Health Center (*Pusat Kesehatan Masyarakat or PUSKESMAS*)) are limited in the West Malaka Subdistrict; there are only six Integrated Service Post-Centre of *POSYANDU* and six Community Health Center of *PUSKESMAS* in six villages, namely in the villages of Lasaen, Umatoos, Besikama, Raimataus, Rabasahain, and Rabasaherain.

Currently in 2017, the local government constructs the infrastructure of hospital in the West Malaka Subdistricts. The logical reason is that the 16 villages of the West Malaka Subdistrict do not have health centers and clinics where community members can gain access to the medical services of *POSYANDU* and *PUSKESMAS*. The hospital in the West Malaka Subdistrict has been newly established by the local government to satisfy the increasing demand of medical service after recent floods. However, medical personnel in the *POSYANDU* and *PUSKESMAS* are still limited due to the lack of human resources, such as the expertise of nursing and pharmacy. Thus, although the 320 respondents said that they understood a positive impact of the public health facilities in the West Malaka Subdistrict, they are still restricted and influenced by their strong sense of local culture especially traditional care and medical treatment. Community is entrusted to the traditional leaders of women (Female Leaders) whose duty is similar to the obstetrician. Besides, some female leaders who use traditional medicines, such as natural spices and herbs, are still prevalent in some remote communities nowadays. Therefore, a combination of modern and traditional medical practices has obviously become a new medical culture that values the practical use of herbs rather than medicine in some ways but both of them are acceptable in the community. It can be concluded that both the medical staff who acknowledges the modern and traditional way of medical care are not mutually exclusive given that they both provide health care services and treatment for the community.

So, the community in the West Malaka Subdistrict stands for the building of a new hospital that can support their needs of their healthcare services, especially in the process of ex-ante or before flooding, during flooding, and after flooding.

Although the availability of medical personnel and public health facilities is very high, they are far out of supply when a flood disaster hits the buildings. Therefore, the medical personnel that has been engaged in cooperation with Disaster Group Volunteers in the phase of evacuation measures are efficient. It can be concluded that at the stage prior to the floods, during, and after the floods, the function and the role of medical personnel and public health facilities are not totally effective and efficient and sufficient in the long run.

(d) Community Traditional Market

The community in the West Malaka Subdistrict has a local or traditional market which is located in Umatoos Village. To support the process of the realization of flood management and the process of community resilience, the traditional market is one of the physical assets that is not only a place for trade and to increase their earnings, but also it serves as a platform for social and cultural activities that create a bond in the community. In general, the sellers in the traditional market are mainly aged above 50 and below 15 (Fig. 6.6).

During the disaster management process, traders in the traditional market have already known each other, but come from different villages in the West Malaka Subdsitrict. Thus, the stages of flood management before, during, and after the flood, information is transferred through daily conversation. In this stage, resilience is represented as a community's marketable resources such as food crops, fish and shrimp ponds, and farm products. Communities sell these marketable resources in the traditional market and earn some income. As one of the physical assets, traditional market supports the flood disaster management process and resilience process through implementation of the social-cultural element and economics. The sustainability of this physical asset is seen as the physical factor that has been affected by floods that come every rainy season. However, the sustainability of this type of asset

Fig. 6.6 Situation in the traditional market (an old woman, 58 years old with her granddaughter, 8 years old were selling their fishes) in the West Malaka Subdistrict. *Source* Author's Research Fieldwork, July 9, 2016, to February 9, 2017

is maintained and preserved for years by the community as an element of social and cultural cohesion. Thus, the traditional market has been known as one of the physical assets that reflects an intimacy as well as socio-cultural familiarity in the disaster management phase and long-term recovery phase.

(e) **Irrigation and Drainage Systems Provided by the Community Versus Indonesian Government**

Irrigation and drainage systems for the community in West Malaka are only funded by the Indonesian government during their initial reconstruction. They can better support the growth of plants and agricultural development in society. Thus, the growth of plants, rice, and maize can meet the harvesting needs of the public for everyday food and for economic value in society. Based on the interviews with the 320 respondents, 160 respondents voiced-out their opinion that the availability of irrigation and drainage is in the early stages from the Indonesian government, and the view is that it brings a smooth process of maintaining their food security. However, some of the respondent commented that the construction of irrigation and drainage initiated by the Indonesian government is too slow. History demonstrates their concern.

During the flood period which lasted from 1939 to 2016, people experienced a food crisis because of the lack of physical assets, irrigation, and good drainage support from the Indonesian government. Irrigation and good drainage are also crucial to sustain effective and efficient food security in the communities and to support local community programs as well as activities that enable the locals to live comfortably.

This issue is rooted in the fact that the government is very slow in handling the case of flood disaster funding in terms of the regulation pattern and policy implementation which dealt with funds distribution for infrastructure (i.e., physical assets). Thus, sustainable use of irrigation and drainage is still determined by the Indonesian government policy. The community has independently attempted to make an emergency irrigation. They simply start getting water from the river and form their narrow irrigation and drainage system. According to the interviews, all 320 respondents claim that their independent efforts to build and to utilize the irrigation and drainage are really helpful for supporting their food security. However, they did not put the water taken into their own irrigation and drainage, but directly sprayed into the fields or crops they planted. This practice was provoked since the visit of the Indonesian local and national governments during to their village. Despite having limited irrigation and drainage, the West Malaka Subdistrict shows their intelligence in defending their food security. Overall, the lack of irrigation and drainage construction by the government leads to interference overflowing of the Benanain River, (i.e., flooding) and the stage of construction of irrigation and drainage is a good initial stage, but it is too late in the process to support food security, ecosystems, and communities as well as safety.

(f) **Electrical Grids Provided by the Indonesian Government**

Electrical grids have been implemented by the government of the Malaka District for the community in the West Malaka Subdistrict. Based on the interviews with the 320 respondents, the data shows that there were 150 respondents who have electrical grids provided by the government, but 170 respondents do not have electrical grids;

they still use kerosene and lanterns. People who do not have electrical grids only earn less than 250,000 rupiah ($25). This financial problem is seen as the reason why the full community is unable to gain access to electrical grids and advance technology. It is only mainly felt by the people in Betun City which is known as the capital city of Malaka and located at Central Malaka. The advance of technology in question is a low representation of human indices in term of interacting with the internet for the benefit of society today's modern life. It is evident that information can be easily obtained by the use of a simple information tool, namely cell phones (handphones) which can be purchased for a relatively cheap price. I assume that the community in the West Malaka Subdistrict has a lot of positive values which enable us to influence other people in modern society. Although the community has the chance to take some benefits of advanced technology for their daily living, some of the communities in West Malaka Subdistrict are in their opposition and practices their local or traditional ways (i.e., physical assets). On the one hand, financial problems restrict the ability for the community to improve their knowledge, access to advanced knowledge, and to advance their technology, but on the other hand, financial problems have led to one of obstacles that negatively support the lack of priorities of the use of media communication and technology as a means of information and communication for disaster management and recovery process in the community. An assessment of the long-term benefits of this type of physical asset (i.e., electrical grids) is certainly that can be very useful for lighting every household in West Malaka. However, when a flood hits every house, electrical grids do not function well. In conclusion, the usefulness of electrical grids in flood disaster management response and recovery processes are seemingly adequate and efficient but they are not effective during the floods.

(g) **Disaster Shelter Provided by the Indonesian Government**

Based on the interviews with the 320 respondents, there was no disaster shelter provided by the Indonesian government for the community in the West Malaka Subdistrict. This is due to the Indonesian government's policy priorities related to the management of floods in the district of West Malaka. All the physical assets that must be implemented by the Indonesian government, including providing shelter for the affected community, was precisely the idea was derived from the community, built, and driven from the community, and for the community. It contradicts the role of the local executive and legislative boards which are superior in the community, and they act slowly for the community in the West Malaka Subdistrict. The valuable strength lies in the people themselves. A strength reflects the irregular system and also it runs in the regulation or statute law, policy, and implementation of regulations and policies for the benefit of the affected community by the floods. At the stage of before the flood, during, and after the flood, the community's independence but with the help of the Disaster Group of Volunteers allows the community to assist themselves either in starting from the stage of preparedness, early mitigation, and evacuation to the stage of safety from flooding, or in the final phase of post-flooding. In this final stage of post-flooding, a large part of their physical assets for optimal recovery were running out, but the community's physical assets are still not effective, efficient, and sufficient in the assessment of this type of physical asset for the long-term period.

(h) **Access to the Schools and to the Public Roads**

Access to schools and education facilities (i.e., sufficient schools building, laboratories equipment, books in the library, football field for sport subject, schools yield, toilettes, table and chairs, blackboard, and erasers) are vital in supporting teaching and learning. However, it was not felt and experienced by the community in the West Malaka Subdistrict. Based on the interviews with the 320 respondents, access to education facilities at the school was disturbed due to the effects of a flood catastrophe. Going to school is not easy and both students and teachers must walk on foot through the public roads which were inundated by mud from the flood. Additionally, 120 respondents described that the inundated water along the public roads disrupted learning activities at the schools. Students and teachers at the school suffer disruption in the teaching and learning processes. Therefore, flood disaster not only impairs the access to school but also to the educational resources such as books and student worksheets (*Lembar Kerja Sekolah*). The mud-rock which flows during floods drifts into classrooms, staff rooms, and sports fields damaging and interfering with these educational resources.

Additionally, school events and extracurricular activities were also affected by a flood. Laboratory facilities were destroyed as well. It is apparent that whenever a flood hits the area of the West Malaka Subdistrict, accessibility to schools and learning facilities becomes a dilemma for many years. The needed role of the Indonesian government regards in this case, infrastructure, and reconstruction of physical assets. That is, were the Indonesian government to support the community in infrastructure and reconstruction of the schools' various physical assets, the community would be able to recover their every day-activities, like attending school.

6.1.4 Social Assets

1. **Introduction and Understanding of the Social Assets in Building Community Resilience**

Regarding community resilience, social capital is a vital capital to support the other capitals of resilience. I define social capital as a social resource that individual, household, group community, and community unifies a diversity of social characteristics in order to bind the community for a resilient community. Social capital is a crucial element in determining the nature of resilience at the community level (Aldrich 2012: 12). For instance, disasters may sometimes enhance social capital because they activate or give rise to neighborhood associations and collective organizations that can be used to disseminate vital information, provide community members with a voice, and afford leverage to assist in taking control of rebuilding efforts (Aldrich 2012: 13). Social capital has often been described as the "glue" that binds people in society together (Frankenberger et al. 2013: 13). Social capital can be described as the quantity and quality of social resources (e.g., networks, membership in groups, social relations, and access to wider institutions in society) upon which people draw in pursuit of livelihoods (Frankenberger and Garrett 1998 qtd. in Frankenberger et al. 2013: 14).

2. **Building Community Resilience Through the Sustainable Use of Social Assets**

Aldrich (2012: 12) provides a useful analytical approach for assessing the influence of social capital on community resilience by identifying three distinct but interrelated forms of social capital: bonding social capital, bridging social capital, and linking social capital. The community in the West Malaka Subdistrict has also applied these forms of social capital (bonding, bridging, and linking) through their system of building resiliency.

Building community resilience through the sustainable use of social assets (i.e., bonding, bridging, and linking) has led to an effective way of empowering a community's social resilience. The community in the West Malaka Subdistrict has strong social assets of resilience that assist them to effectively address the flood disaster problem.

Communities have implemented their social assets (i.e., bonding, bridging, and linking) in coping with flood disasters. *Bonding Social Capital* is seen as an emphatical sense of tying up a togetherness of the community in the form of a mutual assistance to solve problems together. The community in the West Malaka Subdistrict recognizes bonding social capital in their local language, *Tetum Malaka* as "*Hakawak*" or in the Indonesian language, *Bahasa Indonsia* as *"Gotong Royong."* The *hakawak* has been adopted and nurtured since 1939, before flooding began. I then interpreted literally the work bonding as *hakawak* where the community has practiced it into all of aspects in their life such as social, cultural, economic, religious, and disaster group activities.

For social capital's understanding in the communities' life, It became apparent that the community agrees that they had implemented the *hakawak* system in their phase of disaster risk management and the phase of resilience. It binds other aspects of their life through the unity of a group in response to the disaster. Contextually, before flooding occurs or in the ex-ante period, the Disaster Group Volunteer enhances community awareness of flooding and supports the community's phase of preparedness. This traditionally has been one of the key roles of the customary leader and religious leaders (i.e., pastoral roles to inform related-disaster-information like the early warning system (EWS) in the Church and to implement the role as a preventive group to protect the community). This real practice that has been implemented and driven by the community in the West Malaka Subdistrict is a real example of *Bridging Social Capital.*

During the flood disaster, the Disaster Group Volunteer plays the role of a volunteer and an activist in evacuation of the community to a safe place or a disaster shelter.

A community is not active during a flood except for their efforts to secure themselves further and to follow the cues or alerting information from the Disaster Group

Volunteer, customary leader, and/or religious leader. Once flooding has stopped, the community reorganizes their homes which have been damaged by the flooding and prepares for their daily necessities in order to continue on with their normal lives again. In the study area, it took several months (approximately three to six months) to recover from the flood. Long period does not mean the long term of recovery. Time does not answer the quality of recovery process, but intensity and quality of recovery process could answer the target time to finish the recovery process. The community has been linked with the Disaster Group Volunteer, customer leaders, and religious leaders to take further recovery steps in cleaning up mud brought by the flood which has become embedded in and around community infrastructures. It is mentioned as Linking Social Capital. Thus, bonding, bridging, and linking systems of social capital that had been implemented by the community in the West Malaka Subdistrict have been implemented in long-term actions. The long-term assessment for improvement of this *hakawak* could for the community to be able to sustain and improve it as a strong resource.

6.1.5 Economic Assets

1. **Introduction and Understanding of the Economic Assets in Building Community Resilience**

Economic assets can be understood as the economic resources which the household, the group community, and the community endeavors to create their own financial resources that can support their resiliency system (i.e., coping with floods). It is also so-called as "economic resilience" which is really close with the term "livelihood." In this study, economic resilience is perceived by the community affected by flood disaster as the collected-financial resources which further could be a result of incomes and reduce several risks of economic consequences due to floods such as reparation of house's building both interior and exterior, fulfillment of food stocks and drinks, medical care and treatment, and fee of cleaning up the mud in the agriculture field and at homes. Economic resilience creates some alternatives solutions to fulfill a single livelihood of the people. It values several monetary resources to support the community's coping mechanism. The community in the West Malaka Subdistrict understands the "economic resilience" requires the multi-monetary resources that they are struggling to find solutions for addressing their economic resiliency.

Their economic resiliency embraces a coping mechanism (i.e., result of their efforts to get economic resources to deal with flooding). Additionally, the community belongs to a rural livelihood which requires a lot of effort to be survived to fulfill their necessities. They have exceeded their monetary resources from selling their natural assets (i.e., result of agriculture; corns, maize, tubers, and handicrafts; traditional weaving) in the traditional market to obtain income and keep the sustainable use of their economic assets. An example from the economies of a developing country remains overwhelmingly dependent on agriculture, either directly (as

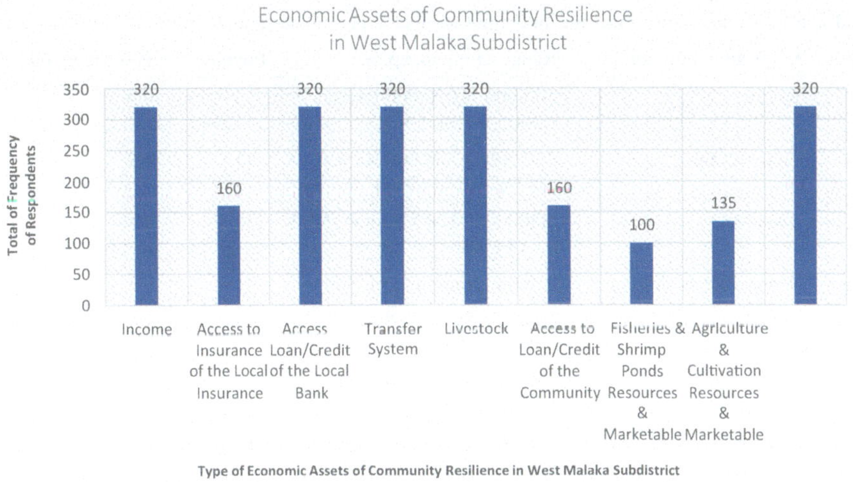

Fig. 6.7 Economic assets of community resilience in the West Malaka Subdistrict. *Source* Author's Research Fieldwork, July 9, 2016, to February 9, 2017

a source of direct employment and income) or indirectly, as a source of employment and income in food handling and agro-processing (Gordon et al. 2000). Markets are also social institutions in which some actors are able to enforce mechanisms of control, which favor access for specific groups and exclude others (Gordon et al. 2000). Another example Linnell (2003) reveals that concerning the role of natural resources in sustainable rural livelihoods of the western Balkans, he found that unemployment and poverty are widespread. Like rural communities across Europe, these villages are struggling along the way toward a sustainable future. It is therefore hard to achieve a sustainable future in terms of achieving a welfare life through a strong of economic resources. It was also experienced by the community in the West Malaka Subdistrict that they struggle for their lives and livelihoods. I thus concluded that the community has their own monetary resources that are effortlessly sought by them (i.e., maximizing their human assets; education, skills, abilities, and hardworking) to cover and manage their resiliency system (Fig. 6.7).

2. **Building Community Resilience Through the Sustainable Use of the Economic Assets**

Income determines how much expenditure is required and how much effort is made by the people to access a variety of assets in their lives. As experienced by the community in the West Malaka Subdistrict, they had access to local insurance, loans from the local bank, and other types of economic assets. Their access to these economic assets is made possible due to their income. A more detailed description of each community's types of income and range of income per month within their efforts and accessibility are discussed in Table 6.1.

Table 6.1 A community's income and range of income per month (p.m.)

No.	Job (source of income)	Range of income p.m.	Frequency	Percentage (%)
1	Teacher	500,000–1,000,000 Rupiah (50 US $–100 US $)	3	0.9
2	Civil servant	750,000–1,500,000 Rupiah (75 US $–150 US $)	13	4.2
3	Household	50,000–150,000 Rupiah (5 US $ –15 US $)	156	49
4	Non-public worker	250,000 Rupiah (25 US $)	1	0.3
5	Fisherman/woman	100,000–500,000 Rupiah (10 US $–50 US $)	5	1.6
6	Farmer	150,000–500,000 Rupiah (15 US $–50 US $)	123	38
7	Entrepreneur	500,000–1,000,0000 Rupiah (50 US $–100 US $)	13	4.2
8	Honorary employer	250,000 Rupiah (25 US $)	1	0.3
9	House builder (workman)	1,000,000 Rupiah (100 US $)	2	0.6
10	Stonesman	500,000 Rupiah (50 US $)	1	0.3
11	Car driver	950,000 Rupiah (75 US $)	1	0.3
12	Motorcycle service	7500,000 Rupiah (75 US $)	1	0.3
Total			320	100

Source Author's Research Fieldwork, July 9, 2016–February 9, 2017

All respondents interviewed have an income that meets their daily needs. This type of income also affects the fluctuation of the quality performance and community work present in their livelihood. The majority of people in the West Malaka Subdistrict are households shown in Table 6.1. There were 150 respondents who lived at home and have a second job as farmers and indigenous cloth weavers. Yet, their agriculture products and clothes are not sold in the market, but they sell the goods through their family-owned business that is easy and simple. However, women typically have to stay at home to manage their household matters such as seeking natural fuel (i.e., wood) from the forest for cooking food and water. Nevertheless, the results of operations that they receive are enough to meet the daily necessities of their family that are between 50,000 and 150,000 rupiah (US $5–$15 US).

Additionally, they gain revenue from the enthusiasts who visit and buy their crops such as corn and tubers.

Traditional cloth is a community-owned creative weaving product that has been produced. Some examples are scarves, handkerchiefs, the plate and bag of betel nut, gloves, tablecloths, refrigerator cloths, some custom dresses, and uniforms. Due to the income produced through traditional weaving, the farmer, entrepreneur, and fisherman/fisherwoman earned more money than other members of the community. The higher income for these workers is also due to the large scale and scope of their markets which promotes crops from fields and rice from paddy fields. The single-business group (i.e., the farmers who are also weavers) was established by the community in order to earn great sums of revenue and gain profits through selling fish and shrimp in the traditional markets.

In the study, there were 123 respondents (38%) whose employment is as farmers and sellers in the traditional markets which have an income between 150,000 500,000 rupiah ($15 US–$50 US) per month. Furthermore, the degree of a presentation followed by entrepreneur, there were 13 respondents (4.2%), which run an independent business with an income between 500,000–1,000,000 rupiah ($50 US–$100 US) a month. Meanwhile, there were five respondents (1.6%) who earn their income through fishing and selling to the traditional markets.

Revenues obtained that obtained by the fisherman/woman was ranged between range between 100,000–500,000 rupiah (10 US–$50 US) per month. Working as civil servants, 13 respondents (4.2%) said that they worked at the district offices with a regular income of 750,000–1,500,000 rupiah ($75 US–$150 US) a month. Meanwhile, three respondents (0.9%) are permanent teachers and part-time teachers in senior high schools and there were two permanent teachers and one part-time teacher with an income of 500,000–1,000,000 ($50 US–$100 US) per month.

There is only one respondent or 0.3% with a fairly minimal revenue that is 250,000 rupiah (25 US $) per month. There is the same representation of 0.3%of the non-public worker, car driver, motorcycle service (motorcycle taxi), and stones man; based on the interview data obtained, there is one respondent for each of these professions. However, these four different professions receive various amounts of monthly income. For example, the non-public worker has a monthly income that is 250,000 rupiah ($25 US) which is equal to the income of an honorary employer and precisely under the rate of a standard household which is equal to the income of honorary employer and precisely under the rate of household income.

Meanwhile, car drivers have a higher income than the other three professions, namely 950,000 rupiah ($95 US) per month, followed by the motorcycle rider and taxi driver profession with a monthly income of 750,000 rupiah per month ($75 US). Meanwhile, another respondent that stated that they work as stones man has a monthly income of 500,000 rupiah ($50 US). Additionally, the house builder's (workman) income amounted to 1,000,000 rupiah ($100 US) where there are only two respondents or 0.6% who have this type of profession.

However, when the flood hits part of a community in the West Malaka Subdistrict, the monthly income earned by some professionals will not be sufficient enough for them to cope with the damage of both the interior and exterior parts of their home. The average cost of the damage to the interior and exterior home ranges between > 2,000,000–3,500,000 Rupiah (> $200–350 US). However, according to the income ranges mentioned above, there are only four professions that have half of the income for repairing their damaged homes both in the interior and exterior, where each month is reached one million rupiah ($100 US) for its cost. The four professions that are able to repair their damaged houses are the teacher, civil servant, house builder (workman), and the entrepreneur. This complicated situation unfortunately happens in the community and people have to make good use of their creative capital to work on reconstruction. A community's independent efforts are still far from meeting the needs of fully repairing the damaged homes and other physical capital and economic capital. However, the community has been valuing their lives that were held up due to the power of culture, morality, and religion in their community. These make them remain robust and strong to survive and to live. Phase by phase and with optimism, the community has made efforts to implement resilience solutions for the flood disaster. One of the efforts of the community was their connection to some public insurance programs (i.e., the cooperative savings and loans).

There were 160 respondents who were connected with the loan/credit of the community along with this community located in Betun City, the capital city of the Malaka District. Furthermore, 160 respondents have also accessed national health insurance (*BPJS*). The presence of a local bank, *Bank Rakyat Indonesia (BRI),* has assisted the community in having access to savings and loans.

Individuals living in the West Malaka Subdistrict have to walk seven to ten kilometers on foot to reach the local bank, *Bank Rakyat Indonesia (BRI),* in Betun City. This local bank helps people for the post-flood recovery. All respondents explained that the local bank allowed them to have a passbook plus (*Taplus BRI*) with a minimum of 250,000 rupiah ($25 US$) in every month of saving. People were allowed to further use the loan and credit system at the local bank. The community made economic efforts to keep their economic capital both for their daily necessities and flood costs. Thus, local banks gently help the community to facilitate the process of post-flood recovery.

The local Bank, *Bank Rakyat Indonesia (BRI)* exists as a temporary option for people, because the main assistance for post-flood-disaster recovery in West Malaka Subdistrict is flood insurance on disaster risk management. Flood insurance for the community is needed to increase the capacity of their resilience system and to be recovered in any type of business they carried out. The community was conscious and has an independent spirit to strive from their capitals and to generate their revenue for the purpose of economic capitals on resilience. The process of striving from their capitals was full efforts done by them through a coping mechanism. In this context, coping mechanism deals with community's strategies or mechanisms to save their money in the preparedness phase or in the period before a flood disaster happens as to assist them after a flood disaster ends. It helps their economic resilience in handling flood disasters.

From the data obtained, there were 320 respondents who own livestock such as cattle, cows, horses, and pigs. Maintenance of these animals does not require large costs due to the nature capital (i.e., green grass, foliage, and water) possessed by the community. This natural capital can support the maintenance of the community livestock, generate new income, and boost economic recovery. The same data has also shown that all respondents choose to ask for money from their relatives and family who are working in the big cities such as in Batam City and Surabaya City and the neighboring country such as Malaysia. This is one of the ways that the community knows how to restore their economic capital. The community is also quite resourceful in that members can save their money from their economic efforts to prepare for flooding, turning economic income into an early coping mechanism.

Three-hundred-and-twenty respondents explained that they keep their money at home and in the local bank, *Bank Rakyat Indonesia (BRI),* in Betun City, the capital city of the Malaka District. It is a kind of coping mechanism that the community uses for the fulfillment of the post-flood recovery.

Natural assets (i.e., resources of fisheries and shrimp ponds) are a marketable business for the people's economic capital in the context of resilience in the West Malaka Subdistrict. Fisheries and shrimp pond businesses also generate regular income. Thoroughly, a community's endeavors to utilize fisheries and shrimp ponds are a coping mechanism because the workers sell them in the traditional market. Coping mechanism is the mechanism or strategy that community implemented in order to prepare their food stocks in the period before a flood disaster happens or in the ex-ante and in the period of after a flood disaster happens or in the ex-post. It is a step of community strategy to meet the day-to-day needs and to strengthen their coping mechanism for early flood recovery or before the flood disaster (ex-ante). All in all, the community has undertaken a doubled effort in economic capital to utilize it as a coping mechanism to deal with flood disasters.

According to the data, there were 100 respondents who were trying to optimize the economic capital through farming. In fact, people whose profession is a fisherman/fisherwoman have their main job as teachers or farmers. People like those mentioned with two or more sources of income are able to strengthen their recovery process with the variety of income possibilities. Similarly, the economic capital (i.e., results of the field cultivation) is marketable by local trading in the community's traditional market. There were 135 respondents who have so far sought this type of economic capital in supporting their coping mechanisms not only for flood recovery but also as a general economic capital which fulfills their families' needs.

6.2 The Additional Applicable Assets of Community Resilience

In another case study in Pakistan, a result of research on community meetings with future strategy agenda by The United Nations Development Programme, UNDP (2013: 4–5) has taken several community livelihood resource information. Information on the livelihood assets of individuals, households, and communities (sometimes collected in a separate "livelihoods assessment") can be classified according to the six capital assets introduced in Chap. 1, pages 4–5: human, natural, physical, political, financial, and social assets. The indicative types of information (UNDP 2013: 4–5) are:

1. Human capital assets consist of skill levels and educational attainment, the extent to which local skills match the needs of existing or emerging local businesses, knowledge base, and local use of science and technology, livelihood aspirations and access to food, education, and health care.

2. Natural capital assets consist of environmental conditions in which people live and work and natural resource management issues affecting livelihoods (e.g., land tenure, disputed natural resources).
3. Physical capital assets consist of access to adequate housing, access to safe water and sanitation and access to and use of infrastructure (roads, bridges, schools, hospitals, agriculture systems, etc.).
4. Financial capital assets consist of access to finance, including microcredit and coping and adaptive strategies for responding to financial shock (e.g., drawing down savings, borrowing from friends, relatives, or money-lenders, and selling or trading assets).
5. Social capital assets consist of participation in social groups and networks that provide support (financial or psychosocial), e.g., community-based organizations, NGOs, faith-based organizations, women's organizations and extended family networks and perceived levels of safety and security.
6. Political/Governance Capital assets consist of access to and participation in government decision-making processes, gender relations and power structures influencing control over decision making within households, communities and the wider society and access to judicial systems and security sector institutions.

6.2.1 Cultural Assets

1. **Introduction and Understanding of the Cultural Assets in Building Community Resilience**

In this context of study, I define the cultural assets as the customary resources that every individual, household, group community, and community endeavors to unify a diversity among the people as well as to encourage people in order to obey the customary or traditional rules. Traditional systems tend to function best in the event of idiosyncratic shocks and stresses, while formal systems are more effective in the context of covariate shocks (Frankenberger et al. 2013; Morduch and Sharma 2002). Frankenberger et al. (2013: 11) assert that traditional systems can break down for a variety of reasons related to the dynamics of social capital: competition versus cooperation, strength of norms around reciprocity, moral or ethical codes, etc. Moral hazard is another problem: insuring risk may encourage people to engage in risky behavior, which burdens other group members when those behaviors result in losses (Morduch and Sharma 2002: 22).

2. **Building Resilience Through the Sustainable Use of the Cultural Assets**

Frankenberger et al. (2013: 11) added that there is some evidence of increasing cooperation between customary institutions/practices and state efforts in resilience-promoting actions. For example, the government of Kenya's drought strategy names "policy recognition of indigenous knowledge and practice" as one of its priorities under rangeland management to increase drought resilience and adaptive capacity. I then emphasize that community resilience reflects an empowered community which has a pattern-ability to adapt to flooding and as well as has a moral hazard to not destroy or ruin the environment. Rather, the community behaviors that keep from harming the environment and do good for others are reflected in the system mutual assistance "*hakawak*" and an act of good behavior toward the creator of the universe.

Bria (2004: 8) states that environmental spirituals are rooted in a vision of the universe and the place occupied by humans in nature. Hereinafter described, the processing environment needs to be implemented as possible like God Himself, is to welcome the life, honor, and respect the rules and rhythm, as well help the universe to stay lush and wonderful. In The Holy Bible (Genesis 2: 15), Adam and Eve were placed in the garden to "cultivate and guard, not to ruin it." (The Holy Bible New International Version 2011).

The second type of custom is in the context of cultural capital and religious capital; it is the local wisdom of the community as distinguished by the role of customary leaders. Furthermore, BRIA (2004: 9) mentions empowering customary institutions and preserving the customs in a place are the manifestation of keeping the natural and cultural environment. One of the cultural values that needs revitalization is to enlighten the public so as not to destroy nature, but to be a nature conservationist or otherwise to maintain the ecological balance.

In connection with the environment is the existence of prohibitions or customary law "*Badu*" in terms of deforestation, forest clearing, and logging which damage the environment, that in turn have an impact on the lack of land cover, causing catastrophic flooding. With the establishment of customary law, there are penalties for those who violate it. Local wisdom of community resilience lies in the spiritual power of the environment, as the living guider in terms of local wisdom of community (traditional house village to invoke the power through *"Sirih Pinang"* for the Almighty).

Based on the interviews with the 320 respondents, there are six types of cultural assets of resilience. They are Customary Leaders' Role for the Community Protection Measures: Regulation Number 1 Year 2012, Routine Traditional Rituals of Preventing Systems, Routine Traditional Rituals of Mitigation Systems, Routine Traditional Rituals of Resiliency Systems, Customary Leaders' Sacral House for Worshiping, and The Symbol of Resilience; "Betel nut" or *"Sirih Pinang"* or *"Fuik Bua"*. These types of cultural assets of resilience are discussed below.

(a) Customary Leaders' Roles for the Community Protection Measures Based on Indonesian Government Regulation Number 1 Year 2013

Indonesia has the law or regulation of customary institutions that the institution customary can set the position and roles of the community, rule number 22-year 1999 (Illicit Discharge Elimination Program, IDEP 2010: 22). Based on the recorded data of the interviews with the 320 households in the West Malaka Subdistrict, all of them have applied the customary regulation into the system, so-called in their local language as "*Manaek/hakawak*" which means in Indonesian "*Gotong Royong*." These terms translate to English as "Mutual Assistance System." This concept is based on community culture values. The Indonesian Constitution in 1945 also defines the entity of customary regulation where the state acknowledges and respects the entity of customary law of community and their traditional rights.

The community in the West Malaka Subdistrict has a local perception and thus perceives their area as a "female land" which the majority leaders for the community are woman. As a "female land" or *"Ra'i Feto"* in community local language (i.e., Belu language: *Tetum*), the role of customary leaders is thus vital in supporting ethical and moral behavior in dealing with a flood disaster. Specifically, the implemented customary regulation has a significant benefit for the affected community of flood disaster in the West Malaka Subdistrict by applying and supporting their protection measures for facing a flood disaster.

The protection measures consist of two sides of assistance from the customary leaders' roles, namely tangible assistance, and intangible assistance. Tangible assistance deals with direct protection from the customary leaders through four applicative components:

- Direct action of spiritual guidance, announcement of flood alerting system for each individual (household).
- Direct action of evacuation guidance for the community to the safety shelter.
- Direct action of delivering ideas and suggestions in a discussion forum related to integrated management of community resilience and disaster risk management held by the public institutions and private institutions.
- General advocacy in terms of cultural leadership for all of the community in the West Malaka Subdistrict; An active role in cooperating, coordinating, and communicating with the religious institutions (i.e., Catholic Church) for the case of flood disaster and any other natural and social phenomenon in the West Malaka Subdistrict.

(b) **Routine Traditional Rituals of Flood Prevention, Mitigation, and Resilience Systems in Customary Leaders' Sacral House Through the Symbol of Resilience, Betel Nut or "*Sirih Pinang*"**

Traditional rituals carried out by indigenous leaders (indigenous female leaders) are the routine ritual which has been occurring since before the 1939 flood. This traditional ritual is an application of culture that is believed to bring good and strengthen the image and presence of cultural values. It is done to accompany the life of the community. This is a primitive element of society that is still awake to this day. A strong image and value of this traditional ritual also can be applied not only to worship ancestors and pray to the Lord but it is also implemented for the sake of preparedness, mitigation, and evacuation as well as resilience-strong-signs and post-recovery of flood disaster, particularly, in the case of community resilience. The symbol of *betel nut* (*betel nut* in the community's local language, Belu language (*Tetum*) is *fuik bua and* in Indonesian language is *sirih pinang*)) is used as the force of nature that connects people with their ancestors who have already helped, maintained, and protected them from misfortunes in their lives. In the West Malaka Subdistrict, it is believed that inclusion and protection of the community is achieved by the Almighty God, which is their major faith, Catholicism. They are also faithful to their ancestors who are believed to accompany and protect people from trouble. Traditional communities recognize this ritual as a manifestation of respect and a sense of hope in the goodness of life. From these cultural practices and beliefs, the value of kindness is expected when floods occur, and people would not experience negative effects such as pain caused by floods. Also, it is believed that through this belief, the community would be kept safe and comfortable although floods damage public facilities, public houses, and their possessions. In this community, a ritual is always done before a flood is predicted to happen with the capitulation of a *betel nut* to the ancestors and offering prayers for their protection. In this ritual, the *betel nut* is first offered and prayed upon and then eaten by the person leading the prayer to the ancestors and the remainder of it is stored in a small bag as a symbol of worship to the ancestors. This is a unique ritual that is trusted by the local people to bring safety to their lives. The *betel nut* is known as the ancestors themselves in where the community believes it can accompany and protect them everywhere, anywhere, anytime, and from anyone who may attack them and under certain circumstances, there are special accompanying and protecting spirits from the ancestors.

For contextual problem of flooding, the *betel nut* is also believed to accompany and protect the community from flooding. When a flood has happened, namely at the stage of mitigation and early evacuation, the customary leaders use the *betel nut* as a symbol of community resilience. It is to be the guardian of indigenous communities. However, a community rescuing or evacuating themselves is not always the most important thing. According to the interviews with the heads of tribal people (female leader) which one of the 320 respondents, she said that when floods occur, she was still inside her house that made of wood and leaves. Further, she explained that when the water reached inside her house, she prayed to her ancestor, but the flood water was not able to reach her house and the bag of the betel nut, which was hanging 65 cm from the wooden floor of her house.

In this case, a large flood disturbed and damaged several people's homes. However, miraculously there was no damage to the traditional leader's house. When the flood was over, the traditional leader still completed her role as a guide or protector for the people; prayer and worship were led through the *betel nut*. The people are guided by the traditional leader to praise in thanksgiving for the good and the bad things of the recovery processes that have been encountered by the people. As this example has demonstrated, the role of traditional leader (female leader) cannot be separated from the power and the role of religious leaders (i.e., priests, sisters/nuns, and brothers in the Catholic Church).

The community acknowledges and trusts the Catholic faith and also in conjunction with the principles of their culture that have preserved and maintained the role of traditional leaders (female leader) as well as their traditional hereditary ritual. The assessment of the sustainability through the role of customary leader (female leader) is concluded to be very strong for the community recovery process in terms of spiritual power. This is because without a strong spiritual force by the female leader, people would not be able to realize the process of their resilience in the face of flooding. This unique thing was found by me during my research fieldwork for seven months in the West Malaka Subdistrict.

6.2.2 Religion Assets

1. **Introduction and Understanding of the Religion Assets in Building Community Resilience**

Understanding religious capital is a sensitive matter because several research studies have tried to conceive an idea of religious spiritual and religious capital, but they did not include all elements in the universe. Thus, in this study, I divided religion capital into two definitions. Firstly, religion capital can be seen as a spiritual and or a faith medium of several certain religions, which is generally also known or adhere as, of individual, ordinary people/communities faith as vertically in people spiritual forms, trust, and their self-defense to the creator of the universe that creates a set of faith-resources (intangible and tangible things), and as horizontally within people spiritual forms, trust, and their self-defense could serve and or solve their own cases. Secondly, religion capital can be seen as a spiritual and or a faith medium of several certain religions that vertically people are belonging and believing in the creator of the universe, which is particularly adhering to the servants of the creator of the universe (i.e., priests, clergy, leader, or servant of several certain religions), as to intangibly and tangibly serve and/or solve the surrounding humanistic cases of people in the universe in a horizontal perspective based on the rules of the religious system of ecclesiastical structure or resources (i.e., capitals they had).

Customs in terms of spiritual/religious views are intended as environmental spiritualism and the role of religious institutions in this case and the church in providing services to the people of the fortitude of faith and steadfast when struggling with trials and the threat of living there. Additionally, other assistance comes in the form of sponsorship funds and the volunteer power of the church (nuns and priests/ecclesiastical structures). For instance, flooding has occurred in the study area. In response, the nuns give assistance as a voluntary service to indirectly or intangibly serve the people who are sick (treat them) due to flooding. The church also may provide mass ministry/worship to the community, and the church's voice is known as one of information sources in addition to the preparedness disaster team and the head of village in each village. Church assistance in terms of financial or directly or tangible assistance is sourced from the Catholic Relief Service (CRS), which is the central institution in channeling funds and agencies/institutions in the country and abroad to help with the process of recovery after flooding. The role of the church is very helpful in the recovery process and aftermath of a disaster.

Yet, people need basic solidarity not only from humanistic assistantship from the government and other institutions but also the community requires strong support in devotional matters. The idea of this is supported by Solnit (2009: 3), who explained in her book entitled, "A Paradise Built in Hell: The Extraordinary Communities that Arise in Disaster" that frequently, and more often than not, ordinary people respond to disaster with self-sacrifice, humanity, kindness, and basic solidarity. The devotional and religious solidarity appears in religious coping systems. Chatters et al. (2009: 9) describe in other problematic contexts that it is critical to explore how religious concerns shape major areas of one's life, including the use of specifically religious strategies to cope with significant life problems. They had examined demographic predictors of attitudes regarding religious coping (i.e., prayer during stressful times and look to God for support, strength, and guidance) within a national sample of African American, Caribbean Blacks, and non-Hispanic Whites (National Survey of American Life), Chatters et al. (2009: 9).

In another example, the religious solidarity in the United States is proven in the presence and actions of the "non-profit organizations, including faith-based and community organizations, which play a vital role in both preparing for disaster, and in ensuring an inclusive and participatory community-wide recovery from a disaster," (Brown 2015: 5). A specific example is the Georgia Emergency Management Agency (GEMA) which formed "Praise and Preparedness" to provide congregations with resources to help reduce the impact of a large-scale emergency. The conception for this group happened when a dozen tornadoes touched down in Bartow County in 2013, several faith-based organizations jumped in and played a key role in post-disaster recovery. The GEMA officials realized these groups were a key avenue for disaster preparation that they had previously overlooked. "Praise and Preparedness" strives to help churches on three levels. First, they aim to help prepare a physical facility in case an emergency takes place during a service. Secondly, they want to help prepare church leaders and volunteers to spread the word about disaster readiness among their congregation through information and readiness toolkits. Finally, the groups seek to provide church leaders and volunteers with the training required to help the community during disaster response and recovery.

In conclusion, the role of faith communities in disaster and building disaster resilient communities is integral to all phases of the disaster life cycle (Holman 2009: 8) in Tearfund International Learning Zone (TILZ)'s Programme. The backbone of mitigation education and preparedness training for all relief and long-term recovery operations is the work of religious organizations collaborating as interfaith or faith-based networks. This action could enlighten and revitalize the values of communities' faith as to cope with flood disaster.

2. **Building Resilience Through the Sustainable Use of the Religion Assets**

Building resilience through the sustainable use of the religion assets is also a substantial relief issue that needs supporting conception and practical actions from local church and other reliefs institutions in the West Malaka Subdistrict, Indonesia. This study uses some foreign organizations' conceptions and practical actions as a mode assessment regarding building disaster resilient communities in the study area. The first conception is mentioned by Holman (2009: 8) in TILZ's Programme, that the work of disaster management can best be done when local churches and their communities work together with specialist NGOs and relief agencies. This conception corresponds to the idea of implementing an integrated process of ecumenical services in the study area. Such ecumenical services are also crucial to assist and enhance community religious assets.

In this religious resilience context, I emphasized that building community resilience should require an integrated process in providing ecumenical services and its process involves local churches, affected community due to natural disaster (i.e., flood disaster), and related NGOs and GOs in Indonesia. Second conception is also supposed by The White Office of Faith-Based and Neighborhood Partnership (2009: 6) on the role of faith communities in disaster and building disaster resilient communities alluded on Integral to all phases of the disaster lifecycle are the support and services provided by congregations and faith-based social service and disaster programs.

The distribution of aid in terms of spiritual things that was provided by the churches in Thailand and the state of Nevada, from the, United States of America (USA) has had a positive impact on increasing the resilience of people to the problem of natural disasters, in this case study of flood disaster. The distribution of aid from the church is a mechanism for channeling the action of spiritual activities in the life of the community that can be carried out effectively and cooperatively to the natural environment and social environment of the community. The church is the owner of effective roles as an effective provider and supplier of spiritual aid and their role is also able to increase community resilience in facing difficult situations such as floods and resilience management. The role of the church (i.e., religion asset) can combine with other elements of the resilience assets (i.e., human, natural, physical, social, economic, cultural, law or legal and political or policy) into a peaceful manifestation despite the crisis and disaster. Furthermore, based on the survey and interviews with 320 households, this study reports that communities have also the tenth roles or ecumenical services were provided by the local church as explained by The Micah Network Global Consultation (2006: 16) in Thailand and The White Office of Faith-Based and Neighborhood Partnership (2009: 6) (see Chap. 2).

For more details, a discussion of the religion assets of community resilience in the West Malaka Subdistrict can be seen in Fig. 6.8.

Based on the conceptual opinion of The Micah Network Global Consultation (2006: 16) in Thailand where they mentioned eight potential roles of the local church in disasters for consultation of the community and the twenty plans role of the church in the United States from the Tearfund International Learning Zone (TILZ)'s Programme in 2009, the community in the West Malaka Subdistrict has also implemented and driven these eight potential roles and twenty plans role of the church into ten potential roles of Catholic Church. Based on interview data with 320 respondents, it can be stated above that there are ten potential roles of Catholic Church (i.e., priests, fraters, and sisters) in the distribution of humanist aids, among others:

(a) *Facilitator of Community Action*

The role of the Catholic Church as a facilitator of community action has a positive impact in terms of expediting resilience action against the flood disaster that the community runs (i.e., the Catholic Church as a venue for the community's training activities and practice of flood disaster simulation and resilience management). The Catholic Church as a facilitator where the information from the preparedness of the flood disaster or prior to the flood occurs, they function as a facilitator that provides disaster tents or shelters, food, and beverages for the community during the flood disaster, and they also function as an effective facilitator in the recovery process (i.e.,

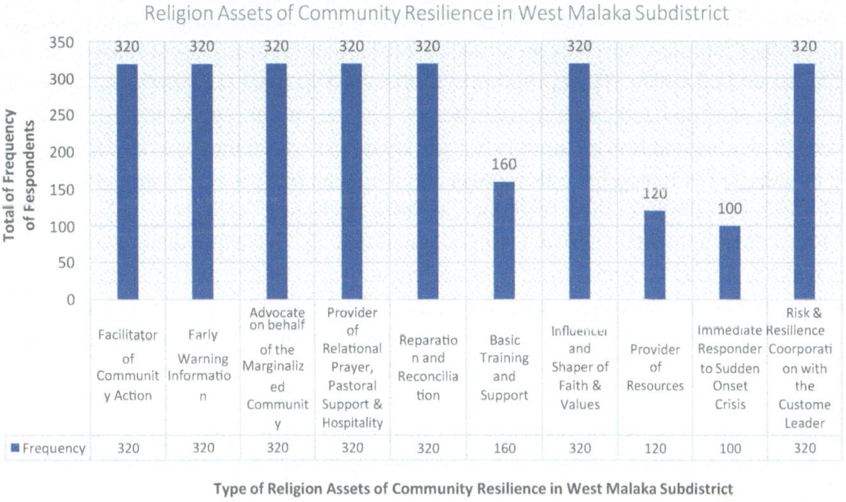

Fig. 6.8 Religion assets of community resilience in West Malaka Subdistrict. *Source* Author's Research Fieldwork, July 9, 2016, to February 9, 2017

spiritual support through a worship and a prayer together). Thus, based on the 320 respondents' assessment, all of them were experienced the realization of help from the Catholic Church as a facilitator of community actions.

(b) *Early Warning Information*

The community considers that the potential role of the Catholic Church as a provider of emergency information especially flood disaster can be felt by the community in the West Malaka. One example is the Catholic Church has ever provided emergency information in the early warning system (i.e., before a flood disaster happens or in the period of ex-ante). The Catholic Church has a strong connection with the community in delivering the community's ideas and thoughts in an emergency situation (i.e., the flood disaster preparedness information). The flood disaster preparedness information was weekly informed by the Catholic Church (i.e., priests, sisters, and brothers) through Sunday church worship. Therefore, community in the West Malaka admitted that this potential role of the Catholic Church is effective for handling a flood disaster problem. From the above data can show that all 320 respondents have the same opinion, the Catholic Church as an effective facilitator and has the power of spirit to encourage community to continue their own strategies actions of resilience to flood disaster.

(c) *Advocate on Behalf of the Marginalized Community*

The Catholic Church has also potential role as an advocate on behalf of the marginalized community where the Catholic Church provides their guidance and advocacy on the resolution of humanitarian disaster reliefs issues. One part of humanitarian disaster reliefs issues is the problem of resilience to a flood disaster that has been experienced by the community. The community's opinion regarding to the potential role of the Catholic Church as an advocate of the marginalized community is so far effective and adequate. The community in the West Malaka considers that they are part of the marginalized community in a very weak economic aspect in terms of their monthly income and less of support for this weakness which they are needed to meet the economic resilience for a flood disaster. This can be realized as something that is difficult for people in their resilience to flood disaster.

From the graphic data above, it can be noted that all 320 respondents were agreed upon the Catholic Church's role as an advocate (i.e., guider and protector) of the community in the improvement of spiritual matters to remain firm and strong in the challenge of faith (i.e., resilience for floods) and the improvement of realistic and implicative matters that are useful for the community resilience such as the community has to be working strategically to collect food stocks before floods or during ex-ante periods for purposes of during floods, and after floods or ex-post floods. Thus, the fulfillment of spiritual and outward needs could be fulfilled in flood disaster though.

(d) *Provider of Relational Prayer, Pastoral Support, and Hospitality*

This type of the Catholic Church's potential role is a type of Catholic Church's relief distribution which was routinely run by priests, fraters, and sisters. This is called as the "ecumenical service or compassionate service." Community appraisal of the Catholic Church as a provider of relational prayer, pastoral support, and hospitality is effectively and sufficiently in support of community inquiries to maximize their resilience assets. It can be proved from the graphic data above that all 320 respondents described that the Catholic Church has a potential role as the provider of relational prayer, pastoral support, and hospitality. The community in the West Malaka requires spiritual power from priests, fraters, and sisters through prayer, pastoral support, and loving service in support of their resiliency power (i.e., building resilience assets) that community is still fighting for to overcome flood problems.

(e) *Reparation and Reconciliation*

The community's point of view of the Catholic Church's potential role was their outlook on the spiritual and outward side of the good that is directly and indirectly connected to the daily lives of the community. The community understands that reparation and reconciliation are an essential part of improving the quality of human resources (i.e., human resilience) and natural resources (i.e., natural resilience). This can be understood as a combination of utilizing the human ability to process a valuable and beneficial natural resource in terms of monetary power. For example, community has the skill in weaving traditional custom fabrics which is typical of the community in the West Malaka, and they sold their traditional weaving in traditional markets for supporting their daily living's necessities and resilience's savings for floods.

The role of the Catholic Church in the context of the community daily life which is seen in terms of reparation and reconciliation can be understood as the outward support of resilience assets (i.e., human resilience and natural resilience). In 1999, the Catholic Church had been reconciled a faith meeting on a basis of a humanist solution for the community in the West Malaka with other local, regional, and national churches to assist and repair other assets in community resilience (i.e., physical resilience; broken public facilities, traditional markets, public road, and school). Thus, the actions and initiatives of the Catholic Church have been so far felt by the community in the West Malaka. Furthermore, the community believes the reparations and reconciliation of the Catholic Church have helped them in voicing their minor voices. As can be seen from the graph data above, all respondents said that there was a realization from the Catholic Church in reparation and reconciliation in order to support the community's resilience power to face flood disaster.

(f) *Basic Training and Support*

The community's assessment of the Catholic Church's potential role regarding basic training and support is in the internal spiritual aspect (i.e., individual spiritual empowerment). A faith training and support from priests, fraters, and sisters have been effectively in assisting communities in terms of maintaining and implementing their positive thoughts that are beneficial to their daily lives, especially in this case, resilience management for resolving flood disaster. Although, only 160 respondents say that the potential role of the Catholic Church has been working within the community, spiritual strengthening and church's support have so far helped each individual as individually in their mental recovery to deal with floods.

(g) *Influencer and Shaper of Faith and Values*

According to the opinion of the 320 respondents, they agreed upon the realization of the Catholic Church's potential role as the influencer and shaper of faith and values of resources. This is related to the moral environment that is expected to be owned by the community for the achievements of the utilization of resilience assets to face a flood disaster. The role of the Catholic Church through priests, sisters, and actions in providing their positive influences (i.e., communicating warmly and lovingly), disaster shelter during a flood disaster as well as supporting the values of all the existence living (natural environment and social environment). Yet, the Catholic Church helped the community in strengthening community's faith. The purpose of this potential role is to bring the community into a circle of a moral environment, which safeguards, protects, and considers the natural environment and the social environment in order to realize or obtain positive impacts (i.e., appreciating the value of God's creation; the natural environment and the value of social services; social environment).

(h) *Provider of Resources*

A resilient community is a strong community regardless of their weaknesses in implementing their resources or resilience assets (i.e., human, natural, physical, social, economic, and cultural). However, community is a single actor of resilience which requires Indonesian government and other related institutions in terms of law or legal assets and political or policy assets to tackle the case of resilience for floods. Community has another resiliency power through a potential role of the Catholic Church as the provider of resources (i.e., religion asset). Provider of resources for the community could be understood as the Catholic Church supported the community resources or resilience assets (i.e., human, natural, physical, social, economic, and cultural) by providing ecumenical services both tangible and intangible. Tangible services deal with several services which are provided by the Catholic Church such as aids for foods, drinks, tents or shelters and church building for training and early warning system or preparedness in the face of flood disaster. Additionally, intangible services deal with several services which are provided by the Catholic Church such as spiritual aids for strengthening people faithfulness to survive during an onset crisis in this case study, flood disaster and its management resilience to cope with and recover from it. As can be seen from the graphic data, there were 120 respondents that agreed upon the potential role of the Catholic Church as the provider of resources.

(i) *Immediate Responder to Sudden Onset Crises*

In order to be a positive influencer for the community, the Catholic Church has proven their effective roles as part that supported community resilience for floods. The role of the Catholic Church that has been felt by the community was the rapid action of church parties to the problems of resilience and flood as well as problems or other conflicts that occurred in the community such as ethnic riots (i.e., riots between tribes and groups), land-sharing conflicts, and violence). So far, the community believes that the main problem that needs to be addressed is to increase their resources to deal with flood disasters. From the graphic data above, it can be concluded that there are 100 respondents that considered that the role of the Catholic Church was insisted for them in supporting their increased resilience to flood disaster management in terms of priority management and immediate action as responder.

(j) *Risk and Resilience Corporation with the Customary Leader*

Potential and effective roles of the Catholic Church are the embodiment of cooperative cooperation with the customary leader to help the community to handle their resilience problem to flood disaster. From the graphic data above, there are 320 respondents explaining that the role of the Catholic Church is adequate in terms of corporation in delivering disaster preparedness messages and information with the customary leaders. An understanding of the role of the Catholic Church in this context is also sufficient because it is assessed by the community in the flood disaster management phase before the flood disaster occurs or ex-ante (i.e., preparedness), during (relocation to a safe place/disaster shelters), and after flood disaster occurs where the Catholic Church communicates and works with the customary leader in expediting disaster management and coping action.

The community in West Malaka Subdistrict admitted that the role of Church since 1939 (i.e., first periodic flooding strikes the community) has taken an important role in assisting the community actions for disaster risk management and resilience before or ex-ante, during and after or ex-post of a flood disaster happens. The process of the church's implementation for the community was adequately helpful in terms of their role as an effective facilitator. The real implementation that was felt and experienced by the community in West Malaka Subdistrict, the Church has been a facilitator, providing early warning system Information (EWS) before flooding occurs or ex-ante period and also pastoral hospitality, basic training and support and gaining a good corporation with the Customer Leader (Female Leader) in terms of risk and resilience phases. The similar actions were implemented by the Church for the community in West Malaka Subdistrict after flooding or in the ex-post period by providing resources such as tents or shelters, medical care, and water, aware to sudden crises that are necessarily fulfilled by the community with lack of economic resources. The pastoral hospitality also plays an ecclesial and *ecumenical* role that influences community faith and values and always embraces a word that is "humanity."

Advocacy on behalf of the marginalized community likewise community in West Malaka Subdistrict can be realized by the Indonesian government and other Institutions through an appropriate assistance both in tangible and intangible actions. Furthermore, the church has significantly helped, supported, and perpetuated the community in West Malaka Subdistrict to achieve a long-term action of resilience, i.e., sustainable development in maintaining and improving the assets. The discussion of pastoral hospitality before flooding occurs or ex-ante, during and after flooding occurred or ex-post was resulted into three different activities namely:

- Before the flood disaster (Ex-ante): Intangible actions such as praying and worships and tangible actions such as early warning system (EWS) in the church.
- During the flood disaster: intangible actions such as praying and worship and tangible actions such as providing tents or disaster shelter, medical care, and strong corporation with other institutions and Customer Leaders.
- After the flood disaster (Ex-post): intangible actions such as praying and worship and tangible actions such as reparation and reconciliation corporates with other Institutions for supporting the recovery processes.

6.2.3 Law or Legal Assets

1. **Introduction and Understanding of the Law or Legal Assets in Building Community Resilience**

Legal or law capital can be assessed by involving the community as a decision-maker. The law or legal capital is defined as the strong component systems in the government regulations and implementations. Here, roles and rules are created and produced, monitored, controlled, and executed by the legislative, executive, judicial, or judicative councils and other related institutions of the country which also implement the trial politics into a certain policy framework that are related to the humanitarian disaster relief case.

Related to the regulation and rules of institutions providing community welfare on the basis of sustained development on the international scale, the United Nations has passed means to involve themselves in the international law of disaster relief including natural disasters (i.e., flood disaster). The United Nations has tried to forefront their efforts to be inclusive across its policy frameworks where through *Reaffirming further the Hyogo Declaration, the Hyogo Framework for Action 2005–2015: Building the Resilience of Nations and Communities to Disasters* and the common statement of the special session on the Indian Ocean disaster: risk reduction for a safer future, as adopted by the World Conference on Disaster Reduction, held in Kobe, Hyogo, Japan, from January 18 to January 22, 2005, and bearing in mind that the implementation of the Hyogo Framework for Action is coming to an end in 2015 (The United Nations International Strategy for Disaster Reduction, UN-ISDR 2007: 1).

Given these efforts of the United Nations (UN), an example case of legislative framework that addresses a flood disaster affected community can be seen in Brazil. The legislative framework most directly addressing disaster risk reduction (DRR) in Brazil is found in its laws on civil defense, and this area of the law has in recent years been frequently amended, often to accommodate changes considered necessary so as to better respond to ongoing disasters (International Federation for Red Cross and Red Crescent, IFRCRC 2013: 8). Likewise, in Indonesia or narrow context, the law or legal capital is still not yet well-designed and implemented for the affected community's capitals for resiliency systems. Civil defense has taken a major step ahead in legislative framework than the government's minor actions.

The term of law related to regulation and rule of institution to provide community welfare in a basis of sustained development is needed by the community to assist their other assets of resilience (i.e., human, natural, physical, social, economic, cultural, and religion). The regulation is part of the law or legal asset of resilience which requires its implementation for the community. The implementation of the law or legal asset is known as the political or policy asset. The Indonesian government regulations (regulation 7 from 2004, and number 42 from the year 1998, and number 46 of 2008 clause 16) concerning the preparation of design patterns and resource management are effective and on target (BARU 2011: 23). However, according to the Watershed Management Bureau of Benanain, explicitly seen in clauses 19, 20, 21, 22 and 23, provincial regulations (NTT number 1 of 2001) constitute a consecutive set of protected areas protected forest area which provide protection against the subordinate (The Watershed Management Bureau of Benanain-Noelmina 2011: 8).

The East Nusa Tenggara Province has also enacted PDAST (*Panduan Daerah Aliran Sungai Terpadu* or Integrated Planning of Watershed Guideline) documents which contain (Regulation number 5 of 2008) planning, implementation, empowerment, guidance and controlling (The Watershed Management Bureau of Benain-Noelmina 2011: 9).

The Benenain River watershed is part of the space that has its own characteristics, and it is located across four districts, namely (*Timor Tengah Selatan, TTS,* or South Central of Timor), (*Timor Tengah Utara, TTU* or North Central of Timor), *Belu,* and West Malaka crossing into the neighboring country of Timor Leste Republic.

The Benenain River watershed was recently characterized as a critical area due to the lack of land cover, as well as floods and droughts that seasonally occur every year. Therefore, it is highly recommended that cooperation initiatives and plans are made in relation to the "Integrated Planning of Watershed Guideline" that involves the participation of multi-stakeholders.

There is no law monitoring human activities in critical areas, especially the land cover area in the upstream portion of the Benanain River watershed. Yet, human interaction in this upstream area can negatively impact the river ecosystem protection. Although disaster risks are by nature beyond human control, it is essential to have a systematic legal framework to control the river (Sawada and Zen 2014: 3). Thus, I may suggest that there is a need for action from the government to create laws to tackle the issue of watershed management. Damayanti et al. (2010) mention that other forms of land such as savannah, farmlands, and forests are very vulnerable to fire, especially when the land had been cleared (qtd. in The Watershed Management Bureau of Benain-Noelmina 2011: 11). The priority areas related to the watershed has not had integrated management implemented. The watershed area needs social corporation and coordination from customary institutions (i.e., the role of leaders of the culture in the community area and pastoral or religious institutions). These leaders in the community area are hierarchically essential in guiding people (i.e., strengthening faith and teachings positive thoughts in order to survive for the lives) from the upstream area, middle-stream area, and downstream area in Malaka-Timor, Indonesia.

To analyze this case, I conceived this capital (law capital) and put it into this analysis to provide an access point between community participation through their self-assessment and self-evaluation in their decision making in the government legal system in tackling three cases:

1. The case of natural and physical factors of controlling, monitoring, and protecting the Benanain River watershed and the dike system management with discussion on the roles and rules of trial politics.
2. The case of human actions (i.e., human activities/intervention) or man-made that have less government responsibility to protect and to give penalties and sanctions in relation to the existence of a customary legal system and in conjunction to requesting a new legal system of the land use and human actions of protecting the watershed resources management in the cross-border area between West Malaka, East Nusa Tenggara Province Indonesia and Timor Leste Republic, and;
3. The last is the case of financial or economic assistantship in terms of microscale insurance for several devastated aspects due to flood disaster, which is supposed to be given by the government for the affected community, not only in the phase of ex-post of aftermath flood disaster but also in the phase of ex-ante or before a flooding disaster has occurred as well as during a flooding disaster.

Specifically, in its ongoing application in the field, the responsibility of the Indonesian government for the implementation of watershed management addresses the question of how to systematically integrate and manage the sectors from the upstream area, middle-stream area to the downstream area. It is important that this is done with consideration of biophysical and socio-economic conditions that exist in a watershed in achieving common goals. This is at the core of an integrated river basin management which applies the principle that embraces "*one river, one plan, and one Integrated management*" (Riwu 2005 qtd. in The Watershed Management Bureau of Benanain-Noelmina 2011: 28). Riwu adds more that, the need for the organization and integrated river basin management requires a comprehensive plan to accommodate the various stakeholders in the Benanain watershed, For its real implication, a discussion and assessment about the three cases dealing with the relationship between the community and government of Indonesia is discussed in the context of "Building Resilience through the Sustainable Use of each of Sub-Law or Legal assets."

From this discussion, I reflected that the essential problem that people complain about is flood management that is still not systematic and integrative so that it can cripple the framework of the resilience system for its long term (i.e., human, natural, physical, social, economic, cultural and religion) that have been formed within the community. The community understands resilience in overcoming the problem of floods. Nevertheless, the community needs realization from the Indonesian government and other related institutions in order to facilitate resilience problem as well as to handle flood disaster that was faced by the community. Discussion requires a critical framework and judgment of the community in the assessment phase of the existing legal assets to follow up its application for the community or in other words, it is named as political or policy asset. Thus, for its implication, the discussion and assessment of the three cases that are dealing with the community and Indonesian government through Building the Resilience through the Sustainable Use of each of the sublaw or legal assets would be effectively, efficiently, and sufficiently or adequately.

2. **Building Resilience Through the Sustainable Use of the Law or Legal Assets**

Legal assets in resilience are the legal umbrella for other resilience assets. Legal assets have a strong foundation for collecting the results of community ideas related to resolving problems of resilience for floods. In addition, building resilience of the legal assets becomes an obstacle when there is no implementation from the Indonesian government to follow up the solution of the resilience problems to floods. Legal assets can play an effective role as a legal umbrella for other assets (i.e., human, natural, physical, social, economic, cultural, and religion) when there is a solvent reciprocal of political or policies assets. The embodiment of legal assets (i.e., regulations and strategic steps for resilience to floods) can be understood as an internal role of resilience actor (i.e., Indonesian government) to realize the implementation of legal assets to the community that affected by flood disaster. The result of the embodiment of legal assets is called "the embodiment of political or policies assets." However, as long as the community is still the sole actor in addressing their resilience

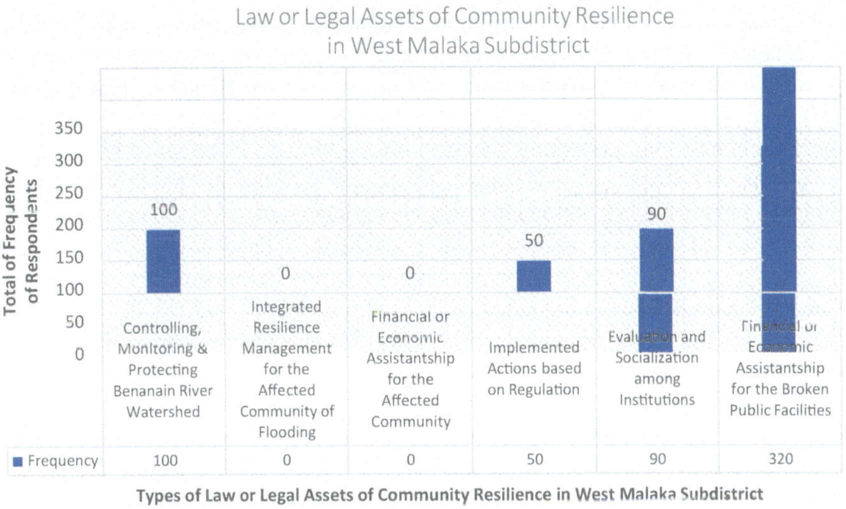

Fig. 6.9 Law or legal assets of community resilience in the West Malaka Subdistrict. *Source* Author's Research Fieldwork, July 9, 2016, to February 9, 2017

to flood problems without the obligation of the Indonesian government to jointly address them, I assume that both these assets, i.e., legal and political or policies cannot be categorized as effective, efficient, and sufficient or in other words, it is not in the long-term phase of sustainable development.

As we can see in Fig. 6.9, the respondents have given their opinion on the use and embodiment of legal assets.

According to Fig. 6.9, respondents give their ideas and aspirations to legal assets. Their point of view regarding the legal assets is more likely to discuss the implementation of economic assets for the improvement of public facilities such as roads, hospitals, and markets where public facilities are always damaged by floods. As can be seen from the graphic data above which was based on the interviews with 320 respondents, all respondents agreed upon the Indonesian government is still weak in considering the major damage (public facilities) caused by flood disaster. Thus, the 320 votes above call for the Indonesian government to earlier focus on the recovery of physical assets from resilience through the improvement of public facilities and then concerns to the other assets. Thus, people can comfortably and safely carry out their daily activities. Meanwhile, socialization and evaluation of related institutions in Indonesia are needed to assist the community in increasing community resilience to overcome the impact of flood disaster. From the graphic data above, there was recorded, of ninety votes said that the need for socialization and evaluation of the Indonesian government together with institutions related to the problem of resilience to flood disaster in order to improve the quality of community resilience that is still based on social, cultural, and religion resilience.

Whereas there are a 100 votes from respondents that were critically analyzing the performance of the Indonesian government and related institutions in handling physical resilience (controlling, protecting, and monitoring the Benanain River watershed). The idea of the community is obviously the quality of physical resilience especially controlling, protecting, and monitoring through an effective repairing of damaged/broken dikes of the Benanain River and a systematic monitoring which enables to reduce the risk of flood disaster. Nevertheless, the Indonesian government has not realized yet their actions on these resilience-related issues. The community assessed that the absence of implementation by the Indonesian government and related agencies has been eliminated or ignored regulations made by the legislative council (i.e., The House Representative *or Dewan Perwakilan Rakyat, DPR*) with the executive council (i.e., The Representative for Consultative Assembly *or Majelis Permusyawaratan Rakyat, MPR*) for the sustainability of an environment that adheres to the origins of natural and human cultivations. Thus, it can be proved based on the graphic data above that shows there are only fifty votes acknowledging the implemented actions based on regulation and no votes from the respondents who rated the integrated resilience management for the community and financial or economic assistantship for the community which was affected by flood disaster. Their opinion that asserts that legal assets have not been implemented systematically yet by the Indonesian government is a critical assessment and therefore it must be handled or realized by the Indonesian government.ich it is exp

(a) **Controlling, Monitoring, and Protecting the Benanain River Watershed**

According to data collected in the research community, over 100 participants concluded that the process or stages of controlling, monitoring, and protecting management for the Benanain watershed can be addressed with these four points of solutions. They four points are namely:

a. The cooperation between the community in West Malaka, the traditional leaders in the disaster territory, religious leaders, and community groups in the disaster territory by the institutions engaged in the field of disaster such as an NGO, Community Association for Disaster Management in East Nusa Tenggara Province, Indonesia.
b. The cooperation established required to proceed in a structured management system and supported by the government because there is a regulation of watershed management for the people who live along the Benanain River.
c. The cooperation that was implemented in a participatory form without losing the executive authority and the legality of the actors in watershed management of the Benanain River.
d. So far, the solution applied here includes the efforts and cooperation of international institutions together with the NGO, CADM. However, the power to settle problems are not resulting from society, groups, community disaster NGO,

CADM, and International institutions. However, it can be supported by all linear-based institutions integration of handling, from forestry ministry, the ministry of culture, ministry of economy, ministry of religion, the ministry of social, ministry of law and the maintenance of the border, the ministry of environment, and agriculture ministries.

(b) **Integrated Resilience Management for the Community Affected by Flood Disaster in West Malaka Subdistrict of East Nusa Tenggara Province, Indonesia**

An integration of resilience management for the community affected by flooding in the West Malaka Subdistrict has been neglected (i.e., flood disaster) where in the process of handling the flood of its beginning occurred in 1939 and it had never been implemented until this research was conducted from July 9, 2016, to February 9, 2017. There was no still implementation from the Indonesian government which deals with an integrated of resilience management as shown in the data. There was no respondent who admitted that the Indonesian government or institutions have applied an integrated resilience management in the West Malaka Subdistrict. I found several solutions (i.e., strong intellectual series of implemented policy (SISIP) through this research which is explained in section c, entitled, "Financial or Economic Assistantship for the Community Affected by Flood Disaster in the West Malaka Subdistrict."

(c) **Financial or Economic Assistantship for the Community Affected by Flood Disaster in the West Malaka Subdistrict of East Nusa Tenggara Province, Indonesia**

The community in the West Malaka Subdistrict is an independent community which tackles flood hazard with their own capacity since the flood has begun in 1939 until 2016. Independent actions encompass individual resilience, social resilience, cultural resilience, physical resilience, and also economic resilience. Based on the interview with the 320 respondents, the community did not receive financial or economic support such as food, clean water, disaster risk management (DRM) insurance, or monetary support for each household who was impacted by the floods. In terms of physical resilience, the Indonesian government did not provide single or some good shelter/s in secure places nor had the Indonesian government implemented irrigation or drainage systems. The strong social capital is positively provoking the role of the community to apply their *hakawak* system or mutual assistance system as to govern the other types of resilience.

Unfortunately, the policy priority of economic assistantship for the community in West Malaka Subdistrict was not strong enough to be implemented in the field. In order to unify the diversity of resilience systems, the community has several purposes or aspirations in terms of resiliency assessment for flood that enables them to take decision making into a strong intellectual series of implemented policy (SISIP). I found some solutions for the case of financial or economic assistantship to help the affected community of flooding in the West Malaka Subdistrict through the strong intellectual series of implemented policy (SISIP) for Disaster Risk Management and Resilience for Sustainable Development. The invented solutions of SISIP are:

- Increasing the capacity of the economy to optimize the power of community resilience through prosperous farmer groups for each household. Considering that people have different jobs but tend to grow crops in the fields and make a regular habit practiced in everyday life. Thus, the culture of farming communities can be developed through policy formation of prosperous farmer groups for the welfare of the people in the West Malaka Subdistrict. Policy prosperous farmer group formation can be supported through participatory steps also from the ministry of agriculture and ministry of the Indonesian economy.
- Increasing the capacity of the economic through group policy insurance for the farmer group's prosperity can be realized through the provision of financial capital for farmers with insurance which is useful when the floods hit people and economic assets (i.e., land and results of degraded land). Assigned capital to the community can be used to measure their recovery and preparation from the ex-ante or the period prior to the flood disaster. This policy supports policy point one, which is the same as a mutually supportive community for the sake of the recovery process in the West Malaka Subdistrict and the implementation of this policy may take a long time (long term) and come phase by phase to achieve the economic welfare.

(d) **The Implemented Actions Based on the Guidelines of an Integrated of the Benanain River Watershed**

These implementation actions are understood as an embodiment of managing flood problems that require community resilience as a strong tool to solve the problems. In support of the flood disaster management process, an assessment of community resilience has increased the community's assets of resilience. There were fifty respondents who agreed upon the implementation of integrated resilience management measures specifically in terms of reconstruction of physical resilience (i.e., dike and irrigation), put in place by the Indonesian government after the Indonesian government action has enacted the rules or regulations of the guidelines integrated watershed (PDAST) at the Benanain River in the Malaka District. A community with limited accessibility to utilize their assets of resilience is supposed to have an integrated guidance system in terms of resilience management. Integrated resilience management can be implemented in practice through the guidance of the traditional/customary leaders together with religious leaders, the Indonesian government, social ministry, environment ministry, economy ministry and, cultural ministry in Indonesia.

(e) **Evaluation and Socialization Among Institutions**

One of the obstacles is the absence of an active response from the Indonesian government regarding the evaluation and socialization among the various government institutions in cooperation with the private institutions. Independent communities have processed their own information through social power that they have achieved to support flood disaster management processes and the recovery processes. Sustainability of the recovery has not been reached yet by the community in the West Malaka Subdistrict. It was not an effective and efficient process that meets community priorities. This case should be met by the local government, national government with participatory-based methods that have been imposed by the Indonesian government. However, it is ignored by the Indonesian government. Assistance was apparently needed through evaluation and dissemination of flood-risk-information among institutions. Non-Government Organizations (NGOs) as the private institutions have played their roles as a facilitator of local moving for the community development and community-based disaster management. It is driven by the NGO, Community Association for Disaster Management (CADM) or Society Disaster Management, with their program that is Resilience for Sustainable Development and Environment (RESUDENVI) in East Nusa Tenggara Province, Indonesia. This NGO is supporting the steps of the community in handling flood disaster management starting from the earliest step, of the early warning system, prevention, mitigation, and community resilience in the West Malaka Subdistrict.

(f) **Financial or Economic Assistantship for the Physical Damage in the West Malaka Subdistrict**

Floods that swept away the community in the West Malaka Subdistrict resulted in physical damage to infrastructure development such as highways, irrigation, drainage, embankment, water dams, schools, markets, media, and telecommunications, internet, the offices of local government, and districts. These are public assets that need to be intensively well-maintained and repaired by society and also the Indonesian government. There are funds from the national budget and local government budgets as national and regional government, but they have not been realized since the beginning of the flood disaster occurred in 1939 in the West Malaka Subdistrict. The embodiment of economic policy in terms of improved public facilitation for the regions affected by the floods was realized by the Indonesian government through the national budget on October 3, 2014, with repairing roads and dike construction of the subvillage of Umamota at Sikun Village in the West Malaka Subdistrict.

6.2.4 Political or Policy Assets

1. **Introduction and Understanding of the Political or Policy Assets in Building Community Resilience**

In this context of study, political asset is defined as a strength of political leaders' roles (i.e., local leaders, regional leaders, and national leaders) that connect to any other of related strength institutions in order to provide a solution for the community's problems as well as a bridge to receive aspirations from the community and solve them in the process of policy implementation (i.e., Indonesian government's policies). As supposed by The Associations of Non-Governmental Organizations, TANGO (2003), political capital consists of power relationships, as well as access to and influence on the political system and governmental processes at local and higher levels (qtd. in Frankenberger et al. 2013: 13). The level of political capital at the community level determines the nature of community participation in the process of policy formulation and implementation (Frankenberger et al. 2013: 13). The community has the rights to accept the assistantship from the Indonesian government. The process of implementing the aspirations of the community therefore requires a strong basis of community participation. It aims at galvanizing the ideas and thoughts of the community to assess how was the sustainable use of political or policy assets that have implemented by the Indonesian government in the community. A discussion of building resilience through the sustainable use of political or policy assets is presented below.

2. **Building Resilience Through the Sustainable Use of Political or Policy Assets**

The implementation of a resilient "political or policy assets" can be realized in the community when there is an integrated solution in terms of flood disaster risk management and resilience. The role of political leaders is required to be involved in a group discussion within the community in order to discover effective, efficient, and sufficient solutions in addressing the resilience assets for floods. Figure 6.10 presents and discusses the role of political or policy assets in the community.

According to Fig. 6.10, there are 120 respondents who describe that the implementation of building resilience's discussion with the community in the West Malaka Subdistrict (i.e., thematic of resilience' assets) was carried out by Indonesian institutions. It indicates that the sustainable use of political or policy assets in terms of the Indonesian institution' roles for creating a thematic discussion about building

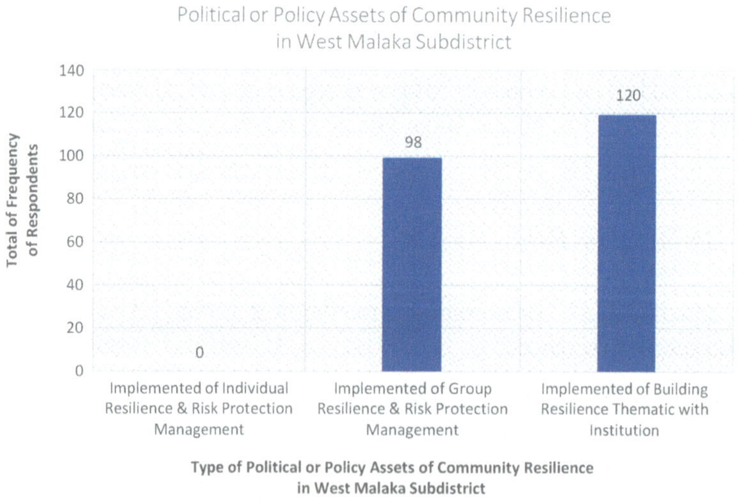

Fig. 6.10 Political or policy assets of community resilience in the West Malaka Subdistrict. *Source* Author's Research Fieldwork, July 9, 2016, to February 9, 2017

community resilience' assets were adequately applied by the Indonesian institutions. Whereas Indonesian institutions have less of concerns in implementing group resilience and risk protection management discussion with the community in which the community perceives these discussions as the most important issues. The community explains that resilience and flood risk protection management issues are an integrated problem which is needed to be solved by the Indonesian government. As it has been shown, there are 98 votes of respondents who agreed that the implementation of group resilience and risk protection management discussion was applied by the Indonesian government. Additionally, there are no votes of respondents who explain that there are thematic discussions about the implementation of individual resilience and risk protection issues with each individual and household.

The community in the West Malaka Subdistrict determines the rule of Indonesian institutions in terms of political or policy assets that are still concerning with a single solution (either it is resilience or flood as well as either it is concentrating for the individual or group, nor both of them), and neglecting a hierarchy solution which considers both resilience and flood disaster risk management solutions. Thus, the political or policy assets in resilience are still less of concerns from the Indonesian government.

(a) **Implementation of Individual Resilience and Risk Protection Management**

There is no sustainability of political or policy assets through implementation of individual resilience and risk protection management which was provided by the Indonesian government. During this time, the community has been independently creative in protecting themselves against the risks of flood disasters. Flood disasters occurring with long flooding cycles are incapable of attracting the Indonesian government in flood disaster mitigation actions, especially the Indonesian government's attention in the special priorities of individuals affected by the floods. Communities experience a long learning process during the flood cycle that always alternates and occurs in their area. They are always trying with independent strategies to reduce the danger of flooding through the assets of the resilience they have.

The efforts that society has taken to date have become a value of their independence in the form of enhancing self-protection against flood hazards, even though it is very risky to live with it. Meanwhile, the protection of assets from resilience is implemented through one of the strong systems of social resilience (i.e., bonding, bridging, and linking) that they have applied, and it has become the symbol of community power that is the system of elements of social assets. Individuals are exposed to risky hazards that threaten various assets of resilience starting from themselves as propulsion of assets from resilience (i.e., human assets like having good physical and spiritual health in the recovery phase), risk to their homes, vehicles, and public or physical facilities, land, availability of clean water, crops and gardens, risk to livestock, in-house properties of economic value, savings or economic assets, risk to daily social activities that support human assets and economic assets such as healthy communities. Community groups have concern about disasters or social assets, risk to the safety and security of customary leaders for the communities. These are known as the cultural assets, risk to cessation of aid mechanisms which are step by step implemented by the community and/or the continuation of the Indonesian government regulations that have not been equally implemented. Indonesian government has not targeted policies that prioritize the interests of the affected community by considering the types of risks from flood hazards that are still occurring in the West Malaka Subdistrict.

The role of the Indonesian government so far is still determined by the priority scale of its handling. Management and applied solutions are structural concrete actions of local actors from executive councils such as the Regent (head of Malaka Subdistrict), head of West Malaka Subdistrict, and head of village. However, the actions and concerns of local actors of the executive councils can be thought to be ineffective, inefficient, and insufficient for flood and resilience problems that require a long-term improvement and several recovery programs. Awareness and self-reliance of the community become a benchmark for the improvement of skills in protecting themselves and the assets of the resilience that they own and have implemented thus far.

(b) **Implementation of Group Resilience and Risk Protection Management**

Independent communities create solutions and increase the assets of their own resilience. Concerning the threat of a flood disaster, people have already developed a good habit of self-protection and protecting their assets. An effective habit is gained through the strong social system that accompanies the life of the community. The social system in question is a social asset consisting of three supporting elements namely bonding, bridging, and linking (as described in the previous part of social assets). Thus, the community has a resilience group to protect themselves from the flood disaster that is supported by the local government. Communities have practiced an adaptation which has a value in disaster management pattern. Implementation of Group Resilience and Risk Protection Management has been driven by the community with a conscience and noble desire which unifies the safety of the environment from the flood disaster. Flood disaster can be seen through the flood period assessment that reduces assets from resilience. However, the cultural value which was generated through the assets of social system has given a good reference for the Indonesian central government that has to pay attention to the environmental safety of the affected communities. The community in the West Malaka Subdistrict also has the rights to receive the assistance from the Indonesian government to deal with the problem caused by the flood disaster. Therefore, the cultural value and social value (cultural asset and social asset) that had been initiatively and effectively driven by the community, have their purpose in sustaining the other assets of resilience. It also aims to sustain their life and brings a sustainable future to their next generation, should they reside in flood-prone areas. Comprised of Indonesian citizens, the community should also have the rights to enjoy a better life assisted by the Indonesian government.

The community acknowledges the implementation of their Group Resilience and Risk Protection Management with 98 votes, showing their concern of overcoming the flood disaster. However, the community will not continue to be independent as a single actor from creating other innovative opportunities as well as beneficial solutions that can support communities in protecting themselves and their resilience assets. The availability of assistance from the Indonesian central government to be structurally run by the local government is part of a major solution that is expected to overcome the flood disaster in the West Malaka Subdistrict.

(c) **Implementation of Building Resilience with Institutions**

The flood disaster management pattern aims at protecting the assets from resilience that the community possesses and implements, and it can be applied also with a thematic learning from private and public institutions. Thematic learning is a frame dialogue that contains some discourses from the Indonesian central government and local government with the community in overcoming the flood disaster. The dominant intervention comes from private institutions such as Non-Government Organization (NGO), like the Community Association for Disaster Management (CADM) in East Nusa Tenggara Province, Indonesia who constantly monitors and observes the types of vulnerability aspects which are experienced by the affected community in the West Malaka Subdistrict. It can be proven by the role of NGOs in tackling the thematic case of resilience for the flood disaster in the West Malaka Subdistrict. There are 120 votes from respondents who said that the private institution, CADM has done the assessment of flood disaster and short-term community resilience response. Even though it was not yet in a long-term or integrative phase of resilience for a sustainable livelihood development, they created a tentative-platform solution for the affected community in the West Malaka Subdistrict. However, CADM has not been well developed and supported by other private institutions and public institutions both in Indonesia and foreign countries in terms of building community resilience and achieving the long term of community resilience. The long term of learning thematic and assessment of resilience was independently built by the community. These issues will be able to find solutions for an integrative assessment of community resilience. As long as the private institutions and public institutions are still discussing the physical aspects of the solution (i.e., water resource management of the Benanain River), the flood disaster management pattern and resilience system of the community are remaining ignored and not being solved by them. It means that the community is still the single actor of resilience to the floods.

6.3 Discussion and Summary

In discussion and summary, a significant part of building community resilience is human consciousness, adaptivity, and positivity in valuing their assets of resilience, in creating and maintaining their assets of resilience, and in preserving their assets of resilience to provide future resilience. Human actions are predominantly movement elements in the making of new coping strategies of an integrated resilience system (also refers to an integrated ecological system), in order to address flood disaster (see Chaps. 1, 2, and 8). Communities in the West Malaka Subdistrict have been consistently driving their actions without considering the scale of the challenge in dealing with flood disasters that threatens them or their livelihoods. Floods come nonetheless and bring problems besides socio-economic issues (less income and lack of social welfare) with which the community must deal. However, while floods bring a problem, they also bring a solution as a flood teaches the community to educate themselves and increase their consciousness of how to deal with floods. The more conscious they are, the more resilience actions they can create to cope with floods. Also, human consciousness is necessary for driving resilience actions to achieve sustainable development. It reflects human moral hazard that also embraces culture; communities value a moral hazard as the cultural aspect of their consciousness, adaptivity, and positivity in managing flood disaster. In managing flood disaster, they owned and implemented their assets of resilience.

This research highlights a solution that contains the communities' actions in striving and driving their resilience assets. Previous researchers have studied the definition of resilience and debated the core idea of how affected people in hazardous areas can achieve sustainable development. All agree that building resilience has become a crucial issue in the twenty-first century that therefore needs to be studied through empirical research to find a solution.

Using surveys and interviews with 320 households in Malaka-Timor has brought to light several findings about the community's existing resources and additional applicative resources. Discussion on the community's *pentagon resources* (five resources) in The Sustainable Livelihoods Framework (SLF), as presented by the FAO and ILO (2009: 11), has shown that a community owns and implements its *pentagon resources* (natural, physical, economic, human, and social) to address flood disaster, along with four further applicable resources (cultural, religion, law or legal, and political or policy). These *pentagon resources* and the four more applicable resources owned and implemented by the community are discussed below.

The active integration of the community is vital and important, as the community which possesses experience of flood disaster understands its own abilities to cope with flood. Communities have gained human capital from their skills (experience) and education (knowledge). Mayunga (2007: 5) and Cohen (1999) qtd. in Okuwa et al. (2016: 2) mention that human capital includes knowledge, skills, and experience. Human capital that is owned and applied by the community helps them nurture their skills and education to face flood disaster. The community learns and understands how they should cope with flood disaster using their ability. Learning and understanding the way of coping while in a difficult situation, such as a flood disaster, increases the human ability to develop other types of resilience, such as natural, physical, social, economic, cultural, religion, law or legal, and political or policy. PRISE & SDPI acknowledge that human capital is the capacity and capability of human beings to contribute to socio-economic growth (Sustainable Development Policy Institute, SDPI and Pathways to Resilience in Semi Arids Economies, PRISE 2015: 16). Human capital enables community resilience assets to be sufficiently developed and/or improved in regard to managing floods. Okuwa asserts that human capital interlaces a set of abilities to achieve resilience and that the dynamic nature of the environment makes it imperative for organizations to embrace human capital development so as to achieve resilience (Okuwa et al. 2016: 1; Schultz 1979: 5).

This chapter concludes that the community owns and implements its seven human assets; skills (experience), education (knowledge), good health, drinking water quality, consuming nutritious foods, shelter provided by the community, and helping mental recovery after flooding. All respondents in Malaka-Timor use these human assets to adjust to the other assets of resilience, natural assets, physical assets, social assets, economic assets, cultural assets, law, or legal assets, and political or policy assets. This *resilience's project* in this book is also supported by Nawrotzki et al. (2014: 12) in their field project which emphasizes a combination of *livelihood capital*, i.e., human, social, financial, and physical, which was utilized between migrants and non-migrants in accessing and utilizing natural assets. Since the community lives in a rural area and near the Benanain River watershed, fulfilling basic needs depends on the natural environment (utilizing natural assets). In addition, with the human assets that the community possesses, it utilizes all its resilience assets to deal with flood disasters. Similar to the example of the community in the West Malaka Subdistrict, another case project from Fragility to Resilience in Africa described in several major global declarations and reports over the past 15 years recognizes the potential of natural resource management to strengthen resilience in fragile situations (African Development Bank Group 2016: 1).

This present research project of this book tells a unique story about seasonal crops in the study area, where the community thought of a flood not as a bane but a boon. The interviewee *Leonardus Jefri* in Fafoe Village in the West Malaka Subdistrict explained: "*…The harvest from our plants after the rainy season (floods) and coinciding in the third harvest period was quite satisfactory to us. Production of corn and tubers and cassava can help our daily supply of food and also food supplies to cope with flooding….*" This example indicates that the community has applied an adaptive pattern in terms of seasonal crops in every rainy season and also used their abilities in planning the optimum periods of planting for the third harvest. This case proves that a community not only has a good adaptation to its natural assets (utilizing soil as a source for planting and consumption) but also has a significant understanding of flood disaster risk management. The community is considering flood disaster management and dealing with the issues of a single flood and is also looking beyond the risk of flood and associated problems (economic loss and lack of social welfare) to use its assets despite the flooding. Thus, they can shift the context of flood disaster management to disaster risk management which requires an integrated approach to building their resilience assets in Malaka. A further explanation is in Chap. 7.

In this chapter, I also present a discussion on the significant relationship between the nine resilience assets. For cultural capital, the traditional system takes its role led by the customary leaders to guide and teach people to obey customary rules regarding environmental protection. Traditional systems tend to function best in the event of irregular shocks and stresses, while formal systems are more effective in the context of predictable or regular shocks (Frankenberger et al. 2013; Morduch and Sharma 2002 qtd. in Frankenberger et al. (2013: 11).

This chapter notes a crucial issue of political assets in the need for the community to be heard by the Indonesian government and considered in the making of policies on assistance to local communities. The political assets implemented in this study will help the community in addressing flood disaster. Building community resilience for flood disaster provides scientific data where the community's aspirations through this research's survey and interviews should encourage the Indonesian legislative and executive councils, ministries, and institutions (universities and research institutions), who are expected to address community flood problems and resilience assets. In addition, researchers in universities and research institutions may also carry out fieldwork and explore entire villages that are affected by natural disaster in different places.

Furthermore, this chapter discusses political/governance capital assets as one of the applicable assets of community resilience, which was added by The United Nations Development Programme, UNDP (2013: 4–5). UNDP explains that political/governance capital assets consist of access to participation in government decision-making processes, gender relations, and power structures influencing control over decision making within households, communities and the wider society and access to judicial systems and security sector institutions. The community has agreed upon its aspirations which it wants to be heard by the Indonesian government so that the government will assist with their case, building resilience for flood disaster. There are two implementation actions that the Indonesian government has taken, and these two actions should be revitalized: implementation of group resilience and risk protection management, and implementation of building resilience with institutions. On the other hand, the communities did not agree on actions on individual resilience and risk protection management as these actions are not undertaken by the Indonesian government but by a private institution in the capital City of the East Nusa Tenggara Province Indonesia (i.e., Kupang City), the NGO Community Association for Disaster Management (CADM). In this case, the dominant role is undertaken by a private institution rather than a government institution in regard to building community resilience for flood disaster.

This political or policy asset data corresponds to law or legal asset data where all 320 respondents in the study area voted for economic assistance to repair the broken dike and public facilities, as well as agreeing that evaluation and socialization among institutions are important issues to be solved. These issues are important for strengthening disaster risk governance, while the community also requires an investment in disaster risk reduction (DRR) for resilience management. Some communities own and apply effective resilience assets while others do not, and all need sufficient resilience derived from their nine assets. Building community resilience through the sustainable use of social assets (i.e., bonding, bridging, and linking) in the study area, as discussed by Aldrich (2012), has led to an effective way of empowering a community's social resilience. The community in the West Malaka Subdistrict has strong social assets of resilience that assist in effectively addressing the flood disaster's problem.

This chapter concludes that the community has a strong system of mutual assistance, or *hakawak*, that the community values and which allows them to embrace their social capital; bonding, bridging, and linking. Social capital has been formed and applied by the community since before the flood disasters began in 1939 with guidance from the customary leaders. This research explains that social capital might be thought of as the single most productive assets among other assets of resilience as measured or viewed by the proactive activities that were carried out in everyday life and in response to flood disaster. These activities involve all the elements of resilience (nine assets) where human assets (skill and experience) are central to community actions.

References

African Development Bank Group (2016) From fragility to resilience: managing natural resources in fragile situations in Africa. A report of AfDB. Environmental Law Institute, Côte d'Ivoire

Ahern J, Galea S (2006) Social context and depression after a disaster: the role of income inequality. J Epidemiol Commun Health 60:766–770

Aldrich DP (2012) Building resilience: social capital in post-disaster recovery. The University of Chicago Press, The United States of America

Allison EH, Hobbs JR (2004) Resilience, adaptive capacity, and the "lock-in trap" of the Western Australian Agricultural Region. In: Ecology and society. Murdoch University, Australia, vol 9, no 1. Retrieved from https://www.ecologyandsociety.org/vol9/iss1/art3/

Associated Programme on Flood Management, APFM (2008) Organizing community participation for flood management: a tool for integrated management. Flood management tool series

Australian Wool Innovation and Meat & Livestock Australia, AWI, MLA (2016) Protect your farm's natural assets. Australia

Bahadur AV, Ibrahim M, Tanner T (2010) The resilience renaissance? Unpacking resilience for tackling climate change and disasters. Strengthening climate resilience discussion paper 1. First published by the Institute of Development Studies of University of Sussex, The United Kingdom

Baru P (2011) Pengelolaan Sumber Daya Air Wilayah Sungai Benenain. Direktorat Bina. Penatagunaan Sumber Daya Air. Atambua, Indonesia

Bashawry A, Stephan g, Moodley K (2014) An overview of the design of disaster relief shelters. Procedia Economics and Finance, Research's Article: 4th International Conference on Building Resilience, Building Resilience, Salford Quays. Elsevier, United Kingdom, vol 18, pp 924–931

Baumann P (2002) Livelihood support programme. An interdepartmental programme for improving support for enhancing the livelihoods of the rural poor. Food and Agriculture Organization of the United Nations (FAO)

Bourdieu P (1986) The forms of capital. The Eltan Burgos School of Economics, New York

Bria F (2004) Along the way to happiness: a perspective of Belu' people about the society's structures, culture, religiosity and tourism in Belu. Caritas Publishing House, Jakarta

Brown I, Thompson D, Bardgett R, Berry P, Crute I, Morison J, Morecroft M, Pinnegar J, Reeder T, Topp K (2016) The United Kingdom climate change risk assessment evidence report: natural environment and natural assets. Adaptation Sub-Committee of the Committee on Climate Change in London. University of York, Stockholm Environment Institute. Stockholm, Sweden

Brown J (2015) Preparedness and recovery for The Georgia Emergency Management Agency (GEMA): churches play a growling role in emergency management. Emergency and Management's News, USA. Retrieved from http://www.govtech.com/em/disaster/Churches-Playing-Growing-Role-Emergency-Management.html

Carney D (1999) Livelihoods approaches compared. DFID, Department for International Development, London

Carroll J, Epstein R, Fiscella K, Volpe E, Diaz K, Omar S (2007) Knowledge and beliefs about health promotion and preventive health care among Somali Women in the United States. Health Care Women Int 28(4):360–380

Chatters LM, Taylor RJ, Bullard KM, Jackson JS (2009)Race and ethnic differences in religious participation: African Americans, Caribbean Blacks, and non-Hispanic whites. Ethnic Racial Stud 32:1143–1163

Cohen (1999) Food security and conflict. Social research: an international quarterly food. Nat Cult 66(1):375–416

Damayanti F, Rusgiarto S, Riveira M, Riwu M, Hertz T (2010) Inter local government cooperation based integrated watershed management: Benenain River, Nusa Tenggara Timur (NTT): Decentralization as Contribution to Good Governance (DecGG) in cooperated with German Technical Cooperation (GTC) and Ministry of Home Affairs. Non-Government's Publication. Kupang, East Nusa Tenggara Province, Indonesia

Don-Nelson MR, Charley O, Lynne (2015) Ecosystem service monitoring for range managers. Sustainable food and farming network. Washington State University

Ellis F, Allison E (2004) Livelihood diversification and natural resource access. Livelihood Support Program, Food and Agriculture Organization of the United Nations (FAO). Overseas Development Group, University of East Anglia, The United Kingdom

Felix D, Feio A, Branco M, Machado J, Jose S (2013) The role of spontaneous construction for post-disaster housing. In: Cruz (ed) Structure and architecture concepts, application and challenges. Taylor and Francis Group. London. ISBN 978- 041566195-9

Fernandez A, Black J, Jones M, Wilson L, Salvador-Carulla L, Astell-Burt T, Black D (2015) Flooding and mental health: a systematic mapping review. A research article. Plos One 10(4):e0119929. Retrieved from https://doi.org/10.1371/journal.pone.0119929

Food and Agriculture Organization and International Labor Organization, FAO, ILO (2009) The livelihood assessment tool-kit. Analyzing and responding to the impact of disasters on the livelihood of people. Policy and Support Branch Communication Division. Geneva and Rome

Frankenberger T, Mueller M, Spangler T, Alexander S (2013) Community resilience: conceptual framework and measurement feed the future learning agenda. Westat, Rockville

GITEC-IGIP GmbH (2016) Improving rural livelihoods, increasing food security and promoting the sustainable management of resources are the core goals. Natural Resources and Rural Development Department. Retrieved from http://www.gitec-consult.eu/index.php/en/business-areas/natural-resources-rural-development

Gordon A, Andrew L, Kate M (2000) The role of natural resources in the livelihoods of the urban poor. Social Sciences Department, Natural Resources Institute—NRI. University of Greenwich

Hawkesbury-Nepean Floodplain Management Steering Committee (2006) Managing Flood Risk through Planning Opportunities—Guidance on Land Use planning In Flood Prone Areas, Hawkesbury-Nepean Floodplain Management Steering Committee, Parramatta

Holman SR (2009) Tearfund International Learning Zone—TILZ religion, global health, and human rights. Oxford University Press, New York

Illicit Discharge Elimination Program, IDEP (2010) Panduan Penanggulangan Bencana Berbasis Masyarakat: Buku Panduan Praktis untuk Sebelum, Saat dan Sesudah Bencana. IDEP Foundation Publication, Bali

International Business Machines, IBM (2009) In the spotlight: the human side of business continuity planning. Global Service: GSW00160-USEN-01, New York. Retrieved from http://www.continuitycentral.com/feature0642.html

International Fund for Agriculture Development, IFAD (2011) Rural poverty report: rural poverty and natural resources, improving access and sustainable management. The Agricultural Development Economics Division, Food and Agricultural Organization of the United Nations, Rome

International Fund for Agriculture Development, IFAD (2012) Environment and natural resource management, resilience livelihoods through the sustainable use of natural assets. The International Fund for Agricultural Development

International Federation for Red Cross and Red Crescent, IFRCS (2011) Participatory approach for safe shelter awareness. Geneva

International Federation for Red Cross and Red Crescent, IFRCRC (2013) Legislative advocacy for disaster risk reduction. International Federation of Red Cross and Red Crescent Societies, IFRCRCS. Legislative Advocacy Guidance Note Series No 1. Geneva

Jennifer E, Palmer W (2009) Building resilience within human capital development global trends versus local necessities. Sir Arthur Lewis Institute of Social and Economic Studies, SALISES, Barbados (2009)

Jha M (2010) Natural and anthropogenic disasters: an overview of vulnerability, preparedness and mitigation. Springerlink, The Netherlands. ISBN: 978-90-481-2498-5

Johnson C (2008) Strategies for the reuse of temporary housing. In: Ruby IA (ed) Urban transformation. Ruby Press, Berlin, pp 323–331

Jones O, Garde-Hansen (2012) Geography and memory: exploration in identity, place and becoming. In: Countryside and Community Research Institute and University of Gloucestershire, Palgrave Macmillan, The United Kingdom

Krishna A, Shrader E (1999) Social capital assessment tool. The World Bank, Washington (1999)

Lee DR, Neves B (2011) Rural poverty and natural resources: improving access and sustainable management. IFAD Rural Poverty Report, The United States of America

Legha RK, Solages M (2003) Child and adolescent, mental health in Haiti, developing long-term mental health services after the 2010 Earthquake. Boston, The United States of America. Retrieved from http://www.sciencedirect.com/science/article/pii/S1056499315000498

Linnell J (2003) The role of natural resources in sustainable rural livelihoods in the Western Balkans. Norwegian Institute for Nature Research—NINA. Trondheim, Norway

Mayunga JS (2007) Understanding and applying the concept of community disaster resilience: a capital-based approach, USA

Miller F, Osbahr H, Boyd E, Thomalla F, Bharwani S, Ziervogel G, Walker B, Birkmann J, van der Leeuw S, Rockstrom J et al (2010) Resilience and vulnerability: complementary or conflicting concepts. Ecol Soc 15(3)

Morduch J, Sharma M (2002) Strengthening public safety nets from the bottom up. Dev Policy Rev 20(5):569–588

Morrone A (2003) Measuring vulnerability and resilience in OECD countries. In: Paper—IARIW-OECD conference on economic issues. ISTAT, Italy

Nawrotzki JR, Hunter LM, Dickinson TW (2014) Rural livelihoods and access to natural capital: differences between migrants and non-migrants in Madagascar. The United States National Library and Medicine, Max Planck Institute for Demographic Research, Germany

Nelson S(2015) River flooding. In: Natural disaster issues EENS 3050–6050 Department of Earth and Environmental Science, Tulane University

Neonbasu G (2013) Kebudayaan: Sebuah Agenda dalam Bingkai Timor dan Sekitarnya. PT gramedia Pustaka Utama Jakarta, Indonesia. ISBN 978-977-22-9343-2

Nunes AR (2016) Assets for health: linking vulnerability, resilience and adaptation to climate change. Tyndall Centre for Climate Change Research, Statistics and Epidemiology Department, Division of Health Sciences, The Warwick Medical School, University of Warwick, The United Kingdom

Okuwa AJ, Nwuche CA, Anyanwu C (2016) Human capital development and organizational resilience in selected manufacturing firms in Rivers State. Department of management, University of Port Harcourt, vol 3, Issue 2, pp 43–50. ISSN 2394-9694

Pierre A, Minn P, Sterlin C, Annoual PC, Jaimes A, Raphaël F, Raikhel E, Whitley R, Rousseau C, Kirmayer LJ (2010) Culture et santé mentale en Haïti : une revue de littérature. Santé mentale au Québec 35(1):13–47. Retrieved from https://www.erudit.org/fr/revues/smq/2010-v35-n1-smq3941/044797ar/ and https://doi.org/10.7202/044797ar

Queensland Government, Department of Environment and Heritage Protection (2012) Pathways to a clean growth economy: Queensland climate transition strategy. Climate Change's Research Project, Queensland

Sawada Y, Fauziah Z (2014) Disaster management in ASEAN. Singapore Institute of International Affairs (SIIA) and Economic Research Institute for ASEAN and East Asia (ERIA). ERIA's Publication, Singapore

Schultz (1979) The economics of being poor. Blackwell Publishers, Cambridge, Massachusetts

Sustainable Development Policy Institute, SDPI & Pathways to Resilience in Semi-Arid Economies, PRISE (2015) Connecting the dots: linking climate change resilience to human capital, Pakistan

Smithers J, Smith B (1997) Human adaptation to climatic variability and change. Glob Environ Chang 7(2):129–146

Solnit R (2009) A paradise built in hell: the extraordinary communities that arise in disaster. Viking, New York

Steinberg F (2007) Housing reconstruction and rehabilitation in Aceh and Nias, Indonesia: rebuilding lives. Habitat Int 31:150–166

Sundaram MM, Jha MM (2004) Natural resources management and livelihood security: survival strategies and sustainable policies. Bhoovigyan Vikas Foundation, New Delhi

The Association of Non-Governmental Organizations, TANGO (2003) Socio-political analysis: critical information needs for addressing underlying causes of poverty and marginalization. Prepared for CARE, Draft

The Holy Bible New International Version (2011) Biblical. Retrieved from Salem web network: https://www.biblestudytools.com/genesis/2.html

The Micah Network Global Consultation (2006) The role of the local Church in disasters and emergencies: a workshop's draft in 11–16 September, Thailand

The Scottish Climate Change Adaptation Programme (2016) Second annual progress report on climate change. The Scottish Government Riaghaltanah-Alba, APS Group Scotland, Edinburgh. ISBN: 978-1-78652-273-3

The United Nations International Strategy for Disaster Reduction, UN-ISDR (2007) Hyogo Framework for Action (HFA), 2005–2015: building the resilience of nations and communities to disasters. The HFA's Research Projects, in the Andean region, Central America, the Caribbean, Asia, Pacific, Africa and Europe

The United Nations Development Programme, UNDP (2013) Livelihoods and economic recovery in a crisis situation. Bureau for Crisis Prevention and Recovery Livelihoods and Economic Recovery Group, New York. ISBN: 978-921-1-126352–7

The Watershed Management Bureau of Benanain, Noelmina (2011) Rencana Pengelolaan Daerah Aliran Sungai Terpadu di DAS Noelmina. Kementrian Kehutanan Direktorat Jenderal Bina Pengelolaan DAS dan Perhutanan Sosial. Kupang, Indonesia

The White House Office of Faith-Based and Neighborhood Partnership (2009) A toolkit: partnership for the good common. The Centre for Faith-Based and Neighborhood Partnership, The United States Department of Health, and Human Services, Washington DC

University of Adelaide (2004) Natural resources management. Building South Australia Programme. University of Adelaide's Working Paper, Adelaide

World Meteorological Organization, WMO and Associated Programme on Flood Management, APFM (2009) Integrated flood management: concept paper. World Meteorological Organization, Geneva, Switzerland

Chapter 7
Community-Based Disaster Risk Management

In this chapter, I introduce the preliminary study of the flood characteristics and community assets and discuss community risk perception for periodic flooding, duration of flood, flood depth, and flood distribution (river flood), as well as community risk perception of how their assets of resilience were impacted by flood disaster and how they coped with it. The central idea of the analysis of flood disasters is periodic flooding, which highlights the essence of building community resilience for flood disaster and is therefore important to be studied and presented in this section. Period flooding in Malaka-Timor, Indonesia, occurred once very twenty years in 1939, 1959, 1979, and 1999, and then twice a year in 2000, 2001, 2002, 2003, 2004, 2005, 2006, 2007, 2008, 2009, 2010, 2011, 2012, 2013, 2014, and 2015. The fieldwork in Malaka-Timor, Indonesia, was conducted in 2016 (see Table 6.1). There was no recorded data of building community resilience for flood disaster in Malaka provided by the Indonesian government, non-government institutions in Indonesia or international institutions. This research project of this book pioneers the knowledge of community disaster risk management using the community's own risk perception for building resilience, and this approach is applicable in both the national scale in Indonesia and the international scale.

In the context of disaster risk management and policies, some previous authors, researchers, and practitioners from Indonesia and other countries have discussed the role of institutions and disaster risk governance in Indonesia. These researchers evaluated Indonesia's disaster risk index compared to the global disaster risk index using the mixed method, where the data was provided by the institutions in Indonesia and international institutions rather than assessing community risk perception in survey and interviews (i.e., survey and interviews with the affected households in the prone area of natural disaster) and in the field discussions (i.e., Focus Group Discussion or FGD and Stakeholder Meetings or SMEs).

In the context of building resilience for natural disaster, in this case flood disaster as discussed in Chap. 6, this research presents a comprehensive analysis of the content of building resilience (i.e., nine resources or *nine livelihood resources*) in coping

A. D. S. da Costa, *Building Resilience for Flood Disaster in Malaka-Timor, Indonesia*, Springer Natural Hazards, https://doi.org/10.1007/978-3-031-63272-3_7

with flood disaster using community risk perception. The case of flood disaster risk management is thus a crucial issue that needs an assessment and a measurement of how to build community resilience (see Chaps. 6 and 9). On this analysis, I claimed that flood risk management entails community risk perception toward their experience and understanding to handle floods. The use of community risk perception is due to the reasoning idea that lies on the assumption that a community who experiences flood disaster has an understanding in how to manage a flood disaster. It then highlights in this analysis that a community who experiences flood disaster has a "*lesson-content of understanding*" or "*knowledge*" which provides it with essential information for dealing with community-based disaster risk management.

On the international scale, some previous authors, researchers, and practitioners have introduced and discussed building community resilience due to natural disasters, including flood disasters. However, this research found that there is no comprehensive knowledge that considers the human geography perspective, which is crucial in analyzing building resilience for flood disaster.

From the survey and interviews with 320 households in the research location, one of the circumstances that this research notes is the role of the private institutions in Indonesia that are still actively running disaster management activities and building community resilience in coping with natural disasters in place of government institutions. This observation proves that the issue of building community resilience for natural disaster has not been so attractive and/or prioritized by the Indonesian government and government institutions.

Hence, this chapter focuses on the content analysis regarding community-based-flood disaster risk management using community risk perception. This book tries also to be at the forefront of the analysis of building community resilience for flood disaster in the research location with the help of related data and information on community disaster management and community coping system from Non-Government Organizations (i.e., Community Association for Disaster Management (CADM) or *Perkumpulan Masyarakat Penanganan Bencana* (PMPB) in Kupang City, in the West Timor Island, East Nusa Tenggara Province, Indonesia.

CADM is the sole Non-Government Organization which focuses its mission on the issue of humanitarian disaster relief with fieldwork such as community disaster risk management, climate change adaptation and reducing poverty, and disease of the community in East Nusa Tenggara Province, Indonesia.

Building community resilience for flood disaster in the West Malaka Subdistrict of Malaka District, in the West Timor Island, East Nusa Tenggara Province, Indonesia, pioneers this research which requires the support of international institutions with humanitarian missions, such as the United Nations. The United Nations intends its efforts to be inclusive across its policy frameworks through: *Reaffirming further the Hyogo Declaration, the Hyogo Framework for Action 2005–2015: "Building the Resilience of Nations and Communities to Disasters and the common statement of the special session on the Indian Ocean disaster: risk reduction for a safer future, as adopted by the World Conference on Disaster Reduction, held in Kobe, Hyogo, Japan, from January 18 to January 22, 2005, and bearing in mind that the implementation of the Hyogo Framework for Action is coming to an end in 2015.*" (The United Nations

International Strategy for Disaster Reduction (UN-ISDR) 2007: 1). This strategy is discussed in Chap. 1.

This book reinforces the idea of translating community-based disaster risk management into practical actions, in that this study agrees with the international policies of the UN-ISDR that promote a solution for tackling the case of humanitarian disaster relief in the world including Indonesia. UN-ISDR declares that "*they have passed their efforts through reaffirming its resolution 46/182 of 19 December 1991, to which contains the guiding principles for the strengthening of the coordination of emergency humanitarian assistance of the United Nations system, as well as all its resolutions on international cooperation on humanitarian assistance in the field of natural disasters, from relief to development, and recalling the resolutions of the humanitarian segments of the substantive sessions of the Economic and Social Council*" (UN- ISDR, 2007: 1). This point is also discussed in Chap. 1.

Flood disaster risk management based on community perception is one of the crucial issues in humanitarian disaster relief that requires international cooperation, as UN-ISDR mentioned in its declaration and resolution 46/182 of 19 December 1991 about "*humanitarian assistance in the field of natural disasters, from relief to development, and recalling the resolutions of the humanitarian segments of the substantive sessions of the Economic and Social Council.*" To understand flood disaster risk management, this study explores the strength of the community's resilience system in coping with flood disasters. In gathering such information related to flood disaster risk management in the research location, surveys and interviews and in-depth field were used. This chapter focuses on the results of the survey and interviews gathered from the 320 households in the research location and discusses how the community understands its risk management in coping with flood disaster.

The use of the field survey and interviews is compatible with understanding the community's problems about how they build resilience for flood disaster and implement their flood disaster risk management based on their own risk perception. These interviews and participative discussions are incorporated with descriptive qualitative research that has been obtained in information related to community-based disaster risk management in the research location. Such fieldwork was supported by the Indonesian disaster management regulation; in clause 26 (1:e) chapter V, number 24 from 2007 mentions that "*every person has a right to participate in disaster mitigation decision making*" (Community-Based Disaster Risk Management, CBRDM 2005: 39). This topic was also discussed in Chap. 1.

The idea of this disaster management regulation might be understood as a support or encouragement from the Indonesian government to researchers in the academic and non-academic, non-government, and government institutions in Indonesia to implement practical action in gathering information related to disaster risk management. The regulation aims to solve the issues around natural disasters in Indonesia. This study asserts that risk perception in field discussions can be considered as community-based disaster risk management that helps communities tackle their flood disaster risk management and improves their resilience management. This approach allows communities to understand and build knowledge of disaster. Risk management in the area prone to natural disaster, and it might be used for current and subsequent studies in regard to social-economic characteristics of the research site.

This book has considered the analysis of building community resilience according to two objectives. The first objective was discussed in Chap. 6 (i.e., building community resilience for flood disaster). This chapter discusses the second objective (i.e., community disaster risk management) where a community owns and applies its resilience system. The data referred to in this chapter is based on community risk perception documented through survey and interviews with the 320 households in Malaka-Timor, Indonesia.

7.1 Disaster Risk Assessment and Discussion

A detailed risk assessment should be carried out by the community disaster committee led by a local leader (Cutter et al. 2008: 19). According to UN-ISDR (2007), within the context of UN-ISDR, the Hyogo Framework for Action (HFA) 2005–2015 states that Building the Resilience of Nations and Communities to Disasters is the key planning document that is guiding international efforts to reduce the risk and impact of natural disasters (qtd. in The World Bank 2008: 20).

The UN-ISDR mentions the HFA's five strategic priorities for action (UN-ISDR, 2007: 9 qtd. in The World Bank 2008: 20) are to:

1. Ensure that disaster risk reduction is a national and local priority with a strong institutional basis for implementation.
2. Identify, assess, and monitor disaster risks, and enhance early warning.
3. Use knowledge, innovation, and education to build a culture of safety and resilience at all levels.
4. Reduce the underlying risk factors.
5. Strengthen disaster preparedness for effective response at all levels.

These five strategic priorities have been implemented by the community in the West Malaka Subdistrict. In order to implement a plan, the most appropriate approach or key method that can be useful to gathering key information from key persons needs to be put in place using a research-based community approach with survey and interviews and FGD and SMEs.

This chapter presents a discussion of flood characteristics and community resilience using the interview method. The information related to flood hazard was contextually experienced by the community in the West Malaka Subdistrict who provided key information as they are central to this research. Ensuring disaster risk reduction involves a community that has perceived its risk and implemented assets of resilience to overcome the risk of flooding. The community has a strong local or ethnic priority (i.e., a strong culture) which has formed a social capital in terms of bonding, bridging, and linking, as explained in Chap. 5 which describes the social capital of the community in the West Malaka Subdistrict.

This particular community has a unique local wisdom that can function as an effective shaper or effective influencer to be practiced by the community. The collaboration and cooperation of the customary leader, religious leader, and the volunteer disaster group, which was founded by the community's capacity of awareness, is the vital resource to assist and guide a community. Therefore, in order to run the phase of identification, assessment, and monitoring, and to implement the assets of resilience that a community has (whether before a flood disaster in the ex-ante period, during a flood or after a flood disaster in the ex-post period), a strong sense of culture is needed. This culture leads a community to have the safety of resilience not only to protect its assets of resilience but also to improve and maintain them in accordance with culture that the customary leader has established for the community.

7.2 Human Risk Perception in Relation to Flood Characteristics (Interviews)

Examining the community perception of flooding is a keyway to recognize the process they have taken in their efforts to manage flooding. Indonesian government regulation (number 76, year 2001), article two states that Indonesian government and rural development organizations should take an active role in the community so that people will continue to take part in the development of life with fellow of villagers (qtd. in The Watershed Management Bureau of Benanain-Noelmina 2011: 29–30).

Recognition that the Indonesian government has issued this regulation made possible the community's actions in preventing flooding and overcoming the problems of recovering their assets due to risk of flood. Communities are therefore very enthusiastic about giving their risk perception through flood-related information.

Additionally, the community who owns and implements their resilience assets has explained their risk perception about how to use resilience assets to deal with flood. The relationship between the assets of resilience that community has perceived as strengthening its resources has supported a solution-based community approach to the negative issue of flood risks. The community was active in providing information related to flood hazard, such as an alerting system in the ex-ante period, communication among households about fleeing to a disaster shelter during floods, galvanizing households to clean mud through physical reconstruction (resilience), and re-establishing their agriculture group in terms of economic resilience. The location of the 320 respondents who participated in the research project of this book can be seen in Fig. 7.1.

Risk perception of floods in the West Malaka Subdistrict involves diverse flood disaster information. The flood disaster information was experienced by the community and also describes the assessment of flood risk itself. Moreover, flood risk which has a destructive potential to intervene in community assets of resilience, does not intervene in the community's sense of positivity and adaptivity in addressing flood disaster. Community continues to own a sense of positivity and adaptivity as a means

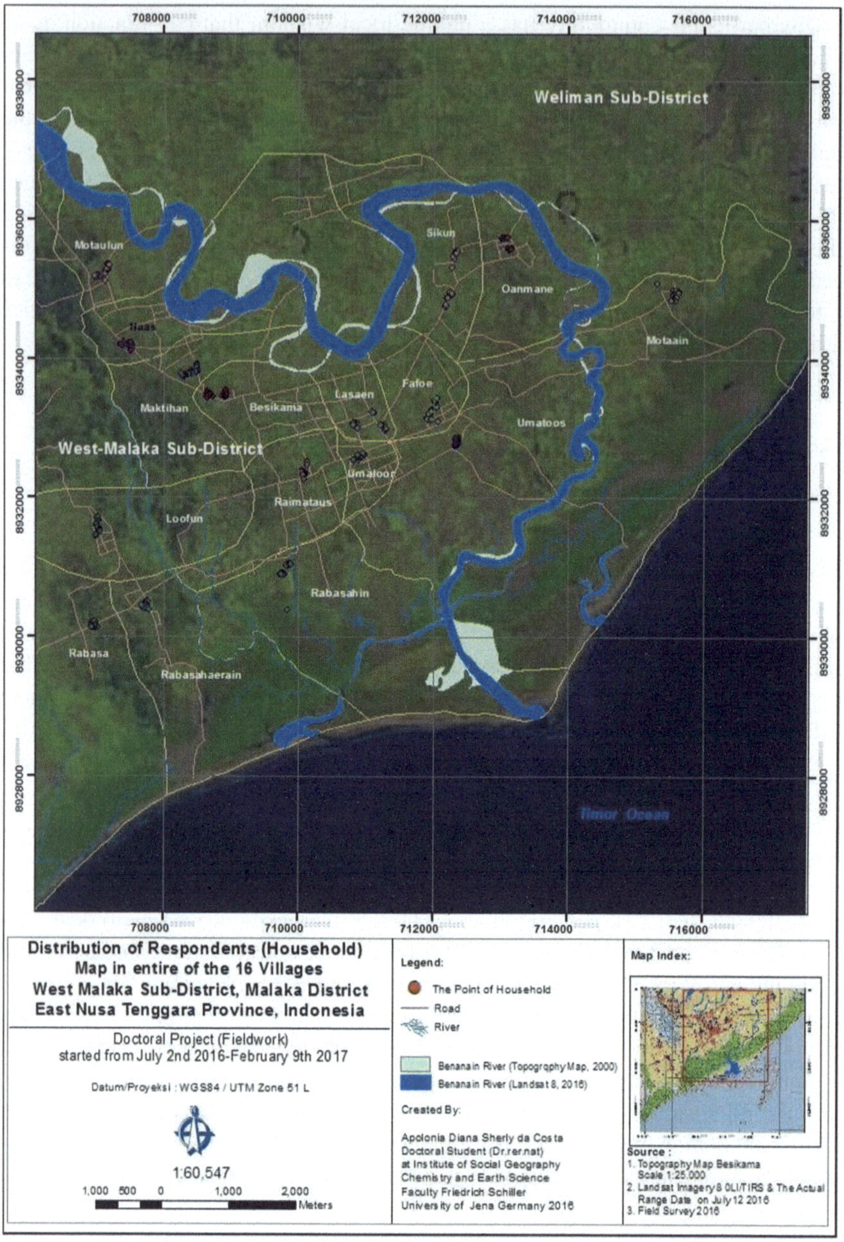

Fig. 7.1 320 distribution of respondents (households) map in all 16 villages of the West Malaka Subdistrict, East Nusa Tenggara Province, Indonesia. *Source* Author's Research Fieldwork, (Survey and Interviews) July 9, 2016, to February 9, 2017

also to increase their coping capacity. Their sense of positivity and adaptivity were sourced from the existence of cultural entity that the community has been maintaining and preserving from their ancestors. For instance, there are some young members of households in the study area who live in the capital city of the East Nusa Tenggara Province, Indonesia (i.e., Kupang City) to pursue their education, while their parents continue to live in the study area. The young people have less experience of managing a flood disaster, but they understand flood information as they are educated people. Nevertheless, the young people are able to handle flood disasters as their parents taught them about the importance of a strong culture. The usefulness of a strong cultural entity helps the people to still value the traditional approach (i.e., traditional rituals and worshipping to their ancestors) being practically implemented to cope with flood disaster such as traditional rituals with customary leaders, with help also from the young people to apply traditional rituals and worship to their ancestors in the phase of constructing the broken dike in Sikun Village. It shows the community's cultural entity that supports them to apply risk management in terms of their daily adjustment to the flood disaster.

The community has also described its abilities to deal with flood disaster management. An understanding of flood disaster management was introduced and discussed through flood characteristics. Flood characteristics consist of four types of characteristics: periodic flooding, flood duration, flood depth, and flood distribution. In order to withstand each type of flood characteristic, resilience is a mandatory coping system to address floods. However, community-based disaster risk management in this chapter tends to analyze the second objective of the study, that is, a perspective of community understanding regarding flood risk management rather than a lot of analysis of community resilience itself. An understanding between the relationship of community-based disaster risk management and community resilience analyzes the way that community assets were being damaged by the flood disaster.

Through risk perception also, community shows that some valuable assets of resilience were used and impacted by the flood disaster while they are still evaluating a process of encountering the flood risk problems rather than just delivering complaint and criticism.

The dilemma that the community faces is how to maintain and strengthen their assets of resilience, so they are always relevant to the challenges they are facing (i.e., risk of flood). It is information that determines the relationship between the community's strength or resource and the risks of flooding. This relationship could be understood as a valuable process for the community to have an essential understanding of their resiliency strategies which enable them to recover from floods. One of the challenges that a community faces is obtaining information about community risk perception in respect of periodic flooding. Daily activities that were undertaken by the impacted community and the type of assets of resilience that they had to implement are described in the discussion that follows. Risk perception explains how they have confronted the types of risk which are described in the discussion on periodic flooding.

7.2.1 Periodic Flooding

Periodic flooding describes the duration of a flood disaster and the way the community perceives and deals with it. Periodic flooding is also known as the return period of flooding and is the time period during which it is likely that a particular magnitude flood will occur (Frazer 2010: 3). All 16 of the villages in West Malaka Subdistrict experienced flooding every twenty years in 1939, 1959, 1979, and 1999. However, activity increased, and floods occurred twice a year in 2000, 2001, 2002, 2003, 2004, 2005, 2006, 2007, 2008, 2009, 2010, 2011, 2012, 2013, 2014, and 2015.

Community perception for this altered cycle of flood disaster is therefore different for each cycle. The perception this research is based upon is the actual date of the research fieldwork in West Malaka Subdistrict, from July 9, 2016, until February 9, 2017.

Based on interviews with 320 respondents in West Malaka Subdistrict, there appears to be no previous statistical data about periodic flooding in their area. There is no institution in Indonesia or overseas that has conducted research about the period of floods and community resilience. The community is witness to history because it experienced the flood disaster.

This study uses statistical data from the community risk perception. The statistical data obtained explains the cycles of floods that the community has observed from 1939 to 2016. Community informs and explains that the accurate information of the periodic flooding is also determined by how long they have been settled in the study area. This means that the longer the duration they live there, the more experience or skill they gain in order to adjust with the flooding. The people settled in the study area since the first flood disaster hit in 1939 are the native people of West Malaka Subdistrict. They are indeed the interviewees' ancestors; thus, the interviewees are also native people, have been settled for many years, and have sufficient information related to flood disaster in the study area. Such information from the native people becomes evidence data, and this data was used to inform and explain periodic flooding in the research location. The discussion is provided in Table 7.1.

Since the beginning of the flood disasters in 1939, people have viewed floods as a natural phenomenon associated with the large river in their area West Malaka Subdistrict, and how it crosses the community settlements. Thus, the risk of flooding was already recognized by the community. Based on the interviews with 320 respondents in West Malaka Subdistrict, the number of flood events has increased. Flood occurred from 1939–1959, 1959–1979, and 1979–1999 only once in each 20-year period. This data is based on group interviews with the leaders of the indigenous people in the Lasaen Village, who spoke of how early the history of flooding and associated difficulties occurred in West Malaka Subdistrict. Figure 7.2 shows the Group Interview with the indigenous leaders and an indigenous woman leader in front of the statue of the Indigenous King in Lasaen Village and Group Interview with village officials and the disaster activists or volunteers in Lasaen Village.

In 1999, the pro-Indonesian refugee community in East Timor chose Indonesia and moved or fled to the border area and southern part of the border area, i.e.,

Table 7.1 Risk perception for periodic flooding in the West Malaka Subdistrict

Periodic flooding	Year of cycle of flood occurrence	Total of flood experience	Frequency of risk perception (Yes)	Percentage (%)
1939	20	Once	320 respondents	100
1959	20	Once	320 respondents	100
1979	20	Once	320 respondents	100
1999	20	Once	320 respondents	100
2000–2010 (Annually)	1	22	320 respondents	100
2010–2016 (Annually)	1	21	320 respondents	100
Total	98	47	320	100

Source Author's Research Fieldwork, July 9, 2016, to February 9, 2017

(a). Group Interview with the Man' Customer Leaders and a Female Leader in front of the Statue "The King of Custom" or Ba'i

(b). Group Interview with the Chief Executive and a Disaster Activist in Lasaen Village

Fig. 7.2 Group interview with the man's customer leaders and a female leader in front of the statue "The King of Custom" or Ba'i in West Malaka Subdistrict and group interview with the chief executive leaders and a disaster activist in Lasaen Village. *Source* Author's Research Fieldwork, July 9, 2016–February 9, 2017

Malaka District and West Malaka Subdistrict. Furthermore, this political disaster in East Timor in 1999 was coupled with a flood disaster that occurred in 1999 to 2000. The worst case was that there were more than 300 refugees from *Ainaro* District of East Timor who fled to the southern part of the border area, Malaka District and West Malaka Subdistrict. The Ainaro refugees were not able to adjust to the new environment; they were living near to the Benanain River watershed and should adapt to flood disaster.

This led to humanitarian disaster relief issues which also connects to socio-cultural anthropology at the same time in terms of individual and social adaptation to the natural environment and socio-cultural environment. In a review of anthropology, displaced people who moved to the southern part of the border came from a society that obviously derived from *Ainaro* District with different or contrasting geographical situations, i.e., living in mountainous areas. This was in contrast to the way of life and social adaptation to the natural environment of the West Malaka Subdistrict community who lived near the Benanain River.

In addition, economic resilience, and physical resilience issues in 1999 were not a serious concern of the Indonesian government. The economic resilience issues provided by the Indonesian government for the refugee people of East Timor in the West Timor Island were the stocks of food, clean water, clothes, and subsidies, while the physical resilience issues were resettlement (i.e., house for the refugees) and sufficient transportation inter-subvillage-village to bind social capital.

The West Malaka Subdistrict is home to a society that is aware of the natural environment and social environment, so the patterns and mechanisms of adaptation were established before the floods in 1939 and continue to the floods of today. Different adaptations are due to differences in geographic location where people settle and adapt and thus are triggered new conflicts in society in practice, and also communities maintaining their cultural values in terms of anthropological history.

Communities have a different adaptive mechanism that means they need assistance and guidance from their customary leaders. However, such assistance and guidance were not implemented by the customary leaders in 1999 and 2000. The customary leaders' role could also help the refugees from East Timor while they were facing a lack of social capital not fulfilled by the Indonesian government in 1999 and 2000.

It is not the fault of either party, but a crisis befell them between natural disasters and bitter political disaster which was felt by both local residents and the migrant community of West Malaka Subdistrict in East Timor's pro-Indonesia integration. In addition, multiple tragedies of political disaster and flood disaster in 1999–2000 were also being noticed by the people in the West Malaka Subdistrict in 2016, as a sudden crisis that their information might not be heard and reached by the customary leaders and also the religious leaders, because the customary leaders and religious leaders in the study area were focusing on the protection of mitigation measures and guidance role for the affected people due to flood disaster.

At that time, the community in West Malaka Subdistrict processed the adaptation pattern to a point of maturity to act as a starting action to maintain the other assets of their resilience in handling the flood disaster. While the lack of management in rescuing the refugee people, and their safety and welfare while flood disaster hit the area and without applicable resilience assets in 1999–2000 is understood, the problems were not appropriately handled by the Indonesian government and international humanitarian mission (i.e., the United Nations High Commissioner for Refugees or UNHCR).

This issue became part of the impact of Indonesian government policies which do not take account of the carrying capacity of a strategic environment for refugees. Reconciliation and reparations have not been fully resolved by the Indonesian government and UNHCR in their duties and responsibilities to solve the problem of East Timor refugees in West Malaka Subdistrict. East Timor refugees were displaced and fled to West Malaka Subdistrict and other Timor Island regions such as South-Central Timor and North Central of Timor, pushing Indonesia into a prolonged conflict for both parties and for both local communities of West Malaka Subdistrict and East Timor refugees.

The data obtained has indicated that more than 300 refugees from *Ainaro* District in East Timor were killed due to floods in West Malaka Subdistrict in 2000. It was a sad tragedy that 300 people who had already experienced a political disaster not resolved by the Indonesian government should have the misfortune to be affected by natural disaster. An update of the Indonesia government's policy is needed in respect of the duties and responsibilities of the Indonesian government together with UNHCR's mission in the reconciliation process, namely:

(a) Providing settlement for East Timor refugees living in Indonesia, especially those who are living in the southern part of the border area and flood hazard area, i.e., West Malaka Subdistrict.
(b) The fulfillment of the rights of financial or monetary issues of the East Timor refugees by providing the rights for the farmers to farm for the sake of food security and by providing the accessibility and assistantship to education. (*Interviewee, Rony Seran: November 11, 2016*)

In addition, a right that has not been met by the Indonesian government is the people's right to get insurance for disaster risk management (DRM) to support the assets of resilience that they have implemented in support of the process of handling the flood disaster.

The East Timor refugees who were not affected by the floods continued to depend on the power of the natural resources near to the Benanain River. However, this violated the balance of ecosystems that have an impact on the flood disaster. The refugees were cutting down trees in the forest near the Benanain River to get access to natural assets for their physical resistance, i.e., build their houses. However, the balance of the ecosystem has been destroyed, leading to natural disaster, i.e., flood. Flood increasingly threatens people during the rainy season with the lack of physical assets, such as no drainage system, broken dikes, and irrigation, and firmly reinforces the destruction of various assets of community life in the aftermath of flood disaster for the people in West Malaka Subdistrict.

Moreover, from 2000 until early 2010, floods swept the community in West Malaka Subdistrict. There were 22 flood events in this period. Every year starting from 2000, two floods occurred each year, up to the beginning of 2010. Meanwhile, at the end of 2010 through 2016, there were 21 occurrences of floods. flood events increased, and the most flood events recorded occurred in each year from the end of 2010–2016. Communities perceived that a coping mechanism is the way of learning process to understand the routine of flood events and to implement their assets of resilience.

Risk perception can be understood by the community as the period of learning the process to evaluate and alleviate a good coping mechanism on each of the assets of community resilience. The process of learning for flood disaster management and building community resilience is determined by the effective, efficient, and sufficient resources that they have perceived in order to deal with flood disasters. The need for effective, efficient, and sufficient resources is important for the community to greater value the process of learning through flood disaster occurrence. The more flood disaster experienced by the community the more they value the process of obtaining information to be able to withstand flood disaster.

7.2.2 *Duration of Flood*

Based on the interview with the 320 respondents in West Malaka Subdistrict, the longest duration of flood that entered the 100 respondents' houses was 25–30 days (a month) or with a percentage of 31.3%. These 100 respondents are from the 16 villages in West Malaka Subdistrict.

The respondents in all 16 villages of West Malaka Subdistrict have similar flood experiences and flood responses. The more they experienced flooding, the more they recognized and invented strategic ways of responding to the flood. One approach was to solve the risk of flood by building an 80 cm high flood barrier; floods experienced by the community were used to determine the level or size of the barrier.

There are 75 respondents (or 23.4%) who have experienced inundation in their homes. Their perception is that this kind of flood characteristic is a temporary nuisance even though some possessions were destroyed by the floods. The duration of floods from 20 to 25 days made them find a strategic way of alleviating water from entering their houses, through building a barrier.

Similarly, 60 (or 18.7%) respondents perceived the duration of flood as a usual thing to happen in destroying their assets both inside and outside of their house. Even though the duration of flood is relatively long at 15–20 days, they clearly described a mindset of thinking of flood not as a dangerous thing to be avoided, but as something to be adapted to. Some forms of strategic or adapted ways are therefore required by the community to prevent water from entering their homes, such as through building a wall or barrier. Meanwhile, respondents who experienced flood disaster from 0 to 15 days took a similar preventive action as those who experienced a duration of 15–20 days, but they used different materials for the barrier. They used wood 30–50 cm high to prevent the water entering their house. Detailed information about the duration of flood is presented in Table 7.2.

The above data shows that the public perception of the duration of the flood is different in terms of mitigation and preventive action. However, 16 villages in West Malaka Subdistrict have a common experience and response to flooding. The opinions expressed by people are based on their experiences in dealing with flooding in their area. Patterns of adaptation in the preventive measures are already known, with a very good level of awareness. According to Table 7.2, a graphic of risk perception for flood duration is presented in Fig. 7.3.

From the graph of the community perception of the duration of risk of flooding above, it cannot be judged how the duration of flooding experienced by people can determine mitigation in flood management. The same processes apply whether the duration is short or long, and people who experience short duration flood apply the same strategy or mitigation to avoid water entering their houses. As already presented

Table 7.2 Risk perception for flood duration in the West Malaka Subdistrict

Flood duration	Frequency	Percentage (%)
0–5 days	10	3.1
5–10 days	25	7.8
10–15 days	50	15.7
15–20 days	60	18.7
20–25 days	75	23.4
25–30 days (1 month)	100	31.3
Total	320	100

Source Author's research fieldwork, July 9, 2016, to February 9, 2017

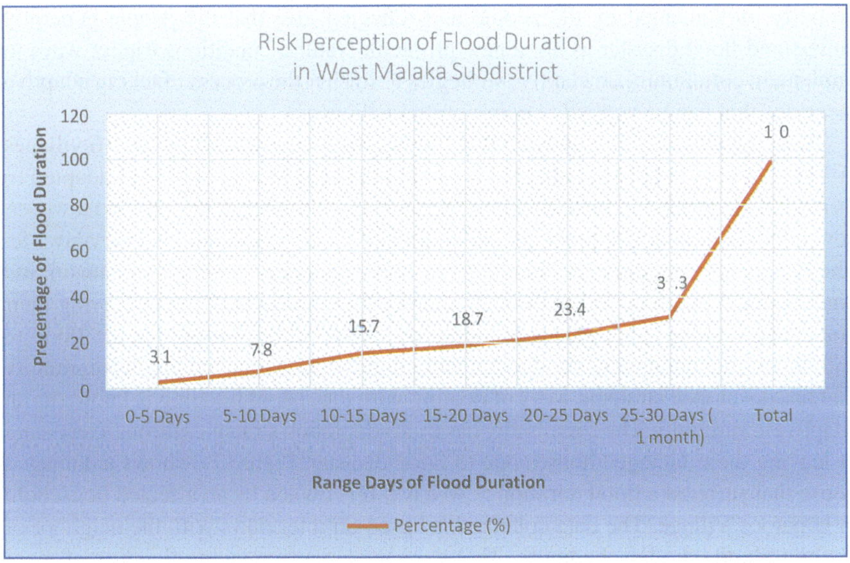

Fig. 7.3 Graphic of risk perception of flood duration in West Malaka Subdistrict. *Source* Author's research fieldwork, July 9, 2016, to February 9, 2017

in Table 7.3 above, it can be reported that respondents with a small percentage 3.1% (0–5 days), have a mitigation strategy with respondents who have a percentage of 7.8% (5–10 days) and 15.7% (10–15 days). It indicates that people have an equal understanding and knowledge in constructing or raising their door wall that is made of wood materials.

The size of the barrier differs between three different categories of day range. People who suffered a flood duration of 0–5 days erected a barrier with a level of 30 cm. People who have a duration of 5–10 days established a flood barrier of 40 cm, and people who suffered flood duration of 10–15 days had an adaptation strategy by building a barrier of 50 cm. The long duration of flood disaster indicates

Table 7.3 Risk perception of flood depth in 16 villages of West Malaka Subdistrict

Flood depth (cm)	Frequency	Percentage (%)
0–50	10	3.1
51–100	25	7.8
101–150	50	15.7
151–200	60	18.7
201–250	75	23.4
251–300	100	31.3

Source Author's research fieldwork, July 9, 2016, to February 9, 2017

community mitigation efforts for facing the flood disaster. Flood mitigation was properly implemented by the people and reflected also that the people generally understand flood disaster as not a new problem. Natural conditions trigger ways to implement community or adaptive strategies to further the process of taking adaptive measures that can be applied to community resilience.

Thus, the research data showed that the flooding experienced by the 16 villages in West Malaka Subdistrict affected a community that was already well-adapted to its natural environment, including flood hazards that threatened their lives. However, new mindset or adaptive practices are ingrained in the community's everyday life. The danger of long-duration flooding has influenced their coping mechanism and their physical building, either through altering their house building or leaving their house and seeking other places of safety or shelter while waiting for the flood to end. Where the duration of the flood is more than 30 days (one month), a community that has developed strategic ways of dealing with the risk of flooding is known as an adaptive community. The community has tried adaptation strategies through repairing or leaving their damaged houses due to flood disaster. Figure 7.3 shows a damaged house that suffered a flood duration > 30 days. It is owned by an affected household in Umatoos Village. The data collected is based on interviews with the neighbor of the household who left the house. This woman also experiencing flooding of more than 30 days but did not leave her damaged house (Fig. 7.4).

7.2.3 Flood Depth

One of the key issues in terms of susceptibility for the community in West Malaka Subdistrict is flood depth and its relation to flood duration. Flood inundation depth determines the duration of the flood. The deeper the inundation of a flood, the longer it takes for the flood's mud to reduce. Dewi (2007: 35) adds that the level of flood depth was measured inside the house with reference to the plinth or ground floor. The depth of the flood can be said to disturb public activities in West Malaka Subdistrict, but an adaptive society that uses traditional methods (i.e., raised foundation of the house or wall door) can cope more easily with the danger of flooding. Different

(a). Flood Duration >30 days in Umatoos Village (b). Interview with a Household in Umatoos Village

Fig. 7.4 Flood duration > 30 days and interview with a household in Umatoos Village. *Source* Author's research fieldwork, July 9, 2016, to February 9, 2017

maximum flood heights in the house vary according to the height of the building above sea level (elevation) and flood-proofing measures such as raised entrances and raised foundation (Febrianti 2010: 51). These mitigation measures were also implemented by the communities in the study area. The data described in Table 7.3 flood depth.

The recorded data showed that 100 (31.3%) respondents had experienced inundation in the range of 251–300 cm. However, no-one has ever experienced the danger of flooding of more than 300 cm. If there was flooding of more than 300 cm, the people affected would be the ones to leave their homes and move to a safer place.

There are 75 (23.4%) respondents who had experienced flood depths ranging from 201 to 250 cm. When flooding occurs at this level, people will evacuate to a safer place based on guidance provided by traditional leaders (female leader), religious leaders and community group disaster volunteers.

Inundation data is recorded based on the flood that inundated both inside and outside the houses of the community. People who experience floods of 0–200 cm have an adaptive process not much different from the people who experience a flood depth of 201–300 cm.

The community in West Malaka Subdistrict that has a strategy of establishing or raising a barrier has also taken a similar approach to adapting their houses, i.e., stage houses or, in their local language *Tetum Therik Malaka*, they are called *Uma-Fehan* which is part of the physical resilience represented in their local culture. The level of the adaptive houses ranged from 65 to 250 cm. Figure 7.5 shows the distinctive features of the adaptive houses built by the community with a high level of 65 and 250 cm.

Communities with a strong sense of custom and tradition have implemented practical behavior in various facets of life. One of these behaviors is living with an adaptive way of coping as a conservationist of nature and managing the risks of nature, which is characterized by the community's typical houses in the above pictures. These houses are recognized as part of the physical assets, i.e., infrastructure of resilience.

(a). Interview with a household's adaptive house house's pole is 65 cm.

(b). Interview with a household's adaptive house which the which the house's pole is 250 cm.

Fig. 7.5 Interview with a household's adaptive house which the house's pole is 65 cm and a household's adaptive house in which the house's pole is 250 cm in Lasaen Village. *Source* Author's research fieldwork, July 9, 2016, to February 9, 2017

A good infrastructure resilience can also solve the danger of flooding through a coping mechanism.

Although the risk of flooding threatens all assets and can destroy people's lives and their homes, customary behavior is maintained and preserved. It has become the hallmark of a civilized society in terms of the physical assets of resilience. The value of "customary behavior" is ongoing, even though the coping mechanism is not sustainable because of the danger of flooding that can threaten the community's houses. The community always applies various strategies in terms of coping mechanisms for flood disaster management. Their coping mechanism was effective, but inefficient and insufficient in addressing flood disaster due to the various advantages being run in only a short-term not long-term strategy.

The depth of flooding experienced by people determines a pattern that is formed and structured by the community in the management of flood disaster. Public information about the percentage of water depth is provided in a graphic illustration, which shows the percentage of people who provided information about the depth of the flood water that entered their homes being from 0 to 300 cm (Fig. 7.6).

The community's strategies were undertaken through setting up or elevating their terrace wall or door to avoid the water reaching their homes. Although during the flood they leave their homes to escape to a safer place, this strategy is always carried out to protect property and belongings.

Figure 7.6 shows that the level of water reached 80 cm and entered houses; the householders left their houses due to the floods which swept away their houses. (Fig. 7.7).

After the flood ends, the community returns to their homes and cleans the mud residue as a result of flooding. The effort required is aided by a system of strong social capital, that is, the system of mutual assistance or *hakawak* in the community local language (Belu language, *Tetum*) *or gotong royong* in Indonesian language with the implementation of successive bonding, bridging, and linking social capital.

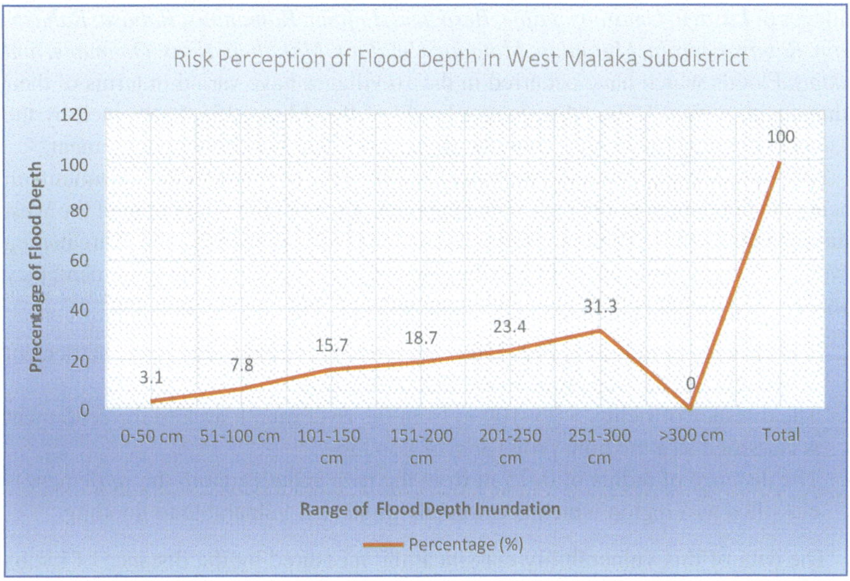

Fig. 7.6 Graphic of risk perception of flood depth in West Malaka Subdistrict. *Source* Author's research fieldwork, July 9, 2016, to February 9, 2017

(a). Water inundated in a Household in Fafoe Village

(b). Water inundated in a Household in Lasaen Village

Fig. 7.7 Water inundated in households' houses in Fafoe Village and Lasaen Village. *Source* Author's research fieldwork, July 9, 2016, to February 9, 2017

7.2.4 Flood Distribution (River Buffer) in the Study Area

In this study, I define flood distribution as the distribution of water due to flooding where the volume of water exceeds capacity. In this case, it is the capacity of Benanain River, capable of distributing the amount of flood water that inundated the entire area. Flood distribution in the area of West Malaka Subdistrict is sourced from the Benanain River which destroys entire villages or areas inside the subdistrict (i.e., the 16 villages of *Lasaen, Umatoos, Fafoe, Besikama, Lofoun, Raimataus, Rabasa, Rabasahain, Rabasahaerain, Motaulun, Motaain, Umaloor, Maktihan, Naas, Oanmane, and Sikun).* Floods which have occurred in the 16 villages have varied in terms of their "threatening levels." The "threatening level" of flood hazard is determined by the distance of radius (in meters) from the Benanain River mouth to the settlement.

I provided the analysis and assessment based on the response of the community in facing the flood disaster through visual data (i.e., River Buffer Class Map of the West Malaka Subdistrict). There are three classes or levels of flood hazard "threatening levels" to a community according to the distance of radius (m) to the settlement; they are:

1. The distance of radius of 100–250 m from the river mouth area to the settlement is classified as an area prone to the risk of flooding.
2. The distance of radius > 25–100 m from the river mouth area to the settlement is classified as a medium prone area of flooding.
3. The distance of radius of 0–25 m from the river mouth area to the settlement is classified as a region which is extremely prone and vulnerable to flooding.

The data of this vulnerability classification, measured by the distance of radius from the river mouth to the settlement has been obtained through the interviews with the 320 respondents in the West Malaka Subdistrict. Community perception about the risk of flooding is acknowledged through their experiences of dealing with flood disasters.

They recognize that if the distance of radius from the Benanain River mouth is farther away from their settlement, the safer it is in terms of the hazard of flooding. On the other hand, the communities also recognize that if the distance of radius from the Benanain River mouth is closer to the community settlement, the more vulnerable they are to the hazard of flooding (Fig. 7.8).

The "River Buffer Class Map of the West Malaka Subdistrict" conveys the spatial distribution of flood danger for all 16 villages. The map shows that all 16 villages were submerged or inundated by a flood; however, the distribution of flooding in several villages is determined by the class of the radius distance from the Benanain River mouth to human settlements. The class is also varied, which can affect people differently in their response to the danger of flooding. The analysis of river buffer shows that:

1. The village which is not prone to flood disaster is Besikama Village with a distance of radius of 100–250 m from the Benanain River's mouth to the settlement. According to the analysis of the buffer data obtained in the community interview in Besikama Village, it can be concluded that classes are not vulnerable to flooding or that they are safe from flooding. In this village, the duration of flooding is very short, beginning with 0–10 days, with an average depth of flooding experienced by every household in the village of between 0 and 100 cm. There was a distribution of twenty households in the Besikama Village that admitted their area is included in a safe or secure location to be. Therefore, Besikama is not prone area by the categories outlined above. Besikama Village

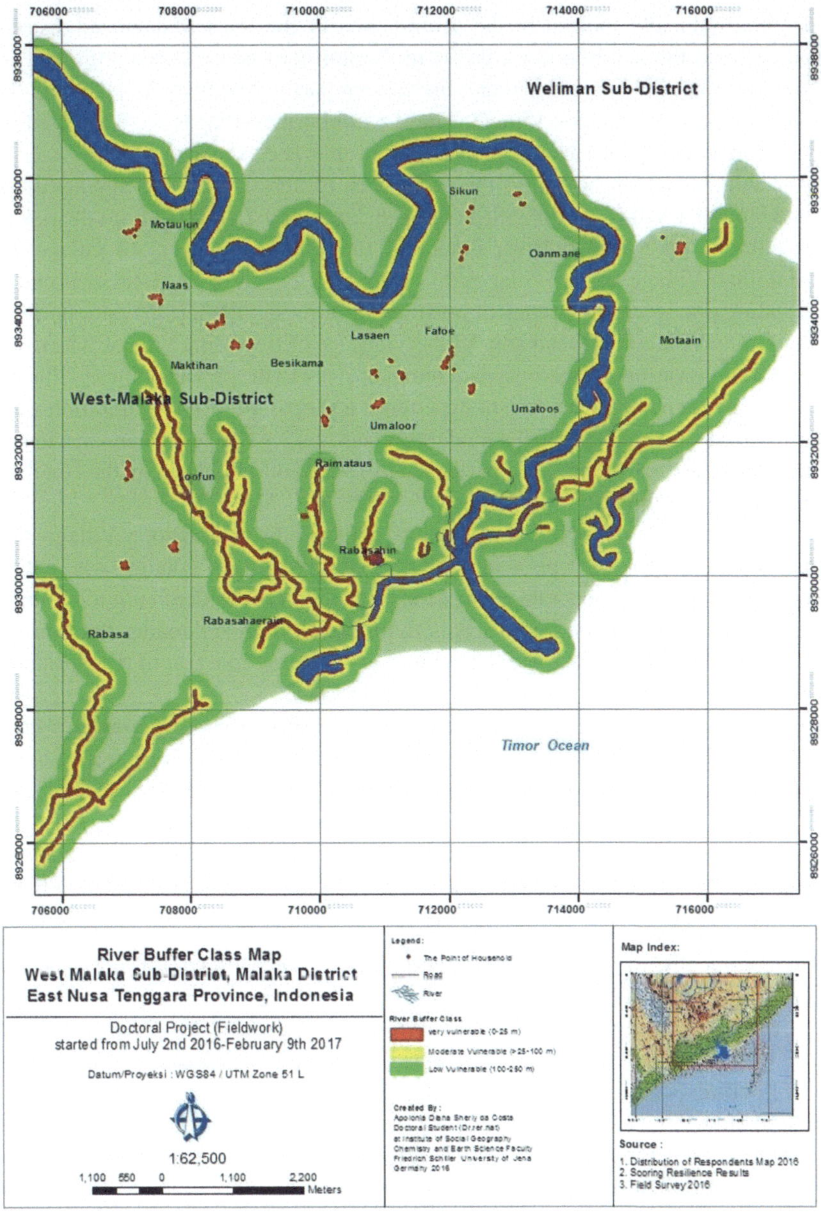

Fig. 7.8 River buffer class map in the West Malaka Subdistrict, Malaka District, East Nusa Tenggara Province, Indonesia. *Source* Author's Research Fieldwork (Survey and Interviews), July 9, 2016, to February 9, 2017

is geographically located in the central area of the West Malaka Subdistrict, adjacent to the neighboring villages in the northern part (Lasaen Village); the southern part is Umaloor Village, the eastern part is Maktihan Village, and the western part is Umatoos Village. These neighboring villages were affected by flooding due to their lack of preventive measures (i.e., social capital).

2. The villages which belong to the prone area to flooding are Maktihan Village and Umaloor Village, each with a radius distance of > 25–100 m from the risk of flooding of the Benanain River's mouth to the settlements. Umaloor Village and Maktihan Village are the villages that belong to a class or level rather prone to flooding due to the geographical positioning of these villages; they are north of Maktihan Village and Besikama Village and east of Besikama Village. The reach of flooding was not dangerous, damaging, or risky to the people of these villages; however, the people who did not experience a short-term flooding duration were also those who belonged to the moderate class (i.e., > 10 days) where an average flood depth of > 100 cm was experienced by every household in both the Umaloor and Maktihan villages.

   There was a distribution of twenty households in Umaloor Village that provided their risk perception about the susceptible area of flooding where their area belonged to a moderate class. Moderate class is considered not too vulnerable, but it is an insecure area. There is, therefore, a different flooding duration and flooding depth than that experienced in a secure place like that of the Besikama Village.

3. The other villages in the West Malaka Subdistrict (namely *Lasaen, Fafoe, Umatoos, Raimataus, Rabasa, Rabasahain, Rabasahaerain, Lofoun, Motaulun, Motaain, Naas, Oanmane and Sikun)* belong to the classes of "vulnerable" and "more vulnerable" to flood hazard with a radius of 0–25 m. In these villages, it is estimated that the highest risk of flooding will be experienced. It is apparent that these villages are susceptible to flooding from the mouth of Benanain River due to the settlements being close distance to the river.

Geographically, the location or position of the Lasaen Village (visually seen on the river buffer map above) can be concluded as belonging to the safety class of flood hazard. However, the results of analysis within a radius of 0–25 m range of flood hazard from the estuary of Benanain River to the people's settlements showed that the risk of flooding is more hazardous than, say, in Umaloor Village, Maktihan Village, and Besikama Village. The other reason is the geographical position of each of the villages to the other villages. Lasaen Village is the closest village, with neighbors to the eastern part of Besikama Village (safe from the risk of flooding), to the western part of Fafoe Village (vulnerable, and highly vulnerable), and to the next southern part of Umaloor Village (moderate vulnerable). Lasaen Village is also the point of the Benanain River's dike and Sikun Village as the location of the ending point of the embankment construction. Benanain River's dike was built in 2001 and swept by floods in 2011, which was one of the biggest floods that occurred in the West Malaka Subdistrict.

As a result, the failure of the dike that was built by the Indonesian government leaves the community in West Malaka Subdistrict in a vulnerable position. In 2014, the Indonesian government's policy was implemented through the construction of a retaining concrete wall. This was an action-implementation by the Indonesian government to repair the physical damage of flood disaster.

Figure 7.9 shows environmental conditions of reconstruction due to flooding, reconstruction of retaining of the building wall construction from the Benanain River, interviews with a farmer living near the Benanain River's wall construction, and the interviews with three teenagers in the border village between Lasaen Village and Sikun Village.

Based on Fig. 7.9, there are four physical and socio-economic conditions of the community that can be analyzed in the framework of physical resilience, social resilience, and economic resilience. Picture (a) and (b) shows the physical condition of the Benanain River wall reconstruction in Sikun Village, in the West Malaka Subdistrict, which is situated at the breaking point of the dike that collapsed in 2011.

(a). Environmental Condition of Benanain River Reconstruction in Sikun Village

(b). Building Reconstruction of Benanain River Wall's Wall's Reconstruction in Sikun Village

(c). Interview with a farmer who is living nearby River Wall's Reconstruction in Sikun Village

(d). Group Interview with the three teenagers in the Benanain border Village between Lasaen Village and Sikun Village

Fig. 7.9 Environmental condition of Benanain River wall's reconstruction and building its reconstruction in Sikun Village, West Malaka Subdistrict, East Nusa Tenggara Province, Indonesia. *Source* Author's Research Fieldwork (Survey and Interviews), July 9, 2016, to February 9, 2017

The Benanain River wall reconstruction was built by the Indonesian government in October 2014 from the Regional Development Budget (*Anggaran Pendapatan Belanja Daerah* or APBD) of East Nusa Tenggara Province, Indonesia which is funded by the State Budget of Indonesia (*Anggaran Pendapatan Belanja Negara* or APBN) (*Interviewee; guide keeper: Rony Seran*).

According to the local community, although the Benanain River wall reconstruction seems to be helpful in blocking the regular floods every rainy season, the discussion with an interviewee, Maria (47), seen in the picture (c), delivers a different opinion. Maria explained that:

> "According to me, there are advantages and disadvantages of the Benanain River wall's reconstruction being constructed by the Indonesian government. The advantage is the people are not too worried when the flood comes, because there is the wall's reconstruction, but the powerful force of flood sometimes is able to destroy the wall construction." She added, "whereas, the people in Sikun Village, who are mainly farmers, only hope for their natural assets of growing corn, tubers, yams, vegetables and consuming these food crops in every day." In addition, she explained more that, "these food crops are also sold by the people in Sikun Village in the traditional market so as to increase their income (economy asset)." She continued to talk about her life at the time of interview: "I am now carrying cattle feed over my head. My husband and I have 8 cattle we take care of, and every day from 07:00 a.m. to 13:00 p.m., I have to walk 1.5 km from my house to a place that is filled with grasses and leafage that grow in some places and come back to my home around 13.00 p.m. I cut the grass and leafage and put them on a sack and carry it over my head. From the result I get, I can help my husband who is taking care of our household needs at home too during 07.00a.m. to 13.00 p.m. The difficult part that we face is the result of taking care of cattle, as a valuable asset for our savings. We used these assets to support our two children's schooling at the primary and senior high school levels, considering we are only farmers and not a government officer or a civil servant. Our monthly income only ranges from 15 to 20 US $ and that is still not enough to meet the basic needs of our daily lives, but we use that income in adjusting to foods; besides we also plant in our garden. Nevertheless, we enjoy our simple life. My appeal to the Indonesian government is to not only pay attention to physical development such as the wall' reconstruction, which is also important for us in the West Malaka Subdistrict, because despite the mighty strength of human power, the mighty strength of natural power will be a stronger one to destroy it again, like the damage to the wall that happened in 2011 due to flood disaster.
>
> *It might not happen anymore if an effective intention of the Indonesian government would be to sincerely help all aspects of our life and not just a single aspect"*. The statement from the interview with Maria shows that it is not only the physical aspects that the Indonesian government should pay attention to in connection with the Benanain River wall reconstruction, but it also needs to consider the socio-economic life in a lower-income community with monthly income about 15–20 US $.

The life conditions of Maria are also experienced by three teenagers from Lasaen Village *(from left, Mariana: 12 years old and studies at the first year of junior high school, Rita: 14 years old and studies at the second year of junior high school and Benekdita, 15 years old, the third year of junior high school)*, as can be seen in picture (d). The location of the interview is situated on the border site between the Sikun Village and Lasaen Village. It can be seen that the three teenagers are carrying pieces of wood which they collected from the branches of fallen tree-trunks and from the live trees in the forests. An interview with them obtains useful information; it is quite surprising that life in Lasaen Village shows a similar profile to life with the Sikun Village. It is evidenced by the statement of the three teenagers, who convey and confess that, "*People are very dependent on their natural assets.*" This statement is shared by the three teenagers. During the interview at 2:00 p.m., they said:

> We just finished our school and arrived at home, and we have not had lunch yet while walking on foot looking for woods and bringing them to our home for cooking and preparing for our lunch with our families. The distance we've walked on foot from our home is about 1.2 km to 1.7 km with a time of 20–30 minutes to the forest where every field is overgrown with trees and cut logs and trees that have fallen or fallen trees. We can collect the wood within 1–2 hours. For us, "Woods for Foods" which means "without woods no foods we can eat." We collect the wood every morning in every day before going to school. However, because of the mud from the heavy floods, we postponed this work until the afternoon after coming back from school. We are very happy to help our parents and we enjoy our lives. My dream is to be a teacher.

From the results of the third point made by the three teenagers, it can be concluded that people in Lasaen Village have a pattern of life that is familiar with nature. Natural resilience is a mainstay of society to meet their daily needs, such as cooking, food, and drink. Natural resilience also supports economic resilience that cannot be sufficiently fulfilled by their society due to lack of monthly income. The community's income is received on a daily basis rather than monthly and the amount per day is uncertain. The community has its own perception of the livelihood resources that, "*what is gained from nature today is thought out for today, and tomorrow has its own condition, energy, and time to think again for the next day.*" They are optimistic amidst pessimistic economic resilience against flood disaster. Their optimism is evidenced by their social capital, which establishes cooperative work within their family with divided household chores. They also have a friendly attitude in delivering their opinions about their lives and the aspects of life and the resilience they apply every day.

7.3 Community Assets on Resilience for Flood Risk and Flood Risk Profile

7.3.1 Lasaen Village as One of the Highest Prone in the Study Area

(a) **The Community in Lasaen Village**

The community in Lasaen Village is characterized by its socio-cultural distinction, which includes their "strength of mutual assistance" and their use of traditional clothes that they dress in for their daily lives and activities. They also have good communication with the local chief or leader of the village, i.e., Head of Lasaen Village, who was interviewed and gave useful information relating to the flood response and community resilience in Lasaen Village (Fig. 7.10).

a). Interview with one of Villagers (household) wearing female traditional cloth Village

b). Interview with the Head of Lasaen Village who was in Lasaen Village's Office, Lasaen Village in Lasaen

Fig. 7.10 Interview with one of villagers (household) who was wearing female traditional cloth and Interview with the Head of Lasaen Village in Lasaen Village's Office, Lasaen Village. *Source* Author's Research Fieldwork (Survey and Interviews), July 9, 2016, to February 9, 2017

Figure 7.10a shows an older woman (*Marthinha, 55 years old*) who is a farmer. According to her, "*The life condition in Lasaen Village is far away from social welfare. Life of the fortunate here is lacking, we are still grateful and enjoying our life; life as a gift, though.*" This short statement shows that the community takes the initiative in the economic, social, and cultural aspects of its life, which can be a form of valuing their lives.

Even though their level of resilience effort is insufficient, they persevere in addressing the flood problem. In dealing with the interviewee's (Marthinha) short statement, the Head of Lasaen Village offers an opinion. He acknowledged that:

> "The tendency of the community is to take their own initiative as the living conditions are far away from the social welfare that should be provided by the Indonesian government." He emphasized, "the Indonesian national and local governments do not support the realization of the prosperous life for remote villages in Indonesia, and international organizations also need to do further analysis and provide a platform of socialization training and workshops on the importance of social welfare to the remote people in the remote villages, especially in the hazardous area in Indonesia including our place here. Socialization is one important thing that functions as a way to understand our social capital and therefore we can mitigate floods. We know the local strategies of flood disaster management and resilience actions that we've applied during the floods that destroyed our village and other villages in the West Malaka Subdistrict. However, that is not enough because cooperation is the best way apart from our independent efforts which may also have several weaknesses in our skills and knowledge; we do not know and master them, how it would be more effective in overcoming flood problems and increasing our resilience. We have insufficient income, but we are indeed able to independently meet our basic needs through the utilization of natural assets that make us tough enough and remain optimistic to live life. However, the most fundamental thing is the balance of our lives depending on our independent positive behavior overcoming our problems without the intervention of the Indonesian national and local governments. The remote villages that are facing the natural disaster problems are not just us, there are many other villages in Indonesia too, but the need of implementing the socialization of social welfare is the thing that we need too now, to support our social capital that we're bonding from the family, bridging our social life among people, and linking our brothership to local

government, local Non-Government Organizations (NGOs) in Indonesia and International organizations in foreign countries who also want to hear our aspirations so far."

The Head of Lasaen Village concludes that the intervention of the Indonesian government and international organizations in terms of socialization is very important to realize a balanced life in various aspects of the community's social welfare such as improving the community's *social capital: bonding, bridging, and linking* to establish more effective socialization and resilience action that people need to be able to deal with floods. The economic aspects of the community would be improved and fulfilled if the social welfare aspects could be solved by the Indonesian government and other local, national, and international organizations as early as possible (see also, discussion on social capital in Chap 6).

(b) **Assets of Resilience in Lasaen Village**

Assets are the results of community livelihood resources. The community has made a lot of effort to fulfill its daily necessities through utilizing natural assets, both to increase their income and to increase their food security, i.e., food stock. The efforts were thoroughly owned by them and strengthened their actions to be more persistent in handling flood disasters and in increasing their assets of resilience. The pictures showing community assets in Lasaen Village are presented in Fig. 7.11.

The six pictures of the assets of resilience which were owned by the community in Lasaen Village show their wealth of resources, i.e., assets of resilience. The community had several assets to support their resilience power to deal with flooding. Picture (a) shows the community assets in terms of agriculture, i.e., maize. In West Malaka Subdistrict, maize is known as the staple food as it replaces rice. The production of rice is limited as the risk of flooding has impacted paddy field yield. The community has its craft of traditional weaving that helps them to increase their income and is also enjoyed as a hobby in their spare time (Pictures b and c). However, there are no appropriate physical assets, i.e., infrastructure of irrigation and drainage, and this is an obstacle for the community in West Malaka Subdistrict in consuming rice. Also, the price of rice in the city center of Malaka District, i.e., Betun City, is high for a community with limited monetary assets for purchasing rice. Maize is a good option as a staple food, and the main source of foods or flavors for cooking a variety of traditional cakes such as corn pudding, as well as daily foods such as boiled maize mixed with cassava, tubers, moringa leaf, pumpkin leaf, and pumpkin. Pumpkin leaves can be consumed by the community as an extra meal every day. Picture (e) shows the interview with the owner of the garden with pumpkin leaf. She said she planted pumpkin leaves beside her house as one of the staples of food every day. Meanwhile, picture (d) depicts the community that owns the well. This situation shows that water is the main source of fulfilling the needs of the community. In addition, the community collects physical assets by means of livestock. The community also has a goat pen (picture f). These physical assets are managed and maintained by the community for its resilience preparation (i.e., coping mechanism) to cope with flooding.

The community in Lasaen Village implements its coping mechanism in terms of the number of meals per day (e.g., they limit their meal, which is usually three

Fig. 7.11 Community assets of resilience in Lasaen Village. *Source* Author's research fieldwork (Survey and Interviews), July 9, 2016, to February 9, 2017

times per day, and it becomes once or twice per day). This indicates that they have insufficient income (>200.000 rupiah or 20 $) per month) to meet their basic needs. However, the community still utilizes natural assets to meet its basic needs. The community has communal wells, with one well for 5–10 houses in a neighborhood. The existence of the wells is limited due to economic factors, i.e., inadequate income of the community to build one well per household. Although the National Program of Rural Community Self-Empowerment in 2008 was designed for the construction of wells for each household in Lasaen Village, it has not yet met the needs of every household in every village in West Malaka Subdistrict. Traditionally, these values have been developed and well-maintained by the community in Lasaen Village. Cultural and social strength are able to overcome the various problems in every aspect of community life, including the risk of flooding. Meanwhile, Figure F presents the community economic asset on resilience, i.e., livestock owned by one of the households in Lasaen Village, West Malaka Subdistrict.

The research showed that cattle have become a supporting part of the economic assets, and they also support the process of community resilience in Lasaen Village. Livestock was a good option for increasing the community's economic resilience. Developing a market for cows is a coping mechanism that the community applied in Lasaen Village. Cows produce a profit through their meat that can be consumed or sold in the market. It is an economic coping mechanism that can increase community income in Lasaen Village. Also, cows have multifunction as the driving engine for the community's paddy fields.

(c) **Flood Risk Profile**

Lasaen Village is one of the villages in West Malaka Subdistrict that has experienced flood disaster from 1939 to 2016. The risk of flood disaster affects the nine assets of resilience, through damage to physical assets of houses and their environment which disturbs the community's daily activities in further handling flood disaster. The floods come twice a year and require as a consequence economic funds for the process of repairing, reconstruction, and recovery of physical assets (Fig. 7.12).

Fig. 7.12 Flood risk profile in Lasaen Village. *Source* Author's research fieldwork (Survey and Interviews) July 9, 2016, to February 9, 2017

Based on the data from field observation in the 16 villages in West Malaka Subdistrict, it can be reported that Lasaen Village belongs to the category of hazard-prone villages and the physical assets owned by the community are susceptible to damage, i.e., public facilities, Head of Lasaen Village's office (Pictures a and b) and the community's houses (Pictures c and d). As Figure B above showed, the community experiences a life-threatening level of physical risk when the flood reaches 70 cm and the Lasaen Village office's yard is filled with mud from the flood and a fallen tree. Meanwhile, natural damage occurred when a tree fell because of the flood that hit Lasaen Village.

The community has taken adaptive steps through increasing the physical capacity of resilience, by building the houses with wood materials, including foundations, poles, and walls. The community also built a stage house with 1 m wooden poles to prevent water from entering. This adaptive step has become a routine activity and can be regarded as a good tradition every time a flood hits the community area.

(d) **Flood Distribution to the 20 Households (Respondents) in Lasaen Village**

The community has a unique adaptive pattern that is reflected through their culture, and it is preserved through physical resilience, i.e., stage house. In addition, they implemented an adaptive pattern of prevention measures such as building a high pole and heightened a wall door of their stage house which protects them from the flood water that reaches their stage house (approximately 1 m) and the wall door (approximately > 60 cm). People are skilled in building their own solutions for their physical assets to increase their coping mechanism in terms of physical resilience for managing flood disaster; floods that have been experienced by the community, increasing their adaptive pattern or coping mechanism. Communities are currently a single actor of resilience to develop and manage their own solutions. However, they need help from other parties or the other actors of resilience to overcome problems regarding building resilience for floods, i.e., local government, central government, and private institutions in Indonesia. The solution implemented by Lasaen Village is seen from their risk profile; the community with high and very high vulnerability levels is able to apply a good adaptation pattern, though it is not yet a long-term solution. Flood Distribution of the 20 Households (Respondents) in Lasaen Village can be seen in Fig. 7.13.

7.3.2 Umaloor Village as One of the Moderate Vulnerabilities to Flooding in the Study Area

(a) **Community in Umaloor Village**

Umaloor Village has moderate vulnerability to flooding, and the community has applied their assets of resilience to aid survival. The community is seeking its own solutions to handle the flood disaster through improving its income. One of their strategies for survival was selling the results of cultivation, such as tomatoes, bananas,

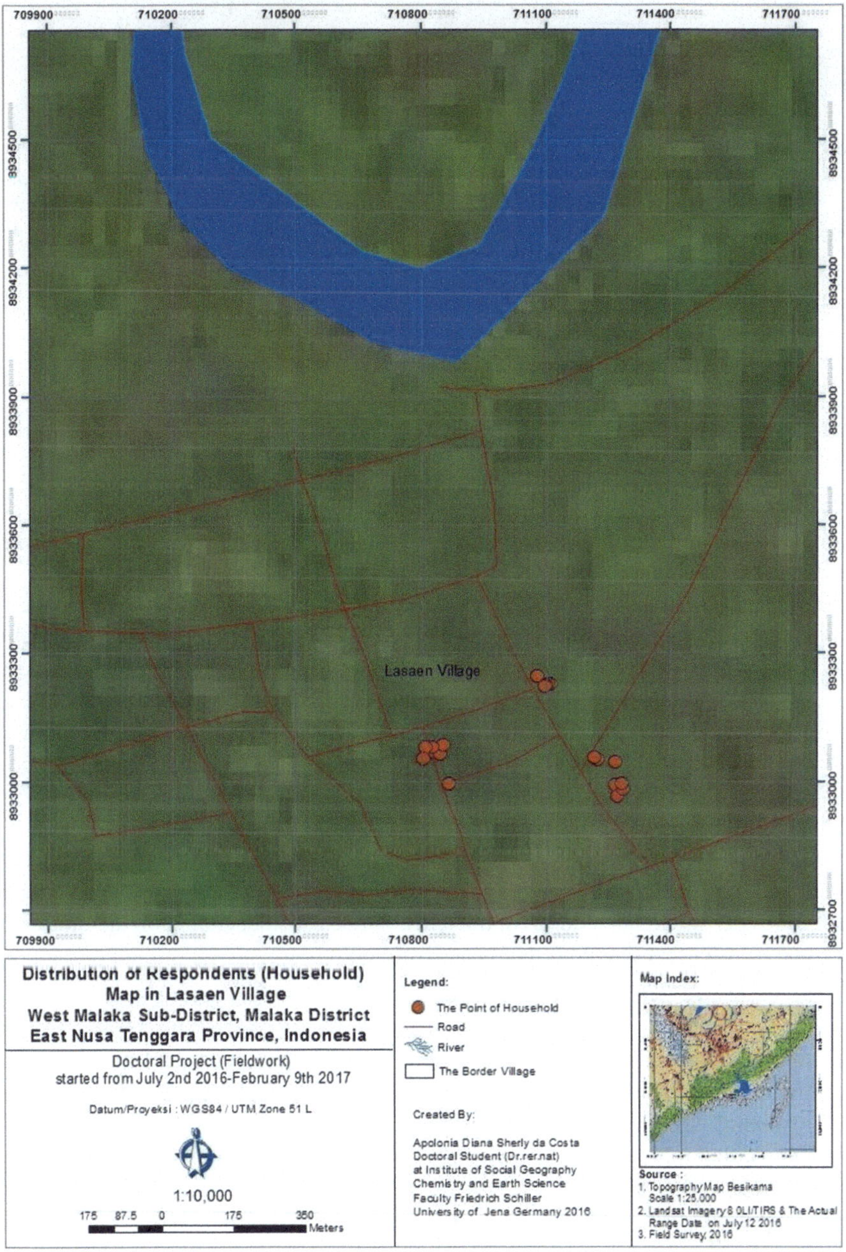

Fig. 7.13 Distribution of respondents (Household) Map in Lasaen Village, West Malaka Subdistrict, Malaka District, East Nusa Tenggara Province, Indonesia. *Source* Author's Research Fieldwork (Survey and Interviews) July 9, 2016, to February 9, 2017

(a). Three female leaders were selling their result of cultivation in the traditional market in Umaloor Village

(b). Situation in the traditional market in Umaloor Village

Fig. 7.14 Situation of the traditional market in Umaloor Village. *Source* Author's research dissertation research fieldwork (Survey and Interviews) July 9, 2016, to February 9, 2017

and betel nuts in their traditional market, which is located in their village, as shown in Fig. 7.14: Situation of the Traditional Market in Umaloor Village.

As can be seen in Picture (a), the three female elders who are wearing their daily clothes, i.e., a sarong, are more than 50 years old. They were pleased to provide information about the position of the traditional market and the reason for selling their results of cultivation. They said that the traditional market is usually stable before a flood disaster and after the flood disaster ends. However, there are no activities during the flood disaster in the traditional market due to water and mud. As a consequence, they sell all of the results of cultivation every evening to get more income for a recovery process after a flood, in the ex-post phase.

The same coping mechanism was used in a situation without flooding to support their needs, but with a different intensity. They generally sell some of their produce four times per week, but a different intensity can be used depending on the intensity of flood disaster that threatens them and livelihoods.

(b) Assets of Resilience in Umaloor Village

Assets are valuable human efforts to support current daily needs and future needs. A human's daily needs can be consistently fulfilled through significant adaptation and resilience efforts, both in short-term and long-term periods. The community in Umaloor Village prepares its resilience efforts and creates additional assets to support their normal assets. They collected their assets with multi-tasking jobs in every day (i.e., taking care of household chores, as a farmer, as a fisherman/fisherwoman, and as a trader). The community is not able to predict how many assets will be damaged once the flood comes over either the short-term or long-term periods; however, their resilience efforts are well-prepared, helping them to adapt to floods.

The community implements its coping mechanism under its own initiative and self-reliant efforts. They have fields to grow crops and fishing gear for fish and shrimp in the Benanain River. These natural assets are the results of crops and fish and shrimp

Fig. 7.15 Community assets of resilience in Umaloor Village. *Source* Author's research fieldwork (Survey and Interviews) July 9, 2016, to February 9, 2017

that the community can sell in the traditional market in Umatoos Village. This market is in a central point that is accessible to the other 15 villages in the West Malaka Subdistrict. Communities implement their economic assets through trading in the market, showing that an economic asset on a microscale can be significant. In Umallor Village, the economic asset on a microscale was in practice implemented by the other 15 villages in order to meet their daily needs and increase their resilience efforts to overcome flood problem. The assets of resilience that the community in Umaloor Village has applied are the results of natural assets, i.e., results of cultivation and fishing that can be sold in the traditional market. Other assets are water accessibility through household wells, stage house, and household's small kiosk (Fig. 7.15).

Based on Fig. 7.15, it can be seen that the community in Umaloor Village has applied some resilience efforts that can be useful in overcoming flood problem. Picture (a) shows a household (grandmother and her granddaughter) from *Umamolin subvillage* that utilizes its natural asset of resilience to support economic resilience. Cooperation that exists in the household is shown in the example of a grandmother and her granddaughter who sell their fish and shrimp in the market; this indicates a good adaptation pattern which is aroused by their awareness. Communities have self-awareness to increase their resilience; picture (b) shows a house of a grandmother with her grandchild. They have pigs which are their economic assets.

The environmental conditions in Umamolin subvillage, in Umaloor Village (picture c and picture d) demonstrate that the quality of the public road and housing infrastructure are poorly managed; this point should be noted by the Indonesian government.

Improvements in intervillages' public road reconstruction in the remote areas in Indonesian, including West Malaka Subdistrict, were undertaken by the Indonesian government, but the public roads in the 16 villages all benefitted. In picture (d), there is a typical stage house that is the community's traditional house and the hallmark of the community's physical resilience. The community has a good level of awareness and takes preventative actions by building stage houses as a form of adaptation and preparedness to handle flood disaster. This action reflects their constant efforts to live their lives every day and increase their resilience in overcoming the flood problem. Like communities in Lasaen Village, communities in Umaloor Village have their own independent businesses that were established from people's homes. Picture (e) shows a household's kiosk in Umaloor Village where local businesses endeavor to improve their economic resilience. The owner of the kiosk business is a middle-aged woman, as can be seen in picture (f); she takes care of household activity (i.e., washing clothes using the water from the well), which she does every day. This example of the household shows their physical asset of resilience (i.e., a well) to access water; not all households have a well. A well can support people's needs such as cooking, eating, drinking, and washing clothes.

(c) **Flood Risk Profile**

Umaloor Village has experienced the risk of flood disaster and learned to create and improve their assets of resilience, especially physical assets, i.e., an adaptive house and building a well for water accessibility. Communities developed their own efforts to cope with flood disasters. Their efforts can be seen in Fig. 7.16.

Figure 7.16 shows the community's flood risk profile in Umaloor Village. Community life in Umaloor Village is affected by the risk of flood which they have to overcome. Picture (a) above shows a stage house with all its building materials made of wood. The stage house is at high risk from flood; once a flood sweeps the stage house, it will be destroyed. Similarly, the high risk of life-threatening damage from semi-walled houses, in pictures (band c) is shown in a house without a pole building

Fig. 7.16 Flood risk profile in Umaloor Village. *Source* Author's research fieldwork (Survey and Interviews), July 9, 2016, to February 9, 2017

which is different to picture (a), a stage house. The highest risk is with a typical house as shown in picture (d); the fatal damage of floods was experienced by the household in that house. The house in picture (d) was made of wood, with some zinc covering the house.

Picture (d) also shows a very unfortunate aspect of life, where a middle-aged woman is holding her daughter who is still a toddler.

It is inescapable that flood risk threatens people's lives in all 16 villages in the West Malaka Subdistrict. However, community resilience endeavors in utilizing their natural resources can improve their livelihoods as part of their coping mechanisms. These coping mechanisms indicate their adaptation in overcoming flood disaster. Communities in Umaloor Village are independently running their resilience of business which is part of everyday life, and which reflects the quality of their lives that is far away from welfare. The risk profile of the communities in Umaloor Village is an example of remote and vulnerable communities which maintain their self-resilience efforts to live their lives and increase their resilience capacity in overcoming the flood disaster.

(d) **Flood Distribution to the 20 Households (Respondents) in Umaloor Village**

The community in Umaloor Village has a unique adaptive pattern through a preserved culture of physical resilience, i.e., stage house. However, most people in Umaloor Village are different from Lasaen Village in the availability of wells owned by each household. Water accessibility is the main element of their recovery process. The adaptive pattern that they adopt is to have a house on stilts with a foundation of more than 60 cm and with a well depth of 2.5 m. People already know the adaptive pattern that must be applied to overcome the flood problem. Although flood disaster sweeps their area, they always apply their physical adaptive pattern. The assistance that they had expected was socialization assistance related to handling the flood hazard in different ways, including access to clean water through the construction of a well for each household, not just in Umaloor Village (where the majority of the community own wells) but also in the 15 other villages in West Malaka Subdistrict where people do not all have wells. Currently, local government, central government, private institutions, and community initiatives have not adopted their adaptive patterns as implemented solutions in addressing flood problems and increasing the capacity of community resilience. The right solution is a participatory approach without neglecting the socio-cultural element that has become the power of the community in the research site. The flood distribution to the 20 Households (Respondents) in Umaloor Village can be seen in Fig. 7.17.

7.3.3 Besikama Village as the Sole Village of the Lowest Vulnerabilities Level in the Study Area

(a) Community in Besikama Village

Besikama Village is known as the capital city of West Malaka Subdistrict. As the capital city, the community has a different view of its environment, i.e., some knowledge of advances in technology. Besikama Village has three internet kiosks with two computers each. The owner of one internet kiosk in Besikama Village was interviewed and said that the kiosk is a start-up small business that provides benefits for the students of junior high school and senior high school who wish to access the internet and computer. Even though the internet kiosk building material is wood, its function is for students to have a good knowledge of using the internet, and also access more technology tools through the internet and computers. The income of this small business is relatively moderate at approximately 800.000 Rupiah or 80 $ per month, and this helps the owner meets his daily necessities and supports his recovery process. The community environment situation in Besikama Village can be seen in this Fig. 7.18.

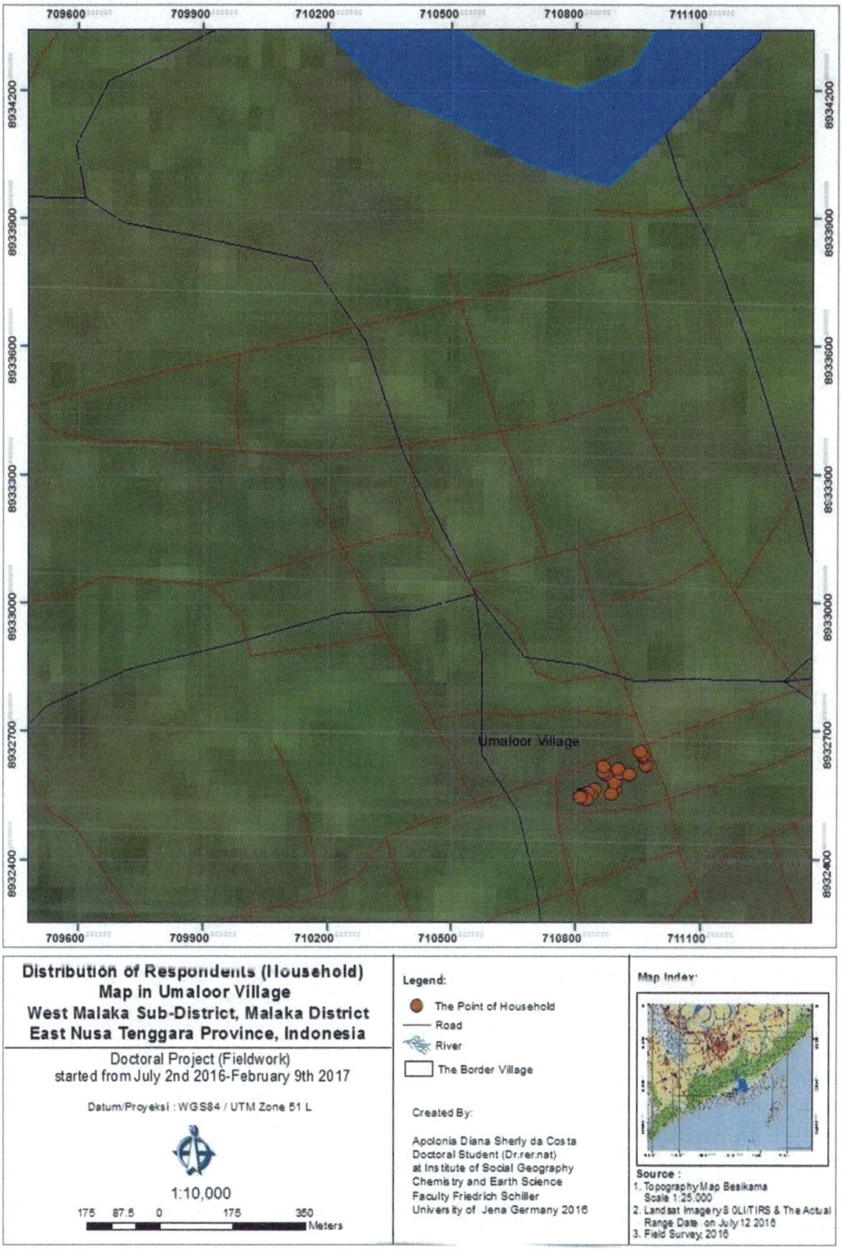

Fig. 7.17 Distribution of respondents (Households) Map in Umaloor Village, West Malaka Subdistrict, Malaka District, East Nusa Tenggara Province, Indonesia. *Source* Author's research fieldwork (survey and interviews), July 9, 2016, to February 9, 2017

(a). Situation In Besikama Village (b). An Internet-kiosk In Besikama Village

Fig. 7.18 Community environment situation in Besikama Village. *Source* Author's research fieldwork (survey and interviews) July 9, 2016, to February 9, 2017

(b) The Assets of Resilience in Besikama Village

The community in Besikama Village has several assets of resilience that can be usefully implemented and support their adaptive pattern or coping mechanism. They have internet kiosks with computer services, satellite dishes (national television cable channels), water closets (WC) and well, stage house and livestock. The pictures that show community assets in Besikama Village are presented in Fig. 7.19.

Based on Fig. 7.19, it can be concluded that the community in Besikama Village has better assets than the other 15 villages in the West Malaka Subdistrict. This is due to the opinion of the community that the lowest (least) vulnerable village in the West Malaka Subdistrict is Besikama Village (see Sect. 7.3.3). Picture (a) shows a small business owned by the local community in Besikama Village. The small business provides the service of some computers and internet access, which are benefits for those who need information and technology, including flood alerting system and flood prevention in the ex-ante period or before flood disaster occurs, during the flood disaster occurrence, and in the ex-post after-period when flood disaster ends. It indicates that the community needs effective and sufficient technology to support them to deal with flood disasters. It is also known for its disaster mobility. Furthermore, this business shows the significant growth of the community's mobility (i.e., information and technology) in handling flood disasters in Besikama Village. The community's mobility disaster in this village is already quite sufficient and needs to be improved in its effectiveness in order to deliver a sufficient provision for disaster mobility that can also be reached by the other 15 villages. Disaster mobility shows an anticipatory practice owned by the community in Besikama Village. It thus shows the physical resilience of the community in terms of helping them to build networking and coordination in the context of information about preparedness and alerting systems to face flood disaster (before the flood occurs or ex-ante, during the flood occurrence and after the flood occurs or ex-post).

Communities in Besikama Village consider that the development of adequate information and technology is important. As can be seen in picture (b), there is a

Fig. 7.19 Assets of resilience in Besikama Village. *Source* Author's research fieldwork (Survey and Interviews July 9, 2016, to February 9, 2017

house which has a full permanent wall and a cable television network (the community calls it *parabola*), some pigs, and some wood for the construction of a compound to save their pigs when the floods come. These assets represent their physical resilience, as presented in picture (b). Pictures c, d, e, and f show an older woman who lives alone and has her own well and a walled building with a WC, as presented in the picture (c). Her stage house has a pole height of 35 cm, as can be seen in picture (d). The woman has pigs, shown in picture (e), and a compound for the pigs, shown in picture (f). The condition of the community in Besikama Village, with its implemented physical resilience assets, does not mean that there is a lack of coordination and networking with the other 15 villages in the West Malaka Subdistrict. This situation in the research site shows a set of collective actions of the rural and affected communities that still prioritize social cohesiveness rather than individualism.

Communities come together in establishing their social links that become a reflection of custom and practice of social life in the West Malaka Subdistrict.

(c) **Flood Risk Profile in Besikama Village**

Besikama Village is the sole village that has the lowest vulnerability to flood disaster in West Malaka Subdistrict. As a safe village, the community has a similar socio-culture that has been embedded in its daily life, i.e., typical of stage houses and housing materials of wood and leaf. The situation in Besikama Village has presented a contradictory picture between two types of the households, i.e., the house in the picture (c) where all the building materials are woods and leaves with the leaf rooftop, and the other house in picture (d) where the building materials are concrete and cement with a zinc roof. These different risk profiles can be seen in different incomes and livelihoods of the community in Besikama Village (Fig. 7.20).

(d) Flood Distribution to the 20 Households (Respondents) and Risk Profiles in Besikama Village (the Remaining 13 Villages' Maps Are Presented in Appendices)

See Fig. 7.21.

Fig. 7.20 Flood risk profile situation in Besikama Village. *Source* Author's research fieldwork (survey and interviews July 9, 2016, to February 9, 2017

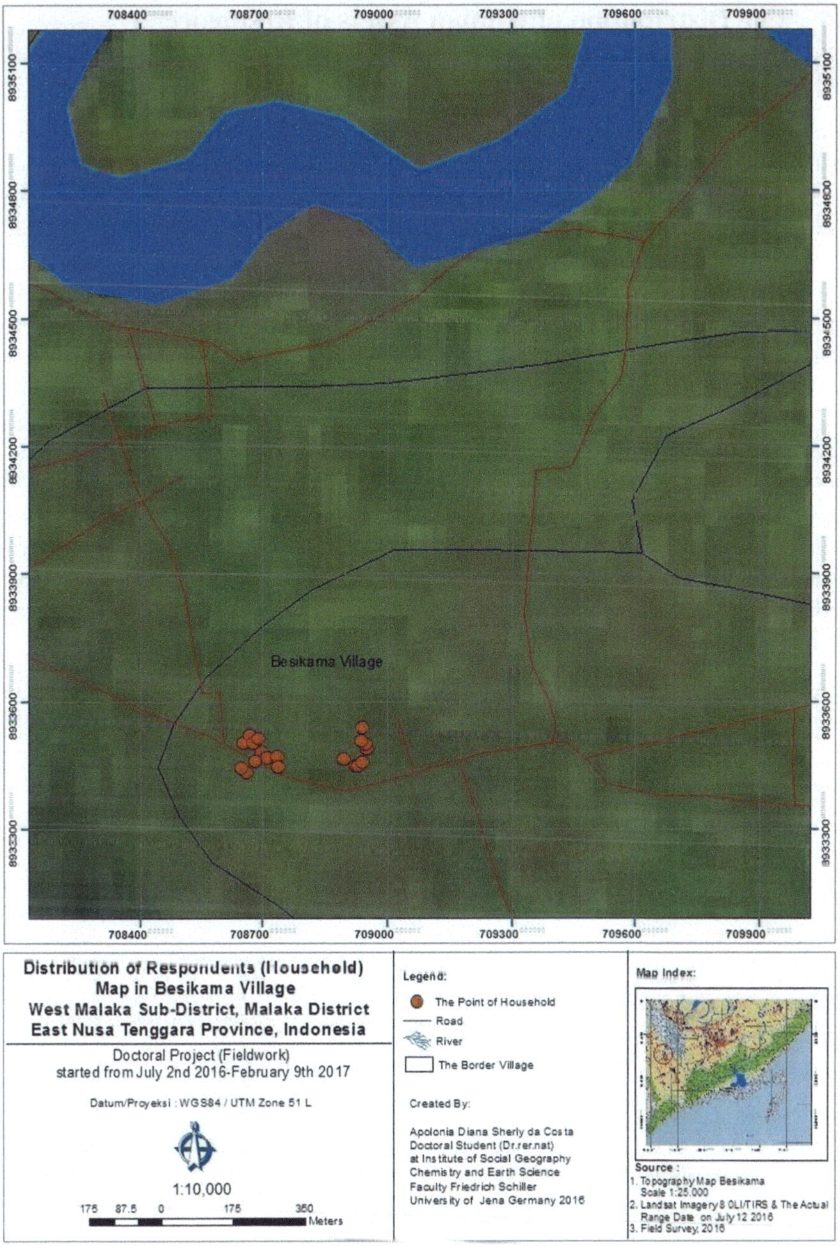

Fig. 7.21 Distribution of respondents (households) map in Besikama Village, West Malaka Subdistrict, Malaka District, East Nusa Tenggara Province, Indonesia. *Source* Author's research fieldwork (survey and interviews), July 9, 2016, to February 9, 2017

7.4 The Distribution of Human Assets of Resilience Impacted by the Flood Risk

Communities have an adaptive capability to overcome problems of flooding that hit their area. However, even with such an adaptive capability as a manifestation of resilience management, a community can still be disrupted by the risk of flood disaster. They had applied their resilience management to handle flood disaster. Their resilience management is represented also as their capabilities of adaptation to withstand floods. One of the resilience management capabilities is human assets. The distribution of human assets of resilience that is impacted by the actual flood in the region is presented in Fig. 7.22.

Based on Fig. 7.22, it can be seen that there are four subsets [i.e., health, consuming nutritional food, drinking water quality, and education (knowledge)] of human assets of resilience that are impacted by flood risk. Communities in the West Malaka Subdistrict have difficulty in maintaining their health due to the danger of flooding; 25 respondents thought that they had experienced health problems after a flood disaster, in every flood incident that has occurred. This is apparent in the presence of several diseases such as flu, fever, diarrhea, and typhus. The flood disaster has had the impact of degrading the human ability to take preventative action before the flood, during the flood, and after the flood.

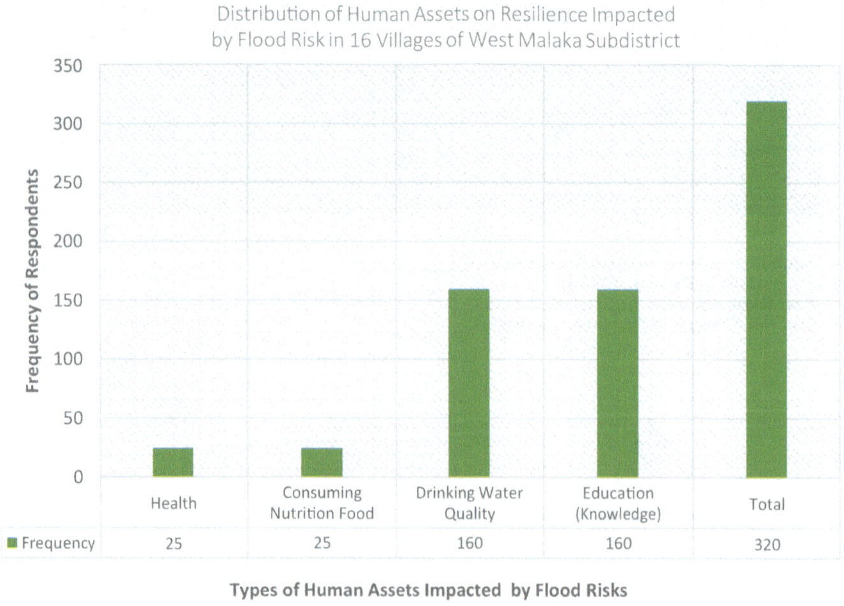

Fig. 7.22 Distribution of human assets of resilience impacted by the Flood Risk in the 16 Villages of West Malaka Subdistrict, Malaka District, East Nusa Tenggara Province, Indonesia. *Source* Author's research fieldwork (survey and interviews) July 9, 2016, to February 9, 2017

Communities have also experienced shortages and depleted availability of nutritious food supplies, which was emphasized by Soekirman (2011: 449a), a nutritionist and a professor at the Institute of Agriculture Bogor. In the publication, the Press Conference of the Nutrition Booklet (2011), the concept of "balanced nutrition" has been changed to the concept of a "Balanced Nutrition Guideline." Understanding what balanced nutrition is equips individuals and families to prevent problems from arising and helps them achieve healthy lifestyles. The Ministry of Health of the Indonesian government has also printed a book on "Balanced Nutrition Guidelines" or "*Pedoman Gizi Seimbang (PGS)*" (2003 and 2005). Every country has a visualization tailored to its specific culture. In Indonesia, the balanced nutrition guidelines are visualized in the form of a "Shaped" or "*Tumpeng*," (the popular and traditional "*Pyramid-Like*") and it is called the "Shaped Balanced Nutrition" *or "Tumpeng Gizi Berimbang (TGS)*". TGS is used as a pictorial representation of the nutrition guidelines (Soekirman 2011: 449b). TGS helps everyone choose foods of the right kind and quantity, according to different needs by age (infant, toddler, adolescent, adult, old age) and according to health conditions (pregnancy, breastfeeding, physical activity, and illness). Unfortunately, the concepts of balanced nutrition and shaped balanced nutrition do not align with the reality that is faced by communities in the West Malaka Subdistrict, where there is difficulty accessing even clean water.

The constraints in accessing clean water are evident in the response of 160 respondents who have struggled with clean water consumption for many years. The source of water for the community is the Benanain River and wells. Every time a flood occurs, the water quality automatically becomes too poor to be consumed, because of the sludge that settles and makes the water brown and undrinkable. Communities seek to address the problems of accessing healthy, nutritious food and beverages from the abilities and experience they have implemented in every period of flooding. The main reason the 160 respondents are facing problems in overcoming their human assets due to the flood disaster is their low education (elementary schools). However, they are aware of their place through implementing cultural resilience and social resilience. Cultural resilience is hereditary in community daily practice while social asset is a unity of bonding, bridging, and linking).

7.5 The Distribution of Natural Assets of Resilience Impacted by the Flood Risk

The community has a resilience system in applying their natural assets through their ability and knowledge during the process of overcoming the flood risk. The subassets of natural assets of resilience that are impacted by the flood risks are river as a source of water consumption, soil as a source of food consumption, land as a source of building settlement, river as a source of fisheries and income, and soil as a source of gardening. A graphic of the distribution of these natural assets of resilience that are impacted by the flood risk in the 16 villages can be seen in Fig. 7.23.

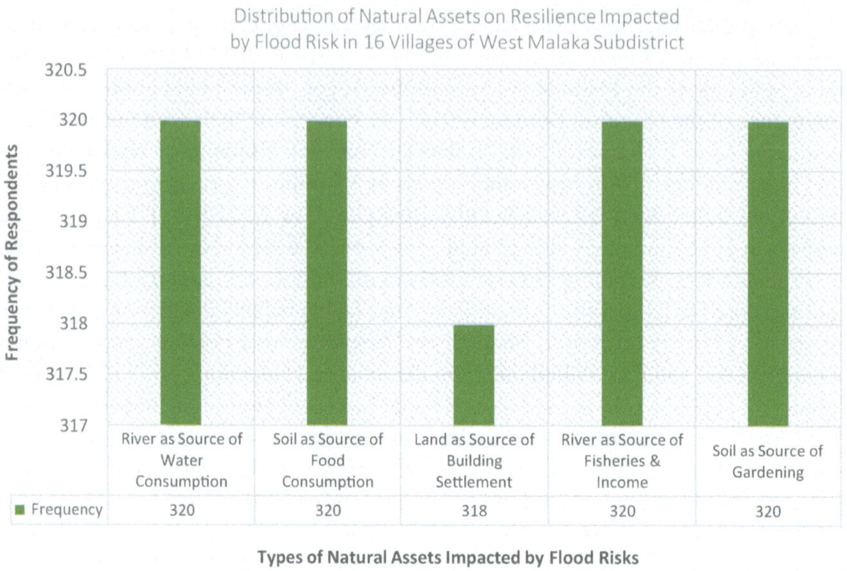

Fig. 7.23 Distribution of natural assets of resilience impacted by the flood risk in 16 villages of the West Malaka Subdistrict, Malaka District, East Nusa Tenggara Province, Indonesia. *Source* Author's research fieldwork (survey and interviews) July 9, 2016, to February 9, 2017

It can be seen from Fig. 7.23 that communities in the region have an adaptive resilience system of their natural assets which are derived from their experience in overcoming flood risk. This same opinion is expressed by all 320 respondents who explain that the river as a source of food consumption, soil as a source of food consumption, river as a source of fisheries and income, and soil as a source of gardening, is a manifestation of their natural assets in the recovery process.

7.6 The Distribution of Physical Assets on Resilience Impacted by the Flood Risk in the Region

The types of subassets that have considerable variability due to the risk of floods experienced by communities are physical assets. There are ten subassets of physical asset: transportation (motorcycle), public internet kiosk, public telephone kiosk, traditional market, irrigation and canal systems, power outage, public school, public road, community health center, and homes. These subassets of physical assets have hampered the process of people recovering power.

However, the adaptive system implemented has been practiced with the appropriate targets, but it is not yet able to provide long-term adaptive solutions. The community believes that the subassets of physical assets affected by the actual flood

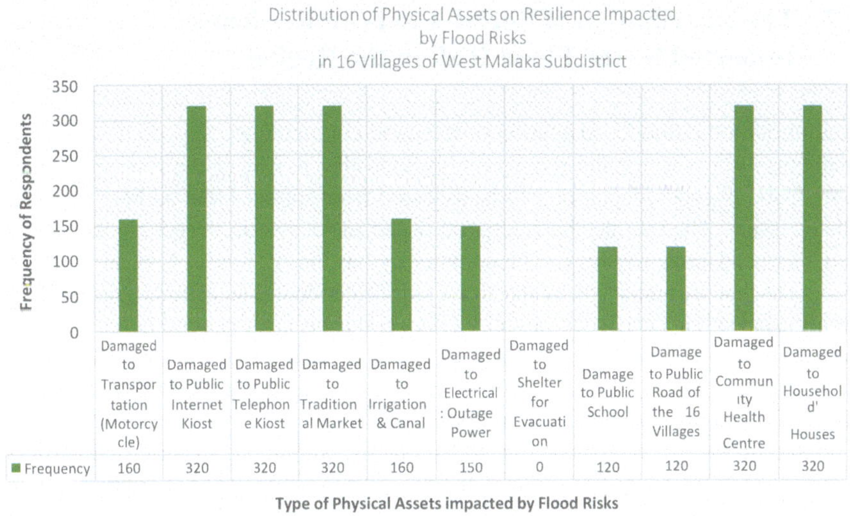

Fig. 7.24 Distribution of physical assets on resilience impacted by the flood risk in 16 villages of West Malaka Subdistrict, Malaka District, East Nusa Tenggara Province, Indonesia. *Source* Author's research fieldwork (Survey and Interviews), July 9, 2016, to February 9, 2017

are among the determinants of resilience management that they will realize in the flood disaster management process. This data is presented in Fig. 7.24.

From the ten physical subassets damaged by the floods, there are five which all respondents experience as having a negative impact, damaged public internet kiosk, damaged public telephone kiosk, damaged traditional market, damaged community health center, and damaged homes. In addition, there are two kinds of physical assets which are impacted by the flood risk, i.e., damage to transportation (motorcycle), and damage to irrigation and canal. The risk of flood disaster also impacted on electrical power. As can be seen in the data, 150 respondents experienced this kind of damage to their physical assets. Meanwhile, there are 120 respondents who highlighted damage to the public school and to the public road for accessibility to the school.

Communities overcome the risk of flood disaster by surviving and moving themselves to a safer place. Considering the physical assets damaged by flood risk that were disturbing community daily activities, they continue applying their economic resilience through collecting more income to support their daily necessities. Moreover, during a flood disaster hitting their house, power was outraged in every household's house for a temporary period. People use kerosene for temporary lights at night, and this is also the primary source of light for people who cannot afford electricity in their homes. Flood brings negative impacts because it disrupts the daily routine of people running their lives. They have done their best to survive and restore their assets.

7.7 The Distribution of Social Assets on Resilience Impacted by the Flood Risk in the Region

Dealing with the flood's adaptation of the local communities, they do have a good adaptation to overcome the risk from the flood that is part of the routine of their daily lives. A good adaptation can be assessed from routines that address the risk-related issues. The routine characteristics of the social system trigger a positive and humanistic adaptive response. The community has the six kinds of social assets of resilience that are impacted by the flood risk: individual connection, household or family connection, disaster community group connection, human rights and discrimination, fear and anxiety, crises of self-confidence or optimism, and applied system of *hakawak or gotong royong* or mutual assistance among the community. These kinds of social assets can be seen in Fig. 7.25.

Based on Fig. 7.25, it can be concluded that the community experienced barriers to social assets due to the risk of flood disaster. The community has an applied system of *hakawak or gotong royong* or mutual assistance, and all 320 respondents said that this kind of social asset can give them a unity of social power to manage the flood disaster. This mutual assistance is manifested after a flood disaster, or ex-post, such as cleaning mud from houses. The system of mutual assistance has had a positive impact on the early warning system, where informing people of the hazard was done by individual connection (180 respondents), household or family connection (200

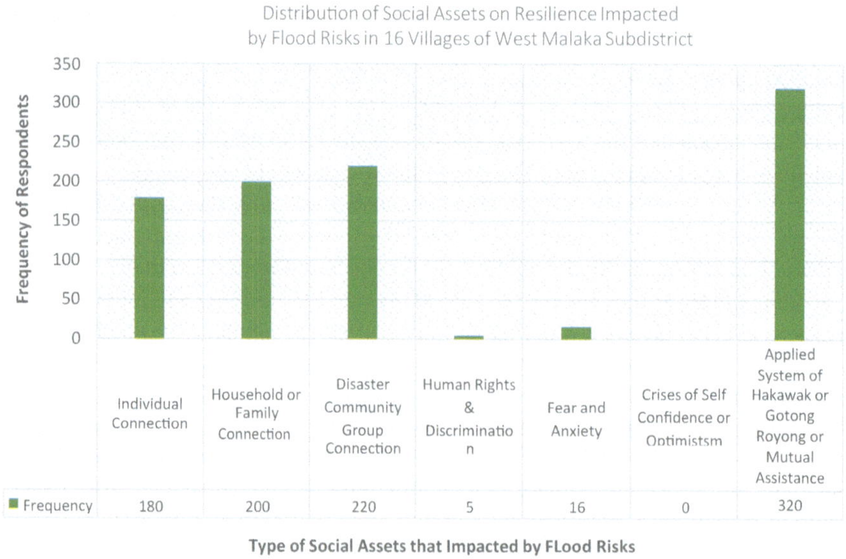

Fig. 7.25 Distribution of human assets of resilience impacted by the flood risk in the 16 villages of West Malaka Subdistrict, Malaka District, East Nusa Tenggara Province, Indonesia. *Source* Author's research fieldwork (survey and interviews) July 9, 2016, to February 9, 2017

respondents), and disaster community group connection (220 respondents). These kinds of social assets are seen as community social adaptations that have the power to address flood disaster issues in the period before the flood happens and after the flood disaster. No activities take place during the flood as people endeavor to keep themselves safe by moving to a shelter or safe place. This social adaptation helps deal with some of the social issues such as human rights and discrimination (5 respondents), fear and anxiety (16 respondents) arising from social conflict and lack of the Indonesian government assistance in the case of flood disaster.

7.8 The Distribution of Economic Assets on Resilience Impacted by the Flood Risk in the Region

Economic assets of resilience can be assessed through understanding the kinds of flood risks that impact the community. There were 9 kinds of economic asset that were impacted by the flood risk, namely, low income (employment), access to insurance of the local insurance, access to loan or credit of the local bank, obstacles on transfer system via colleagues, livestock disruption, i.e., animal deaths, damage to fisheries and shrimp ponds, obstacles on farming mobilities, and limited savings. These kinds of economic assets are presented in a graphic in Fig. 7.26.

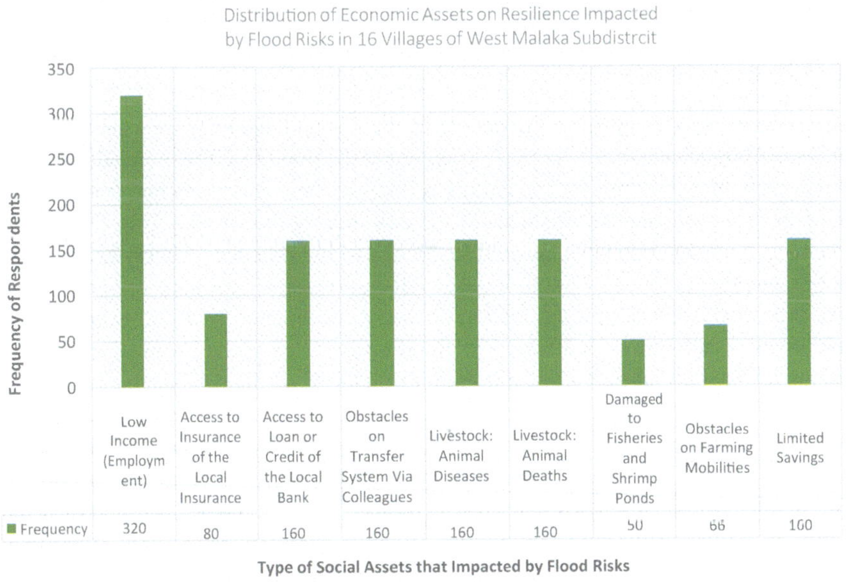

Fig. 7.26 Distribution of economic assets on resilience impacted by the flood risk in 16 villages of West Malaka Subdistrict, Malaka District, East Nusa Tenggara Province, Indonesia. *Source* Author's research fieldwork (survey and interviews) July 9, 2016, to February 9, 2017

From Fig. 7.26, it can be seen that the community has economic risks caused by the flood disaster, notably low income. All 320 respondents explained that their income is reduced due to the flood disaster, which they need to cope with and recover from. The community has good adaptation of economic resilience, but its good adaptation is destroyed due to flood disaster. There were five kinds of economic consequences experienced by 160 respondents to deal with the flood risk; access to loan or credit of the local bank (*Bank Rakyat Indonesia (BRI)* in Betun City—the capital City of Malaka District), transfer system via colleagues was disrupted, animal diseases among the community's livestock, limited ability to save while fulfilling daily necessities, and handling the flood risks that caused damage to other assets such as physical assets and social assets. The community has also been connected to local insurance in the city. The community required a diversity of financial solutions to solve issues related to damage from flood disaster, especially in respect of their livelihood resources, i.e., damaged fisheries and shrimp ponds (50 respondents) and damaged farming equipment (66 respondents).

Independently, the community applied several solutions to handle flood disaster. There is no assistance from the Indonesian government in the case of flood disaster and the community needs to be more resilient in handling the flood disaster itself. The community values its efforts and coping mechanisms as useful strategies for the process of resilience.

7.9 The Distribution of Cultural Assets on Resilience Impacted by the Flood Risk in the Region

The risks of flood disaster to the people in the region have increased their cultural resilience's mechanism, i.e., coping. The greater the flood risk, the more adaptive the practices by the community. The culture reflects the face of society and is seen from the behavior and adaptive practices of the community in everyday life. Communities have a strong system of cultural resilience in the practice of everyday life and in the adaptive practice of flood disaster. All respondents gave their opinion on the five types of assets of cultural resilience that were affected by flood disaster. The five kinds of cultural asset affected by flood disaster are applied-routine traditional rituals for prevention measures, mitigation measures, resiliency system, customary leader's house, and limited accessibility and management of the customary leader with the Church and other institutions. These five kinds of cultural assets were being damaged by flood disaster despite substantial information to understand the way of community coping strategies in terms of cultural coping to deal with flood disaster. Cultural coping deals with community strength which involves their traditional rituals for prevention, mitigation measures, and the mutual assistance system between customary leaders and community in terms of socio-cultural resilience. These cultural copings lead the community to enhance their resiliency system in respect to cultural coping that also needs social coping (i.e., social resilience; bonding, bridging, and linking). Community owns and applies every culture strength it possesses to handle flood disaster. The purpose is to practice their cultural endurance; however, such preventative and protection systems for flood disasters help the people manage the risk of flooding that can inundate people's homes. There is evidence that once a flood disaster impacts the research site, the customary leaders' houses were protected by the female leader, evidenced by a ritual of eating a betel nut and praying to an ancestor and the Lord.

This takes place before a flood disaster happens and after the flood disaster occurs. Such preventative actions implemented by the customary leaders required cooperation with the religious leaders in the Church. Another kind of damage is a limited accessibility from the customary leaders to reach a cooperation of coping with floods with the religious leaders in the Church and other Institutions.

As a flood disaster sweeps the area (during flood disaster), customary leaders have their own single role to protect the community, instead of assistance ship from priests, sisters, and brothers in the Catholic Church. The customary leaders protect the community through a traditional ritual where a betel nut is used as a medium of worshipping to the ancestors. A traditional ritual seeks to ask for help from the ancestors to protect the community's safety from the risk of flood disaster. It also seeks to reconstruct a broken dike that was destroyed by the flood disaster. Thus, the role of customary leader is important for the community's phase of prevention, mitigation measures and the mutual assistance system in terms of socio-cultural resilience. Their role is dominated by the value of cultural advantage that they have experienced (both customary leader and community). Such cultural advantage becomes cultural assets that help the community handle flood disaster, unlike the shortcomings in their cultural assets which require assistance from other institutions.

A strong approach to the community's cultural assets has been nurturing skills and knowledge, i.e., human assets of resilience to survive and reconcile with the flood risks. They continue to carry out five types of cultural assets in the face of floods and are not affected by the flood risk they have to endure because of the strength of the base of cultural assets of resilience and the strong approach by the community.

7.10 The Distribution of Religion Assets on Resilience Impacted by the Flood Risk in the Region

The risks of floods have also disrupted community's religion assets in the study area. In order that the community can cope with their religion assets in a flood disaster, a religious coping process was created and implemented. Religious coping has been created and implemented in the community for more than 20 years. A flood disaster requires religious coping; it is therefore important for the community to establish a caritas system among the householders. A caritas system is a system formed by priests, sisters, and brothers in the Catholic Church that increases community religious coping in terms of flood recovery and its management. Based on the interview with 320 households, they state that there are nine types of religion assets that the local community has experienced in the impact of flood disaster, namely the religious coping actors' role (i.e., priests, sisters, and brothers in the Catholic Church) who act as:

1. A facilitator of disaster community ecumenical activities.
2. An actor in increasing early warning information.
3. An advocate for the marginalized community.
4. A facilitator to fulfill the need for more pastoral care to assist the community.
5. An actor to act for reparation and reconciliation of the infrastructure.
6. A shaper and also facilitator to strengthen the faith of the community.
7. A facilitator to provide financial resources assistance such as water and food for the community when a flood disaster happens.
8. A responder to work quickly for the interest of the community in an onset crises case such as flood disaster.
9. A provider of pastoral services regarding limited resilience cooperation with the customary leaders in the prone area of food disaster (even though disparate networking with customary leaders: ineffective of religious coping).

These nine kinds of religion asset were implemented by the priests, sisters, and brothers of the Catholic Church to accelerate their humanitarian mission, to relieve community efforts, and to serve as hospitality for the community in deepening their faith that was being threatened by unsafe life and conditions (i.e., flood disaster). The advantage of religion assets is understood by the community as a coping religion that also delivers the meaning of life as very precious while in an unsafe condition (see also Chap. 6: Religion Assets).

They believe that a flood disaster that sweeps their area is just a temporary temptation where God can help them. Thus, religious coping helps communities to deal with flood disasters. Nevertheless, a religious coping owned and provided by the religious leaders has effectively manifested those assistantships with the customary leaders after a flood disaster ends or ex-post.

Flood disaster that hits community and their area leads also to the disruption of operational systems of ecumenical or caritas system created by the priests, brothers, sisters, and community in their daily religious life. It is evidenced by none of the community who mention that "action postponed for basic training and support" is effectively applied by the religious institution. Nevertheless, it disrupts their *ecumenical or caritas system* in the short term which means that floods cause ineffective activities on their system during floods while there are effective activities on their system after floods. The need for more institutions to help the community is therefore a crucial issue.

Communities require not only an assistance of religious coping from the role of priests, sisters, and brothers in the Catholic Church, and cultural coping from the customary leaders, but also other humanitarian mission institutions that might be touched to deliver an assistance to the community. Furthermore, there was only one kind of religion asset, i.e., actions for basic training and support, which is actively implemented by the religious leaders without being interrupted by the flood. As the task of religious leaders is to be concerned about ecumenical and charity actions, basic training and support have always been on the agenda to assist the community before the flood disaster happened.

7.11 The Distribution of Law or Legal Assets on Resilience Impacted by the Flood Risk in the Region

Since the community in Malaka is the sole actor of resilience of the flood disaster that plagues the community, regardless of the risk, there will continue to be constraints faced by the community in handling flood problems. There are six kinds of law or legal assets that are impacted by the flood risk as can be seen in Fig. 7.27.

Based on Fig. 7.27, there was one kind of legal asset, i.e., the broken public road, that was impacted by the flood risk and therefore needed assistance from the Indonesian government, as the public road is seen as the main means of accessing the community activities. Every asset of resilience owned and applied by the community explains the need for the Indonesian government's assistance to increase the community's resilience assets. Communities need effective, efficient, and sufficient support from the Indonesian government to tackle their problems (i.e., building resilience

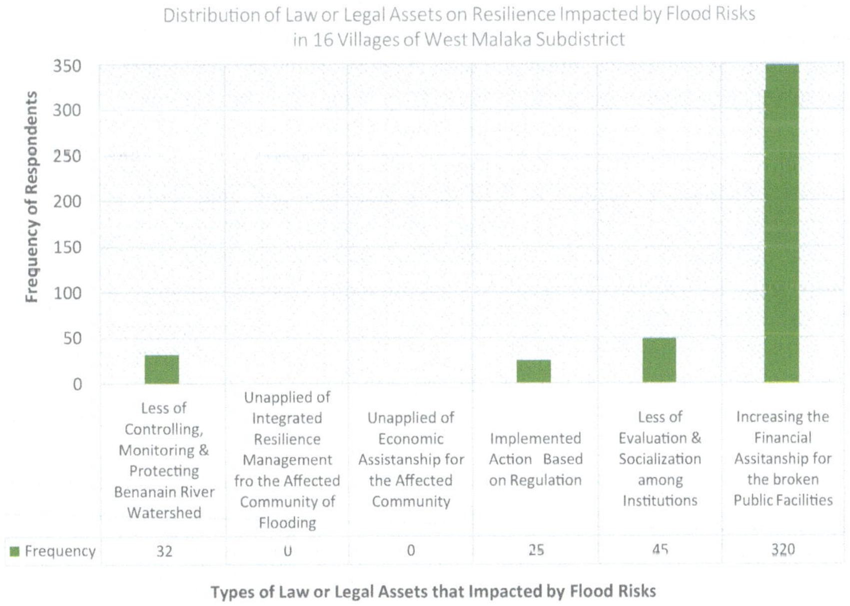

Fig. 7.27 Distribution of law or legal assets on resilience that impacted by the flood risk in 16 villages of West Malaka Subdistrict, Malaka District, East Nusa Tenggara Province, Indonesia. *Source* Author's research fieldwork (survey and interviews), July 9, 2016, to February 9, 2017

for flood disaster). The data shows that there was no intervention in terms of an integration of flood disaster management such as early warning system, prevention, and mitigation measures before a flood disaster happens or ex-ante, during a flood disaster occurrence, and after a flood disaster happens or ex-post. Another example is intervention in terms of building resilience such as human assets, natural assets, physical assets, social assets, economic assets, cultural asset, religion asset, law or legal assets, and political or policy asset. As the task of the Indonesian government is to serve the community in need that experiences the flood disaster, a reflection of this case is still seemingly far from the eye of the Indonesian government.

The Indonesian government's regulation about protecting the protected area near the river was implemented but there were few actions. The role of legislative actors as the creators of the regulation have less of a cooperative action with the executive actors as the most implemented actors-based regulations. They are still running and realizing these kinds of unimplemented law or legal assets for the community in the region.

7.12 The Distribution of Political or Policy Assets on Resilience Impacted by the Flood Risk in the Region

The community in West Malaka Subdistrict describes that the Indonesian government's actions on the prevention and protection measures are ineffective, inefficient, and insufficient to manage flood disaster. They name their own actions as a thematic of flood disaster risk management with Group Volunteer which has been formed into a group of resilience and risk protection management in every village. Their perception of a thematic in respect to their daily activities, and their daily activities in applying assets of resilience is also therefore a mandatory coping action. It indicates that the community is able to create their own thematic that helps them to be more aware about their hazardous area and increase their assets of resilience. It confirms that the community has a theme of policy actions that are driven by themselves with the help of the Group Volunteer. Based on the interviews with 320 households in the region, the community mentions and describes that there is a "distribution of political or policy assets on resilience that are impacted by the flood risk in the region, West Malaka Subdistrict," which is presented in Fig. 7.28.

Based on Fig. 7.28, it can be concluded that policy actions driven by the Indonesian government were assisting the community in the region. There are 60 respondents who reported that the need to implement effective action in building the community's resilience is a serious issue to be overcome by the Indonesian government and other humanitarian mission institutions. Communities observe that such effective action in building their resilience is still in the process of realization which means it is in "rehearsal phase" and has a long way to go to reach a "solution-based-community-approach with the community who experience the flood disaster." The

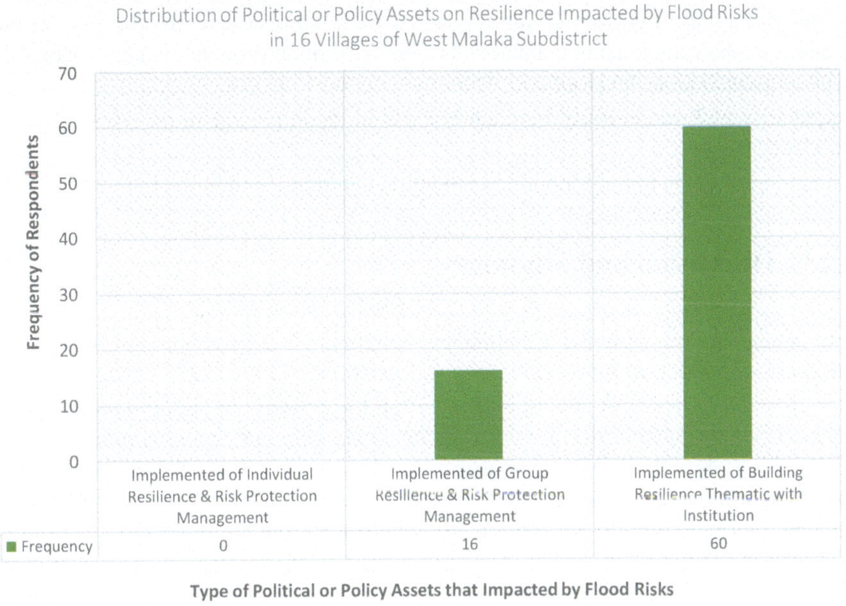

Fig. 7.28 Distribution of political or policy assets on resilience impacted by the flood risk in 16 villages of west Malaka Subdistrict, Malaka District, East Nusa Tenggara Province, Indonesia. *Source* Author's Research Fieldwork (Survey and Interviews), July 9, 2016, to February 9, 2017

active institution is a Non-Government Organization (NGO), Community Association for Disaster Management (CADM) which welcomed discussion on resilience thematic for the flood disaster with other institutions in Indonesia and overseas.

As a single actor of resilience, the community seeks its own solutions to cope with flood disasters. On the social resilience asset implemented, i.e., a disaster group platform for use in serious discussions and the sharing of experiences of flood disaster and coping mechanisms, this has the aim to strengthen community resilience and provide a strong resiliency system. The community also shares information related to the lack of assistance by the Indonesian government (i.e., the role of executive actors; local executive actors in this case, Head of Malaka District, and the role of legislative actors; local representative house or *Dewan Perwakilan Rakyat Daerah*/DPRD in 2014). They delivered criticism about the role of the Indonesian government not acting as responder and facilitator to help the customary leaders (e.g., cultural coping) and the priests, sisters, and brothers in the Catholic Church (e.g., religious coping). Moreover, this policy action was not adequately achieved since it took only short-term action. Communities explain that in order to obtain policy action from the Indonesian government, more attention is required from the scientific or academic institutions (e.g., universities) to conduct a study or research about communities' resilience for handling flood disaster. The issue would then come to the surface and might be heard by other humanitarian missions both in Indonesia and overseas, especially in the northern hemisphere. The assistance needed nowadays by a community is full monetary capital to withstand and recover from flood disasters. In addition, social awareness about increasing capacities to deal with flood disasters is also needed to achieve a sustainable development. Therefore, a coping mechanism could be achieved perpetually and successfully by each household or individual in the West Malaka Subdistrict.

7.13 Discussion and Summary

This study agreed with the UN-ISDR's statement: "...........*its resolutions on international cooperation on humanitarian assistance in the field of natural disasters, from relief to development, and recalling the resolutions of the humanitarian segments of the substantive sessions of the Economic and Social Council*" (UN-ISDR, 2007: 1). UN-ISDR states also in resolution 46/182 of 19 December 1991 that "*humanitarian assistance in the field of natural disasters, from relief to development, and recalling the resolutions of the humanitarian segments of the substantive sessions of the Economic and Social Council.*" This resolution and declaration conform with the Indonesian disaster management regulation; clause 26 (1: e) chapter V, number 24 from 2007 mentions that "*every person has a right to participate in disaster mitigation decision making*" (Community-Based Disaster Risk Management, CBRDM 2005: 39), and also Indonesian government regulation (number 76, year 2001) article two states that the "*Indonesian government and rural development organizations should take an active role in the community so that people will continue to take part in the development of life with fellow of villagers.*" (qtd. in The Watershed Management Bureau of Benanain-Noelmina 2011: 29–30).

In implementation, risk perception supports community-based disaster risk management as a means of discovering solutions based on community experience in using their resilience assets to deal with flood and in practicing flood risk assessment. Cutter et al. (2008: 19) mention that a detailed risk assessment should be carried out by the community disaster committee led by a local leader. In addition, FGD and SMEs' survey and interviews were conducted and guided by the local leader in this study (key informants that connect to customary leaders' internal rules). The results also show a spatial distribution map (river) and buffer map based on community risk perception in the case of vulnerable villages in the study area. Increasing resilience through community-based disaster risk management that reclaims a community action also meets the strategic priorities of the UN-ISDR in the Hyogo Framework for Action or HFA. It thus aims to build community resilience.

This study is in two parts for community-based disaster risk management.

First, this chapter discusses community risk perception about the flood characteristics. Flood characteristics involve periodic flooding, flood duration, flood depth, and flood casualties. The result of periodic flooding was experienced by all 16 villages in the study area which occurred every twenty years in 1939, 1959, 1979, and 1999 and continued with two flood events in a year in 2000, 2001, 2002, 2003, 2004, 2005, 2006, 2007, 2008, 2009, 2010, 2011, 2012, 2013, 2014, and 2015. The data shows that there has been no significant intervention from the humanitarian mission institutions since the first flood in 1939, while the case of the East Timor refugees is still unresolved, leading to twin political disaster and natural disaster at the same time.

Among the community, 31.3% experience floods with the longest duration of 25–30 days. The mitigation measure applied was building an 80 cm high flood barrier in the ex-ante period with no significant actions during the flood. The parts of the community that experience a flood duration of 0–5 days built a barrier with a height of 30 cm. For flood depth, 100 respondents report inundation impact in the range 251–300 cm (max 300 cm and none experienced > 300 cm). This caused people to move to a safe place (shelter) in the city if the flood impact reaches > 300 cm. People applied physical capital in terms of a stage house as a strategy, which is similar to establishing or raising a barrier to prevent water entering their house.

In the ex-post period, people reorganize and manage their assets. They come back to their homes and clean the mud from the flooding with help from their neighbors (bonding), a mini-platform discussion with the Group Disaster Community (linking), which helps them to recall or communicate with their families in other places for remittance (bridging). They also established their social capital "*hakawak*" or mutual assistance. In addition, based on the survey and interviews, discussion with the villages that are prone to flood reveals that there are three levels of hazardous areas: ***Vulnerable, More Vulnerable, and Not Vulnerable*****.** ***More vulnerable areas*** are *Lasaen, Fafoe, Umatoos, Raimataus, Rabasa, Rabasahain, Rabasahaerain, Lofoun, Motaulun, Motaain, Naas, Oanmane and Sikun* with a radius of 0–25 m from the Benanain River to their homes. The high risk of flooding impacted on communities recreating their resilience assets.

Vulnerable villages are Maktihan and Umaloor, where flood reaches a radius of > 25–100 cm based on the distance of the Benanain River's mouth to their homes. These villages' geographical positions, which are north of Maktihan Village and Besikama Village and east of Besikama Village, led them to apply social capital. Besikama Village is the only village categorized as a ***Not Vulnerable village*** with a radius of 100–250 m of the Benanain River's mouth to their homes. Flood sweeps their area and leaves broken and damaged public facilities but does not endanger life. Communities experience floods of 0–10 days with an average depth of flooding between 0 to 100 cm. People overall are adapted to flood and use their assets that were impacted by the flood. This discussion is provided in section two.

Second, this chapter discusses community risk perception about the types of assets that were affected by the actual flooding. The types of assets affected by the actual flooding are human assets, natural assets, physical assets, social assets, economic assets, cultural assets, religion assets, law, or legal assets, and political or policy assets. ***Human assets*** damaged by the actual flood are health, consuming nutritious food, drinking water quality, and education (knowledge). The problem they face is the availability of nutritious food supplies. Their foods are not in line with the concept of "Balanced Nutrition" mentioned by Soekirman (2011: 449b) and flood disaster particularly affects people in vulnerable groups (infant, toddler, elder, pregnancy, breastfeeding).

Soekirman affirms that his idea is that everyone chooses foods of the right kind and quantity, according to different needs by age (infant, toddler, adolescent, adult, old age) and according to health conditions (pregnancy, breastfeeding, physical activity, and illness). These concepts were not experienced in reality by people while they also lacked access to clean water. ***Natural Assets*** damaged by the actual flood are the river as a source of food consumption, soil as a source of food consumption, river as a source of fisheries and income, and soil as a source of gardening; 318 people voted for land as a source of building settlement. Overall, people applied their adaptive resilience system of their natural assets to cope with flood. ***Physical Assets*** damaged by the actual flood are transportation (motorcycle), public internet kiosk, public telephone kiosk, traditional market, irrigation and canal systems, power outage, public school, public road, community health center, and homes. Similar to natural assets, people applied an adaptive system in overcoming the flood but with short-term not long-term solutions. Five respondents experienced negative impact on their public internet kiosk, public telephone kiosk, traditional market, community health center and homes. The same adaptive system was implemented by the people. People say they face constraints to transportation (motorcycle) and to irrigation and canals due to the flood. The impact of flood on 15 respondents' electrical power is also a crucial problem. Moreover, 120 respondents highlighted difficulties accessing the public school and the public road due to mud. Nevertheless, people maintain their action on collecting more income (economic resilience) to meet their daily necessities. ***Social Assets*** damaged by the flood are the people's social bonding, bridging, and linking to their neighbors (to clean flood mud), relatives or families in other places (communication and remittances), the local bank in the city center of the district (loan in the bank), and private and public institutions in the city center of the district and province (communication of asking help for such sudden crises due to actual flood). However, people applied their system of *hakawak or gotong royong* or mutual assistance unity in a partnership to increase social power to cope with flood. This is evidenced by 180 respondents who applied (bonding) to clean the mud, 220 respondents who connected to their family for remittance, and also 200 respondents who connected to the community disaster group. People applied their social capital, representing their effective social adaptation. In addition, social issues such as human rights and discrimination were experienced by five respondents, and fear and anxiety by 16 respondents caused by social conflict and lack of Indonesian government assistance. This study concludes that the community applies its social capital to overcome damage to its social capital due to the flood.

The community applies various adjustments on its ***economic assets*** in order to deal with a flood disaster. Consequences on economic assets made the community trigger its *various adjustment with a variety of job* (a multi-tasking person who works as a farmer, a trader in the traditional market, and as a fisherman/woman) in fulfilling both the community's daily necessities and its resilience assets. Economic assets which are part of community livelihood resources advance a community's efforts and its stabilization in coping with flood. The main idea of livelihood resources is to integrate all of the assets of resilience affected by the current flood disaster in anticipation or preparedness for the next flooding. Nonetheless, community adjustment reflects the economic consequence of a flood disaster and there is no ideal way to reform or recreate their economic assets. Communities encourage themselves in optimizing their social capital, practicing their strong cultural values with the help and guidance of the customary leaders and religious leaders to re-establish their economic assets that were affected by the flood. It shows that a community's efforts absorbed values of social cohesiveness for its social resilience effectiveness in coping with flood. The effectiveness of adjustments depends on a community's knowledge and skills to operate their daily activities, including working as a farmer, a fisherman/woman, trader in the traditional market, and traditional weaving to maximize their resilience assets after a flood occurs. ***Cultural assets*** are valued by the community as its cultural values are sourced from their traditional values. The community has practiced their cultural values for many years, and these values help the community to withstand flood disasters. They have a strong cultural resilience. There are five types of assets of cultural resilience affected by flood disaster; applied-routine traditional rituals for prevention measures, mitigation measures, resiliency system, customary leader's house, and limited accessibility and management of the customary leader with the Church and other institutions. Communities name their cultural assets as a cultural coping mechanism where they implement the traditional rituals, the role of customary leaders and strong social cohesiveness among customary leaders and religious leaders in coping with flood. Cultural coping mechanisms help the community enhance its resiliency system to cope with flood disasters. On the discussion of ***religion assets***, this study assesses religious coping mechanisms that contribute to solving the community's problems including assets that were affected by the flood. The community affirms that there is a caritas system that was established by the religious leaders (priests, sisters, and brothers). Such a caritas system supports the community in both tangible and intangible ways in facing flood disaster, though it is still ineffective once flood strikes and there is less assistance in terms of cooperation and coordination among the religious leader and customary leaders in helping the community.

The caritas system, which was formed by priests, sisters, and brothers of the Catholic Church, has increased community religious coping mechanisms in terms of flood recovery and its management. Based on the interviews, communities said that they experienced nine types of assistance from the religious role which are at risk of being affected by the actual flooding and the future flooding. The nine types of religion assets where religious actors have a role are; an actor in increasing early warning information, (1) a facilitator of disaster community ecumenical activities, (2) an actor in increasing early warning information, (3) an advocate for the marginalized community, (4) a facilitator to fulfill the need for more pastoral care to assist the community, (5) an actor to act for reparation and reconciliation of the infrastructure, (6) a shaper and also a facilitator to strengthen the faith of the community; (7) a facilitator to provide financial resources and assistance such as water and foods for the community during the flood, (8) a responder to work quickly for the interest of the community in an onset of crises such as flood disaster, and (9) a provider of pastoral services regarding limited resilience cooperation with the customary leaders in the area prone to flood disaster (even though disparate networking with customary leaders; ineffective of religious coping). ***Law or Legal Assets*** discusses that the regulation issue is a crucial one for the community in understanding its position as citizens of Indonesia. Based on the interviews, the community shares the opinion about the role of the Indonesian government tending not to prioritize social welfare in rural villages; in this case the community's situation from living in a rural area. They oppose the government prioritizing physical capital in terms of infrastructure for the reconstruction of the broken dike and public facilities (320 respondents) rather than the social welfare of the community's location and its development. In addition, communities are facing challenging issues in dealing with environmental protection, river watershed resources, and an unstable economy living near the Benanain River watershed. The community triggers action in making an independent forum discussion among institutions as they convey their opinion about the need for evaluation and socialization among institutions (45 respondents). Also, 25 respondents said that there are actions based on regulation in terms of environmental protection which exist but have not been implemented. This indicates that communities require a significant prospective environmental management from the Indonesian government highlighting the government's actions for controlling, monitoring, and protecting the Benanain River watershed. Also, the community highlights its case that building resilience and disaster management should be implemented by the government (none voted in this case).

Political or policy assets were impacted by the floods, and the community has agreed on the two types of implementations, namely, implementation of group resilience (16 respondents) and risk protection management, and implementation of building resilience with institutions (60 respondents). None of the respondents mentioned the implementation of individual resilience and risk protection management by the Indonesian government. The community confesses that the only active institution is a Non-Government Organization (NGO), the Community Association for Disaster Management (CADM), that engages on the issue and theme of resilience. The required government intervention is important, as the active roles of customary leaders and religious actors are dominant in taking care of the community's problems (building resilience for flood disaster) and considering how to best enhance the community's resilience. While the community enhances resilience as a single actor of resilience, helped by the customary leaders and also religious leaders, there is a lack of assistance from the Indonesian government (i.e., the role of executive actors: local executive actors in this case, Head of Malaka District, and the role of legislative actors: local representative house or *Dewan Perwakilan Rakyat Daerah*/ DPRD in 2014).

In addition, I added that the government intervention in the wider scale is not able to embrace other humanitarian mission institutions in Indonesia and overseas, especially northern hemisphere countries, which could assist in developing solutions regarding building political or policy resilience for floods in the study area. It highlights the issue of community-based disaster risk management in the study area that turned out to be one of the most crucial issues in humanitarian disaster relief experienced by the people in the West Malaka Subdistrict, in the West Timor Island, Indonesia.

References

Community-Based Disaster Risk Management, CBDRM (2005) Masyarakat Penanggulangan Bencana. Community Based Disaster Risk Management's Article. Kupang, East Nusa Tenggara Province, Indonesia

Cutter S, Barnes L, Berry M, Burton C, Evans E, Tate E, Webb J (2008) Community and regional resilience: perspective from hazards, disasters, and emergency management. Hazard and Vulnerability Research Institute, Department of Geography University of South Carolina. Columbia, South Carolina

Dewi A (2007) Community-based analysis of coping with urban flooding: a case study in Semarang, Indonesia. Gadjah Mada University, Yogyakarta City, Indonesia. M.Sc Thesis. International Institute for Geo-Information and Earth Observation. Enschede

Febrianti F (2010) Flood Risk Perception and Coping Mechanism of a Local Community: A Case Study in Part of Surakarta City, Central Java Province, Indonesia. Gadjah Ma- da University, Yogyakarta City, Indonesia. MSc Thesis. International Institute for Geo- Information and Earth Observation. Enschede

Frazer J (2010) Flood frequency analysis. International Edition University Corporation for Atmospheric Research, The COMET and NOAA Programme, National Weather Service (NWS), The United States of America

Soekirman (2011) Taking the Indonesian nutrition history to leap into betterment of the future generation: development of the indonesian nutrition guidelines. Indonesian Danone Institute Foundation for Food Fortification. Paper, Asia Pac J Clin Nutr 20 (3):447–451. Jakarta, Indonesia (2011)

The United Nations International Strategy for Disaster Reduction, UN-ISDR., Hyogo Framework for Action (HFA), 2005–2015: Building the Resilience of Nations and Communities to Disasters. The HFA's Research Projects, in the Andean region, Central America, the Caribbean, Asia, Pacific, Africa and Europe (2007)

The Watershed Management Bureau of Benanain – Noelmina, Rencana Pengelolaan Daerah Aliran Sungai Terpadu di DAS Noelmina. Kementrian Kehutanan Direktorat Jenderal Bina Pengelolaan DAS dan Perhutanan Sosial. Kupang, Indonesia (2011)

Chapter 8
Sustainable Livelihood Development Model

In this chapter, I explain an aspect of the sustainable livelihood development model's components which bring forward a relationship between building community resilience and the risk factor of flooding, through embodying both as a perceivable social-ecological integrity. Section 8.1 unifies two components of ecological integrity (i.e., building community resilience and risk factor of flooding) into a critical thematic model, delivering an idea of human risk perception in understanding their own built resilience for subduing flood risk management.

This book uses Venn's sustainable development model that consists of and integrates three elements: social equity, ecosystem capacity, and economic security. Building resilience in sustainable development illustrates ecological integrity represented by the community's capacity to cope with flood disasters in an effective, efficient, and sufficient manner.

In regard to the risk of flooding in sustainable development, social equity takes on a key role in the community's livelihood resources development. In Venn's model, social equity is illustrated as "Societal Welfare" (see Fig. 8.3, point 2). Sustainable development values the community's efforts as an enabler of resilience for handling flood disasters; the community drives its assets of resilience in order to reach a sufficient level. Unfortunately, this view is contradicted by the reality, since the community is still the single actor of resilience that drives its livelihood resources in response to flood disasters. The ecological capacity is not balanced with the concern of the Indonesian government.

In this section, the essence of valuing the community's actions and the ecological capacity were not supported by Venn's model; revealing that more research is needed to describe an integrated and balancing system in ecological capacity.

Hence, this chapter postulates three integrated ecological perspectives in this sustainable development model. They are "effective," "efficient," and "sufficient" and they reflect an understanding of the three critical divisions of the three thematic models in this study. Building community resilience for flood risk management

A. D. S. da Costa, *Building Resilience for Flood Disaster in Malaka-Timor, Indonesia*, Springer Natural Hazards, https://doi.org/10.1007/978-3-031-63272-3_8

corresponds to the model which represents human risk perception of the livelihood resources (i.e., the nine assets of resilience). It is understood as an integrated ecological approach.

8.1 Three Components of the Sustainable Livelihood Development Model

The three components model encompasses three critical divisions of three thematic models. They are building community resilience, risk factors of flooding that impact the building system of a resilient community, and community sustainable livelihood development. Each of these components in the model represents human risk perception of their livelihood resources (i.e., the nine assets of resilience) toward an integrated ecological approach. The discussion is presented in Fig. 8.1.

Based on the three components model as displayed above, the research project of this book can be prescribed that there are three different levels of class for measuring a community's implementation of assets of resilience for handling the risks of a flood. The three levels or classes distinguished above are created on a weighted factor of how a resilient community achieves a long-term action. The three components model

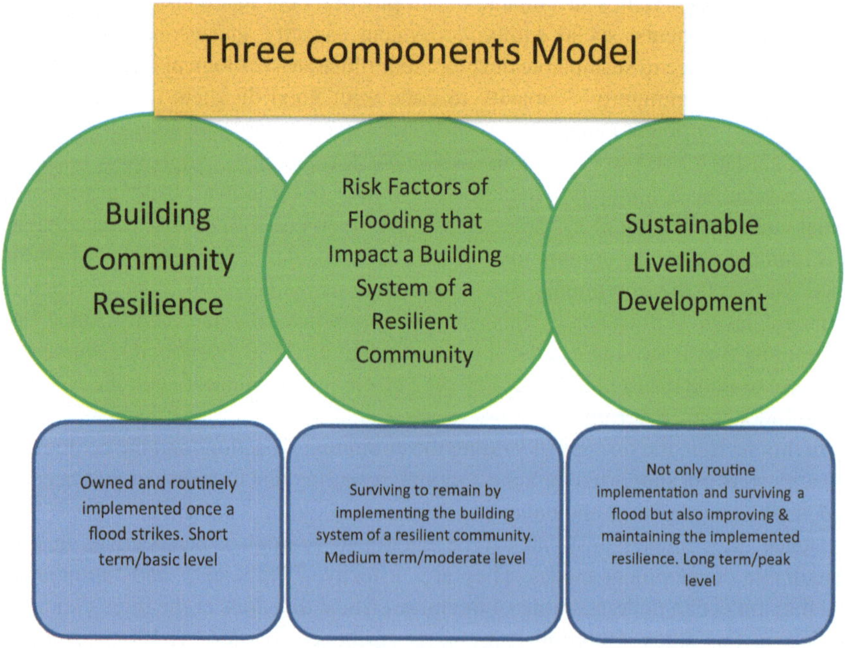

Fig. 8.1 Three components model. *Source* Author's sustainable livelihood development model dissertation research fieldwork, July 9, 2016 to February 9, 2017

describes the listing phases of sustainable livelihood development. The listing phases of the trial components model are:

1. *Building community resilience* is the first phase of an achieving resilient community. At this phase, I asked the people to give their perceptions through the lens of the assets of resilience that they had already established. Furthermore, this phase of resilience explains that the assets owned by the community on a regular basis are also implemented by the community whenever floods hit the assets of resilience.

   The community was able to discover and understand the kinds of assets of resilience that they have to be applied in flood disaster management. Assessing this phase is the assessment of how a system of building assets on resilience can run smoothly without losing one or more than two elements of the balanced system of resilience, which is integrity. The element of *integrity* of the intended balancing system of resilience is a dynamic element that can change the pattern of its coping mechanism caused by the elements of the risk of flooding. The element of integrity of the intended balancing system of resilience is based on three categories of assessment, namely effectiveness, efficiency, and sufficiency from the application of the assets owned by the community. Even though one element or more than two elements of integrity of the intended balancing system (i.e., sufficiency, effectiveness, and/or efficiency) are ignored, the community keeps running and improving on their applied assets of resilience. Thus, it can be judged that the community is at the first level which is the level of formation and the discovery of a system established as the strength of the foundation of improving assets of resilience. However, this phase still does not achieve the target of sustainable livelihood development.
2. *Risk factors of flooding* that impact a building system of a resilient community is known as the second phase of an achievement of building a resilient community. This phase has different processes in which the communities themselves perceive and experience the risk of flooding and are able to manage and maintain their assets of resilience. This phase can be understood as the process where a community continues to value their assets of resilience in spite of flood disaster striking them. Assessing this level was based upon a measurement with scoring 1 point; this is an assessment of the community's ability to build community resilience. There are three risk factors that were assessed; the three elements are not only known as the intended balancing system of resilience or as an integrity system but also as a foundational assessment of the kind of assets of resilience that has applied by the community to cope with a flood disaster.
3. A community that has experienced a flood disaster has taken lessons from the process they have experienced and overcome. In this phase, I am trying to urge the community to see the value of the assets of resilience that they have had and applied through the lens of "risk of flooding" in which the community sees the natural factors in reference to or reviews and analyzes them for assessment of community resilience. A review of this lens explains that the balancing system sees, practices, and assesses any assets of resilience that must be applied in a

balanced way without ignoring one or more of the elements that exist. However, not all of the elements can be run simultaneously by actors of resilience (i.e., community, related institutions, and the Indonesian government). Rather, integrated elements (i.e., effective, efficient, and sufficient) for the case study of flooding are only implicated with an active role from the *complete actors* of resilience not only one actor of resilience. The community as a single actor or mover of resilience has an increased capacity from their driving for assets of resilience in dealing with flood disaster but it is not sufficient. Thus, further analysis of sustainable livelihood development as a union assessment of the building resilience is the first phase, I presented. I also presented the second phase of analyzing a community's sustainable livelihood development which is about to consider the community resilience perception of the risk of flooding. Then, I described the sustainable livelihood development model in point three as a model of community resilience that unifies the integrated elements (i.e., effective, efficient, and sufficient) since all of the actors of resilience are actively working together, utilizing the community's assets of resilience. In my opinion, the achievement of sustainable livelihood development lies in that the elements bind and correlate with each other and are then possessed and applied by the community. The assessment of this category is assessed by the community based on three elements (i.e., effective, efficient, and sufficient) with their explanation of utilizing each asset of resilience as described in Chap. 9. However, at this phase, I analyze a sustainable livelihood development model as presented in Fig. 8.2.

The sustainable livelihood development model presented in Fig. 8.2 has explained each of these three models:

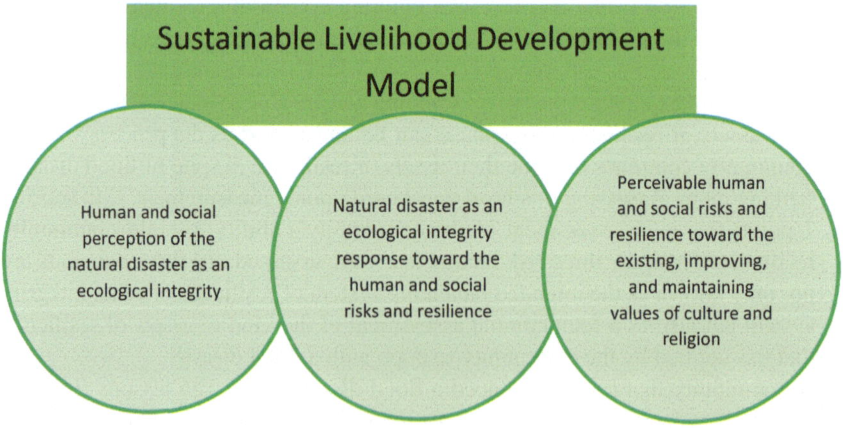

Fig. 8.2 Sustainable livelihood development model. *Source* Author's sustainable livelihood development model dissertation research fieldwork, July 9, 2016 to February 9, 2017

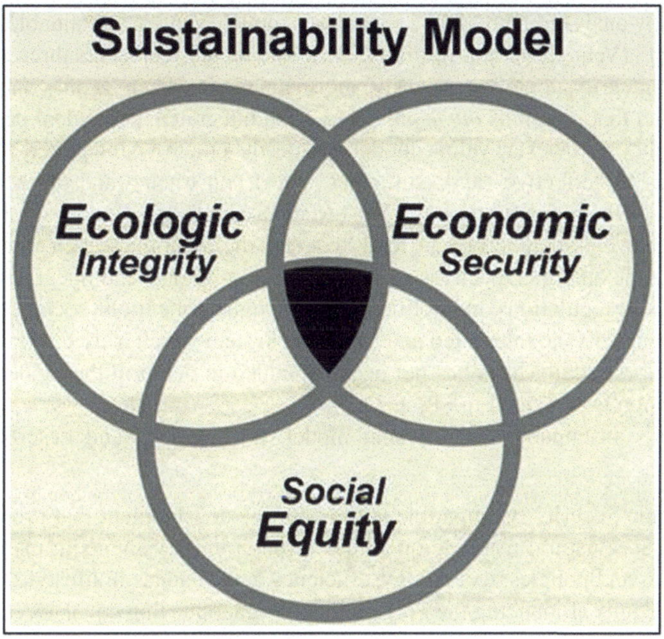

Fig. 8.3 Venn's sustainable development model. *Source* Venn 2002 qtd. in Flint (2013: 34)

1. Human and social perception of the natural disaster as an ecological integrity (human perception toward their *nine livelihood resources* or assets of resilience in an integrated, ecological approach to perceive natural disaster (i.e., flood disaster).
2. Natural disaster as an ecological integrity response toward the human and social risks and resilience (the efforts and capacity of human and social resilience can positively respond and adapt well to the natural disaster (i.e., flood disaster) within the implemented *nine livelihood resources* or assets on resilience (i.e., economic, physical (infrastructure), regulation, and policy actions) in an integrated, ecological approach.
3. Perceivable human and social risks and resilience for existing, improving, and maintaining values of culture and religion (as a means of balancing the system with an integrated-based-practical-ecological method for the revitalization of their *nine livelihood resources* or assets on resilience without neglecting the cultural and religious values).

The three components which were conceived by me after a discussion with the community have led to a model for measuring community resilience in terms of three elements of an ecological integrative outlook. The three levels or classes are: effective, efficient, and sufficient, which are so-called as a unity of sustainable livelihood resources.

I critically analyzed a previous researcher's model, Venn's "sustainable development model" (Venn 2002 qtd. in Flint 2013). The model represents three elements: ecological integrity, economic security, and social equity. There is only one element in this model that mentions *one component* of an integrated, ecological perspective (i.e., efficiency). However, Venn's model ignores the other two integrated ecological perspectives of "effective" and "sufficient" that I constructed and utilized for the assessment of the community in the West Malaka Subdistrict, Indonesia. In Venn's sustainable development model, he tends to assess the economic perspective without considering the other perspectives such as cultural, religious, and integrated discussion of laws or regulations and politics. Even though he mentions social equity and ecological capacity, an integrated and balancing system which is used so far to solve a specific case in daily lives has not been explained in detail in the model by him. The discussion is presented in Fig. 8.3.

In Venn's sustainable development model, he mentions and describes three elements:

1. Economic Security (Compatible with Nature)—development that protects and/or enhances natural resource quantities through improvements in management practices and policies, technology, efficiency, and changes in lifestyle.
2. Social Equity (Balancing the Playing Field)—guaranteeing equal access to jobs, education, natural resources, and services for all people; total societal welfare; access to fair conflict resolution.
3. Ecologic Integrity (Ecosystem Capacity)—understanding natural system processes of landscapes, watersheds, and seas to guide design of sound economic development strategies that preserve these natural systems. (Venn 2002 qtd. in Flint 2013: 34).

From the model defined by Venn, I argue that community has applied the three models of sustainable development (i.e., economic security, social equity, and social-ecological integrity). The community has a good risk perception of resilience. Community risk perception which can be understood as community understanding about resilience and flood risk perception until the climax stage of development, i.e., sustainable livelihood, has been expressed in detail their ideas of resilience that connect to these three components model.

I have found this sustainable development's model and juxtaposed with mine that is applicable and based on the facts of the field-case study of flood disaster with the community as an actor of resilience and experiencing the flood disaster. The community has shared their experiences on the assessment of their resilience and their response to floods and also the assessment phase of their development sustainable livelihood. They already have had well-established community resilience, as well as a mitigation response which is relevant or appropriate to the assets of resilience that they applied during flooding. It can be seen from the perspective of natural elements that interfere (i.e., flood disaster).

However, two components of Venn's sustainable development model cannot be balanced if the carrying capacity of the others is not fused. *Carrying capacity* means sustainable livelihood development. To achieve a sustainable development of livelihood, resources (i.e., *nine resources)* should be owned and implemented by the community. Therefore, the community's system must bring the three components of this model together in a balanced system or an integrated system (i.e., effective, efficient, and sufficient) of a resilient community.

As a single actor of resilience, the community has managed and implemented the nine assets or *nine resources.* Instead, the community had been struggling with their independent capability in the process of improving their assets on resilience. Thus, this causes the sustainable development of livelihood resources that they harness and have applied to be still quite far from a successfully integrated resilience.

I concluded that community as a single actor of resilience requires some help from the other institutions to actively run their assets of resilience. There is inequality that the community experiences in getting assistance from the local and national governments of Indonesia. Therefore, it is supposed that the Indonesian governments and related institutions in the country could help and support community's resilience assets in the face of flooding. If we have seen through the lens of the law or regulations and implementation of the policies of the Indonesian governments, the community has a right to obtain treatments from similar policies (i.e., economic aspects and aspects related socialization and water resources utilization management; in this case, protection areas nearby Benanain River). They live around the river basin and downstream area, and the dependence on natural resources due to the economic crisis situation they face has imposed in addition to the preparation of economic budgets for flood disaster prevention, mitigation and recovery efforts. Thus, the situation is not only about how much variety there is in domestic and foreign interventions for disaster-affected communities in Malaka-Timor, Indonesia, but also about the sustainability of the type of assistance programmed legally and successively in the national policy documents in Senayan, Jakarta that takes into account and considers the physical, human, psychological, social, economic, cultural, religious, legal and local political data of the village/region. I argue and provide a simple illustration, that *the natural disaster of flooding in Malaka-Timor* is part of the disaster for humans/communities themselves, disturbed psychologically, and is also an ecological, social, economic, cultural, religious, belief, legal, and political *disaster* for us all, which needs our solution hand in hand together, in an integrated manner from various parties both from the community in Malaka, village officials and stakeholders and traditional/female leaders, the national government, and the global community. Criticizing this situation, issue, and polemic more deeply from time to time, on this discussion in Timor, Indonesia, I bring a holistic understanding of the interrelationships and continuity between these aspects. A study report on this similar topic in Malaka-Timor, Indonesia, stated that natural disasters such as floods in Malaka-Timor, Indonesia, are part of ecological disasters because natural and human damage results in social, psychological, cultural, religious, economic, and environmental imbalances that trigger the carrying capacity of the environment and people to weaken over the increased environmental overcapacity, i.e., the impact of the hazard/disaster itself (da Costa 2020). Furthermore, she explains more that the imbalance between carrying capacity and risk of threatened environmental capacity requires a comprehensive empirical study for sustainable development assessed from the human side (resilience) and the natural hazard risk side (vulnerability) for the determination of local and national government policy certainty for sustainable protection from natural hazards for future generations. When we talk about sustainability, we're talking about a development model that can meet the needs of the present without compromising the ability of future generations to meet their own. It's a holistic approach that considers the social, environmental, and economic impacts of actions and decisions taken today. From a historical perspective, the concept of sustainability was formulated at the first United Nations Conference on the Environment in 1972, but it has only really taken and shaped since 1987, when the publication of the so-called Brundtland Report ("Our Common Future") clarified the goals of sustainable development (ENEL Group 2023). For this discussion, Müller et al. (2000) add also that the concept of sustainable development has been evolving on the basis of the ideas of World Commission on Environment and Development (1987) that, on a global scale, we need a development that meets the needs of the present without compromising the ability of future generations to meet their own needs. This concept includes the consideration of the carrying capacities of ecological entities as well as an intragenerational (geographical) and an intergenerational (long-term) equity in their paper's discussion section Ecological Demands of Sustainable Development (Müller et al. 2000, p. 15), where I analyzed and discussed the phenomena and issues of this topic to the surface through my onsite empirical research in Malaka-Timor, Indonesia. We need uniformity of thinking about ecological integrity on this issue. Müller et al. (2000) state that ecosystem integrity turns out to be the ecological branch of sustainability (qtd. in Roger et al. 2022). Many definitions of ecosystem integrity exist because ecosystem integrity is not a simple absolute physical property but rather a multidimensional and scale-dependent emergent phenomenon that encompasses important system components and their interactions. The concept has received considerable attention over the past several decades because of the human benefits derived from natural processes and ecosystem states (Rogers et al. 2022). Some communities may be consistently affected by high levels of localized disturbance that allow for the coexistence of species competing for the same limited resource (e.g., intertidal communities), whereas others may experience periodic catastrophic episodes, such as floods, forest fires, or pest outbreaks, that reset the timing of successional cycles (De Leo and Levin 1997). Reflecting on the issue about resilience, this research on building resilience of affected communities facing flood hazards/disasters in Malaka-Timor Indonesia that I conducted formally from July 9, 2016 to February 9, 2017 (informally, I have been staying in Malaka-Timor onsite in the pre-monitoring of spatially affected areas by flood hazards/disasters and engaging with local communities since July 2, 2016) is therefore a strategic step because it opens the gates of moral awareness of disaster hazards, the safety of residents living in flood hazard areas, the rescuing and mitigation systems, and the impact of ecological hazards on sustainable development of environments and resilience capitals/resources/assets, to be researched by the local and national scientists in Indonesia, recognized and provided extra-serious attention by the Indonesian local and national governments, researched by the international scientists, and recognized and understood by the global community at large. It is thus imperative that the research studies on this topic in various hazard-prone areas of Indonesia and other countries in south hemisphere and north hemisphere be conducted by both domestic (Indonesian) and international scientists. Meanwhile, in the context of the Indonesian intervention for this issue, we can see that the Indonesian national government has indeed the latest intervention of the Ministry of Environment and Forestry, Indonesia, a virtual climate discussion (*Pojok Iklim Diskusi*) with the theme "Resilience, National Defense in the Environmental Sector," organized by the Ministry on February 10, 2021, led by the Chairman of the Advisory Council for Climate Change Control, Sarwono Kusumaatmaja (Ministry of Environment and Forestry, Indonesia, in the Information Documentation and Management Officer 2021). Furthermore, from this *Pojok Iklim Diskusi*, the Chairman, Sarwono Kusumaatmaja emphasized that "it is important for Indonesians to build community resilience in adapting to and coping with disasters from an early age," as "it can channel the power of capital of resilience for sustainable recovery for all aspects of safety and protection" (Ministry of Environment and Forestry, Indonesia, in the Information Documentation and Management Officer 2021). We may notice together that there were diverse opinions on this topic, including an international's point of view from the United Nations (UN) sustainable development goals (SDGs). *The UN SDGs which cover a wide range of sustainability issues on the building around several goals is the UN 2030 Agenda has set a plan of action for people, planet and prosperity, signed into law by the UN General Assembly on September 25, 2015. The Agenda includes 17 goals, valid for everyone around the world, articulated along the three dimensions of sustainable development: economic, social and environmental* (ENEL Group 2023). We then reflect on the case of Timor, Indonesia, that the lack or limitations of Indonesian local and national government intervention for a single system or unit of building resilience for natural disaster/climate change is not the reason for natural disaster/climate change management and resilience measures to be eliminated or avoided or ignored, in fact we, you and I especially as geography/ecology scientist (s) and scientists in whatever field you are in, should also work together to harmonize our actions (scientific and humanity) regarding what the next resilience strategy is, and what and how we need to do for the benefit of people and nature; where this our case study example in Malaka-Timor, Indonesia, is a reflection of us thinking and acting together for the good of the future of our generation and this planet.

However, I need to review and emphasize more on the previous important point about lack of regulation to support the resilience actions to cope with flood disasters in Indonesia. I am also aware of this issue that the issue of river watershed protections is still a popular concern, not the people safety and security from the natural hazards/disasters in this country, and how to build/increase their resilience resources. It has been more critically examined and thus uncovered that there was only an integrated water resource management regulation from Indonesian local and national governments regarding river watershed protection in Indonesia (See Chap. 3 "Malaka-Timor, Indonesia", Sect. 3.2 "Physical and Social Characteristic") while there is no regulation governing the integrated community resilience as a single system/unit that has its own regulation and policy despite natural disaster risk/climate change management's regulation and policy. I may dare to express my critical thinking as a scientist from both Timor-Leste (born here) and Timor Indonesia (raised here) about some constructive questions. Where is social justice for the neglected communities in flooded villages and they were poorer economically, where are the decision-makers in various government departments and ministries in Indonesia, there must also be moral concern and contribution from the neighboring country of Timor-Leste, which shares the same historical and cultural background as the *Wewiku-Wehali* story from the same ancestors with Malaka-Timor, they enjoy natural resources, i.e., water (Benanain River), and live side by side on the Southern Border regions. As a scientist who holds multiple roles in the profession as a science explorer, a practical humanitarian observer, a poet, and an environmental/geography philosopher, I am still struggling to this day, this is also a question mark for humanitarians and interdisciplinary scientists, this is actually our task. The efforts of local communities and NGOs in flood hazard areas have surpassed those of the national government until very recently. I reiterated that as a scientist, I aim to reveal the facts and potential solutions for the communities in Malaka-Timor, Indonesia, through my this empirical research onsite using the Focus Group Discussion and Stakeholder Meetings. In these meetings, I therefore aim to work with the communities to help find points of meaningful solutions with also the direction of justice and humanity for their resilience adjustment and sustainable livelihood resources case. The discussion is provided in Chap. 9.

8.2 Discussion and Summary

In this book, ecological integrity in a sustainable livelihood development model considers a community's goals in building its assets of resilience to cope with flood disasters. The discussion of the sustainable livelihood development model in this book relies on the element of a balancing system of resilience. Each category of assessment that this study uses is described, namely effectiveness, efficiency, and sufficiency. These three types of assessment reflect the community's actions in developing its assets of resilience for flood disaster. The community's actions also distribute a set of local abilities to understand its space (i.e., flood disaster) as an integrated ecological capacity. Such an integrated ecological capacity has transformed as an integrated social-ecological capacity that is progressively moved by human progressivity (abilities) in achieving their social justice (environmental and sustainable recovery from social, cultural, economic, and other vulnerabilities) caused by the ongoing and continuous impacts of natural hazards/disasters, in this case floods, based on an understanding of human and spatial philosophy (*Sustainalism*). This progressive movement of the community requires an effective or a significant role of the movement elements in the sustainable development model. In the sustainable development model, community is one of the elements that move a system of capacity in proximity with the other movement elements, e.g., social, cultural, economic, and other resilience capacities which counter the existence of the above vulnerabilities, in this context it is understood as the '*carrying capacity*'. The '*carrying capacity*' is supposed to be moved, not only by the community as the internal actor of resilience but also by the other internal actor of resilience in building community resilience for flood disaster in the study area (i.e., the Indonesian government). Thus, the elements of human vulnerability are reduced, and the elements of human resilience are built or increased, and progressed gradually in a definite sustainable stage within the framework of the national government policy program.

I pressed that the assessment of the community's sustainable development is critical in measuring its ability to cope with flood disasters. This assessment considers not only how the community builds resilience for flood disaster but also how the community implements its understanding of flood risk management through owning and implementing its assets of resilience that were impacted by the flood disaster. This way is applied to survive in independence by implementing their resilience assets to face flood hazards/disasters in the areas where they live, as long as the government has not provided the expected solution until this period. However, they also expressed their aspirations that they would continue to survive in this same way even until the government helped them restore the resilience of their assets which were at risk of being damaged by the impact of the protracted flood disaster.

The second phase of the sustainable development model, "Risk factors of flooding", concludes that communities perceive their own experience of the risk of flooding as the process of valuing the means of resilience in their daily lives. They could handle and sustain their assets of resilience with less prominent assistance from the local/national Indonesian governments as they are the single internal actor or mover of resilience; it is understood that the process of valuing "*community resilience*" in the social movements are not as effective as "*carrying capacity*" in an integrated of ecological point of view. In an integrated ecological point of view, a community's "*carrying capacity*" refers to its balancing system of resilience.

A balancing system of resilience represents community abilities (i.e., human ability) to achieve an integrated system of building resilience. The community requires a "*carrying capacity*" to deal with flood disaster that threatens their life and assets of resilience. The need for an integrated system is therefore substantial to reach a harmonious life in which the excessive gap is caused by the flood disaster rather than human abilities. An integrated system is achieved by the community's conception of understanding its own ecological perspective. Community understands its space with its own ecological capacity's perspective, that means none of the sustainable development becomes reality without the support of all the actors or movers of resilience. It also means that the community's actual abilities in regard to the Indonesian government's abilities should be mutually synchronized to each other so as to build and sustain resilience for coping with flood disaster. Thus, there is a balanced system in an integration of ecological capacity for community sustainable livelihood development as the entire elements movements that move to each other were insisted but were not effective.

The idea of human geography in this study centers on a subjective perspective that considers a community understands its space then connects with understanding of its actions. Furthermore, it is also noted that the most essential analysis was developed as part of this book (see the first paragraph of this chapter: "*an aspect of the sustainable development model's components which brings a relationship between building community resilience and the risk factor of flooding, through embodying both as a perceivable ecological integrity*").

From the point of view and understanding of *sociological perspectives*, this study discusses how a community implements its assets of resilience. Community presents its abilities to cope with flooding as one of the processes to achieve a sustainable livelihood development. Furthermore, resilience is viewed as a process itself to value a process which is driven by *community's persistence abilities* to cope with harmful and/or disturbance events, in this case the actual flooding. A sustainable livelihood development conforms as "Persistently System" of building community resilience for flood disaster. It then becomes a part of the "Social Welfare" that determines how effective it is a social movement applied by the community, as supposed by Venn in his sustainable development model in Point 2, social equity or social justice. Communities as a social mover drive their resilience actions against flood disaster as to achieve social justice. Adding to one more important point that connects to the need to fulfill their social justice which they are still waiting for at this time, their social justice stems from their empirical reasoning/experience in the philosophy of man and space with the horizon of social geographical and social ecological thinking, which I elaborated my understanding and reflection briefly here. My understanding and reflection in deepening my research in this discussion section is that the people of Malaka-Timor, Indonesia, in facing the risk of flood hazards/disasters and applying their resilience assets, they have found a philosophical answer in the study of spatial conflicts from any hazard/disaster including floods, that their subjective reality experience in the truth that is oriented towards objectivity, which they themselves truly experienced and understood, not us who have not experienced it like them. Thus, the aspirations regarding social justice in the participatory discussion that I led were expressed passionately by them.

They emphasized that they do require social justice from the Indonesian government because they are the only ones at the moment acting as a major social mover to drive their community resilience actions intensively to deal with flood disaster, despite the limitations of various aspects of their life's livelihood. In this section of the book, I discuss community's understanding with its actions (i.e., resilience assets) to seek solutions due to harmful and/or disturbance events caused by its space (i.e., flood disaster), (see Theory of Werlen 1992, Society, Space, and Action, in Chap. 1; Sect. 1.3, point one).

References

da Costa ADS (2020) Discussion on the economic resilience and fisheries development in Malaka-Timor, Indonesia. A social-ecological report

De Leo GA, Levin S (1997) The multifaceted aspects of ecosystem integrity. Resil Alliance Conserv Ecol [online]1(1):3. http://www.consecol.org/vol1/iss1/art3/

ENEL Group (2023) The 3 pillars of sustainability: environmental, social and economic. A news. https://www.enel.com/company/stories/articles/2023/06/three-pillars-sustainability

Flint WR (2013) Practice of sustainable community development: a participatory framework for change. Springer, New York

Ministry of Environment and Forestry, Indonesia in the Information Documentation and Management Officer (*Kementerian Lingkungan Hidup dan Kehutanan Indonesia, Pejabat Pengelola Informasi dan Dokumentasi*/PPID) (2021) Ketahanan Nasional Bidang Lingkungan dan Kehutanan. A national news. https://ppid.menlhk.go.id/berita/berita-tapak/5817/ketahanan-nasional-bidang-lingkungan-dan-kehutanan

Müller F, Hoffmann-Kroll R, Wiggering H (2000) Indicating ecosystem integrity—theoretical concepts and environmental requirements. Ecol Model 130:13–23. https://doi.org/10.1016/S0304-3800(00)00210-6

Rogers BM, Mackey B, Shestakova TA, Keith H, Young V, Kormos CF, DellaSala DA, Dean J, Birdsey R, Bush G, Houghton RA and Moomaw WR (2022) Using ecosystem integrity to maximize climate mitigation and minimize risk in international forest policy. Front For Glob Change 5:929281. https://doi.org/10.3389/ffgc.2022.929281

Werlen B (1992) Society, action and space: an alternative human geography. University of Zurich, Switzerland, Routledge International Publishing, Department of Geography

Chapter 9
The Assessment of Building Community Resilience Adjustment for Community Sustainable Development Through Focus Group Discussion (FGD) and Stakeholder Meetings (SMEs)

In this section, I assessed a practical methodology using community-participatory action based on the theory of resilience, provided by Walker et al. (2004). Walker et al. used their theory to determine risk perception that embraces resilience in an adaptive manner. An adaptive manner is applied by the community in the research site as "an adaptive capacity" or "an adjustment" rather than as "an adaptation." It means that "an adjustment" defines more *absorptive capacity and collective actions* than the community perceived and possessed for handling flood disaster. Meanwhile, "an adaptation" is rather a set of less-direct steps the community takes to turn its resilience efforts into practical actions. Even though most previous scholars and practitioners have used the term "adaptation" as a "connecting concept" between resilience and flood disaster management and vice versa, identifying a comprehensive solution through practical research concerns implemented actions and so the term "adjustment" is used in this analysis.

Previous chapters use the term "adaptation" to describe the positive actions of the community as well as the existing adaptation regarding the resilience assets applied in coping with flooding. Furthermore, to find out the results of the absorptive and collective actions of the community through a deep discussion with them, this chapter aims to present "a new adaptivity meaning." The term "a new adaptivity meaning" transforms to "an adjustment" as to form a new concept which involves assessment and measurement in a deep discussion of building community resilience for flood disaster using Focus Group Discussion (FGD) and Stakeholder Meetings (SMEs).

The term "adjustment" is contextualized in the discussion of FGD and SMEs as "an adaptive manner in participatory information for disaster risk management" that enables the community to lead the recovery processes and achieve sustainable development goals. The theory of resilience in this section is introduced as a base solution for targeting the community's decision-making process through its own risk perception. The community's decision-making process positions its understanding of its resilience for flood disaster in a broad way. Understanding of community resilience for flood disaster based on FGD and SMEs enabled the community to

A. D. S. da Costa, *Building Resilience for Flood Disaster in Malaka-Timor, Indonesia*, Springer Natural Hazards, https://doi.org/10.1007/978-3-031-63272-3_9

assess critically its ways of implementing resilience assets for handling floods and was also pivotal for brainstorming and reflection to bring out lessons and practical knowledge which led the community to adjust to flood disaster using its resilience assets.

The community's practical knowledge that came to light in FGD and SMEs is discussed in this section. Each asset of nine types of building resilience that the community possesses to adjust to flood disaster is also discussed; these are human resilience adjustment (HRA), natural resilience adjustment (NRA), physical resilience adjustment (PRA), social resilience adjustment (SRA), economic resilience adjustment (ERA), cultural resilience adjustment (CRA), religion resilience adjustment (RRA), law or legal resilience adjustment (LRA), and political resilience adjustment (PolRA).

The study illustrates each resilience adjustment based on FGD and SMEs in regard to two motions' analysis in order to answer the research questions and research objectives of this thesis. Each motion and its analysis are discussed in each of two sections: Sections 9.1 and 9.2.

9.1 The Assessment of Community Assets for Sustainable Development

Theory of resilience is seen merely as an adjustment rather than as an adaptation to answer the critical case of resilience that links to risk perception (Walker et al. 2004). Risk perception of the danger of flooding is underlying participatory information in disaster risk management. It can bring resilience information that supports the recovery process both for flood disaster risk management and advance recovery stages in order to achieve sustainable development goals. The theory of resilience can be critically deepened to reach the target solution through an assessment of the resilience which is owned by the community. To carry out the assessment phase of community resilience, a resilience assessment of each of the assets owned by the community must be completed. Section 9.2 describes the discussions on Community Resilience Measuring Sustainable Actions on Community Adjustment which also explains in detail each of the assets owned by the community resilience through the FGD and SMEs.

9.2 Measuring Community Resilience Adjustment on Community Sustainable Action

A community's assessment of the adaptation of their resilience assets which they had implemented to achieve sustainable actions can each be put through a phase of asset' assessment in the resilience discussion. As already mentioned above, this is the FGD and SMEs. FGD and Stakeholder Meetings SMEs are a practical methodologically based-community-participatory action that to help to achieve the targets or solutions for the sake of solving a case study in this situation, it is a case study related to the public perception of resilience in how they view the threat of floods to their lives and the livelihood of their systems. The analysis in this chapter leads to a number of aspects of the community. Firstly, the analysis concerns the economic aspects of the community in their daily life practices of their hereditary system of resilience that they apply for the sake of the handling of floods. Additionally, the analysis concerns the community's representation of a system that is holistic or integrative in that all aspects of resilience are analyzed and assessed for solutions in answering their case studies. In the previous study, I conducted resilience research in assessing how a community had applied the five pentagon resources in their everyday lives and what their abilities are in coping with the risk of flooding.

However, during the period of this new research study, I found that some of the findings of livelihood resources that people who were in possession of and had applied, both in their everyday lives and in the process of handling a flood disaster, have been experiencing a stable measure of adaptation in even more increasing ways to adjust with flooding. The findings that were obtained include three more additional livelihood resources that a community had and applied in the realization of the framework analyses of this book, namely the "*nine resources of resilience*." It was therefore interpreted and assessed by the community that they have power of utilizing their assets to support their adjustment of resilience before or ex-ante, during, and after or ex-post the flood disaster.

A critical assessment is characterized by a detailed assessment that reveals a number of things. Firstly, during the particular research period, it reveals to me and the community how to improve their assets on resilience to flooding. Additionally, outside of the research period, critical assessment helps the community think critically about their decisions in tackling their own cases, adjusting to flooding by using and improving their assets of resilience. As presented in Fig. 9.1, I applied the FGD and SMEs with the community of West Malaka Subdistrict in the head of Subdistrict's Office in Besikama Village.

A community that assesses and discovers their implemented assets on resilience to cope with flood disaster. The discovery that community has managed and maintained so far as a form of an applied discussion and as friendly and adaptive actions that they have brought out into a positive direction for the sustainable livelihood resources issues.

In progress, positive direction in the assessment phase is poured into FGD and SMEs. As corresponding to the form of framework or scheme of analysis of FGD and SMEs in Chap. 4, there are two motions or two themes for discussion. This participative research is arranged in two motions in answering questions and objectives of this empirical research on this book, which are:

1. Participative talk on addressing and answering the case study' questions: What and how does community resilience deal with flood disaster using FGD. This

a) Process of assessment and discussion SMEs

b) Community Group's Picture Session for FGD and FGD and SMEs

Fig. 9.1 Process of assessment and discussion of FGD and SMEs in an Aula of Head Subdictrict's Office in Besikama Village of the West Malaka Subdistrict, Malaka District, East Nusa Tenggara Province, Indonesia. *Source* Author's Research Fieldwork on FGD and SMEs, November 19, 2016

first phase of participative talk consists of five submotions to be discussed in FGD; they are:

(a) Introduction and explanation of the nine capitals of resilience.
(b) Listing the existing capitals of resilience (pentagon resources) owned by the community.
(c) Finding the barriers and obstacles accessing the nine capitals.
(d) Discussing and assessing the efficiency, effectiveness, and sufficiency of the nine capitals.
(e) Targeting solutions for decision making of the nine capitals on building resilience.

1. Participative talk on addressing and answering the case study questions: What and how does the risk of flooding disaster impacted to community livelihood (capitals or assets, known as pentagon resources). This second phase of participative talk also consists of five submotions to be discussed in FGD; they are:

(a) Introduction of the key targets of solutions from the community resilience deal with flood disaster.
(b) Listing the risk of flooding disaster impacted community capitals.
(c) Finding the barriers and obstacles in channeling the nine capitals.
(d) Discussing and assessing the risk of flooding disaster and the nine capitals.
(e) Galvanizing the key targets of solutions from the community and stakeholders.

Each of submotion phases in these two motions of FGD is further applicable to processing in SMEs through Discussion and Assessment Processes of before the flood disaster or ex-ante, during the flood disaster, and after the flood disaster or ex-post.

SMEs aim at finding the final solutions as in final decision of motion I: key targets of solutions from the community resilience deal with flood disaster and final decision of motion II: key actions of channeling and galvanizing both the community and stakeholders' solutions. Furthermore, I concluded all of the community' resilience perception on Final Decisions of Motion I & II: Building Community Resilience for the Flood Disaster where these final solutions have answered this research. The community perception is presented in a spatial way (i.e., the scoring map) on each of the assets of resilience that the community had and implemented for achieving a long-term action.

Community adjustment is seen as spatially as community integrated solution in a framework of resilience itself, and FGD and SMEs have successfully brought up the target of solutions. A spatial manner could represent a spatial casualties of community livelihood resources for achieving a sustainable action whether it is effective, efficient, and sufficient actions so-called as a long-term action. These categories of assessment have been depicted in terms of how a sustainable action of the community is in perceiving their assets of resilience to deal with flood disaster.

The assessment by the two motions phase concludes that the community had an inherent knowledge to measure their own livelihood resources for long-term actions. I also analyzed the community building system of resilience through each advocacy assessment into an integrated assessment of resilience.

An integrated assessment of resilience based on FGD and SMEs was further analyzed through a spatial description (i.e., scoring map) each asset of resilience, adjustment, and community sustainable actions.

In order to determine the categories or classes for measuring community resilience, the community was involved and took an initiative to create four divisions of values based on their experience dealing with a flood disaster. The community's initiative was approved by me after a deep discussion with them. The four divisions of values that were determined by the community have been described in their level or scoring of values with specific assets and subassets of resilience that were owned and applied by them. The level or scoring values explained that each of values has its classes from one to four. The four categories or classes of assessment which explain each value and the classes are explained below:

1. Class (Scoring Value 1) means that all of the assets and subassets of resilience owned and applied by the community are effective, efficient, and sufficient for long-term action. This class indicates the long-term action of community in perceiving and implementing the nine assets of resilience or *nine livelihood resources* (i.e., human, natural, physical, social, economic, cultural, religion, law or legal, and political or policy).
2. Class (Scoring Value 2) means that some of the assets and subassets of resilience owned and applied by the community are effective and efficient but not sufficient for long-term action. This class indicates the medium or moderate-term action of community in perceiving and implementing the *nine livelihood resources.*

3. Class (Scoring Value 3) means that few of the assets and subassets of resilience owned and applied by the community are effective and efficient but low sufficient for long-term action. This class indicates the low-long-term action (i.e., nearly moderate-term) of community in perceiving and implementing the *nine livelihood resources.*

Class (Scoring Value 4) means that none of the assets and subassets of resilience owned and applied by the community are productive or effective, efficient nor sufficient for long-term action. This class indicates the short-term action of community in perceiving and implementing the *nine livelihood resources.*

These four categories or classes reveal the community's problems (i.e., flooding). Their building system of resilience is determined and measured by their self-assessment in a group discussion through FGD and SMEs with me as a part of teamwork and also as a guide or supervisor in the discussion. Overall, the assessment and discussion of each asset and subasset of resilience adjustment that was owned and applied by the community are clearly discussed.

9.2.1 Human Resilience Adjustment (HRA): Assessment of Participative Talk on the Building Community Resilience

Human adjustment of community resilience can be understood as the human livelihood resources (i.e., assets which are applied by the community in the context of flood disaster management). A community that has experienced the flood disaster has its own applicative adjustment of livelihood resources (Fig. 9.2).

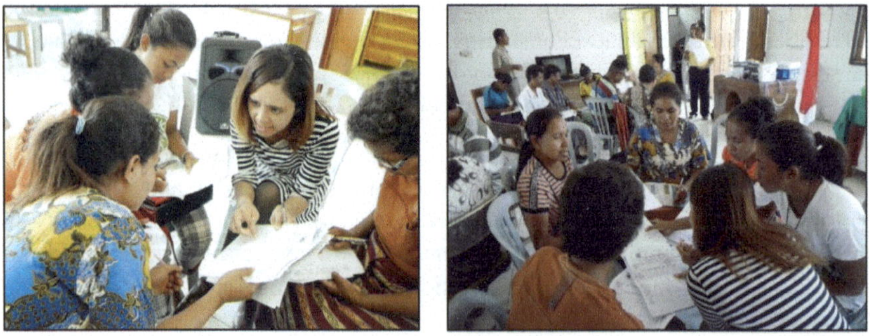

Fig. 9.2 Focus Group Discussion (FGD) and Stakeholder Meetings (SMEs) on human resilience adjustment in Head of West Malaka Subdistrict's Office, Besikama Village, West Malaka Subdistrict, East Nusa Tenggara Province, Indonesia. *Source* Author's Research Fieldwork on FGD and SMEs, November 19, 2016

Participatory Assessment of Sustainability for the Flood Disaster in a Resilient Community. The two pictures above show that the community was actively participating in FGD and SMEs in the process of assessing resilience. The results of discussions with the community can be seen in Chap. 10, Table 10.1. The results of discussions with the community about human assets assessment measures for resilience reveal that the 15 villages in the West Malaka Subdistrict have their own power to leverage their human assets skillfully and appropriately. However, the community also provides an objective assessment where Oanmane Village is still classified into class or level four (Scoring Value 4) meaning that none of the human assets and its subassets of resilience that are owned and applied by the community are productive, effective, efficient nor sufficient for long-term action. This class showed fatal damage of the human assets, especially the health and well recovery of factors of economic recovery and a long physical recovery experienced by the community in Oanmane Village.

Adequate skills and knowledge to think smartly for the recovery phase takes longer than compared to the other 15 villages in the West Malaka Subdistrict. Having well-organized human resources power in order to re-establish human assets that had applied by the community can be seen through the improvement of their physical assets such as damaged houses, adjustments to the recovery for economic assets such as livestock, crops, and the contents of the homes and property which made the issue takes a bit longer in the recovery process.

The community considered that the constraints they are facing is due to the lack of accessibility and evaluation of the other 15 villages with Oanmane Village. Although the effects of flooding have destroyed entire assets of community life and livelihood, the 16 villages in the West Malaka Subdistrict are still quite strong in their social assets and cultural assets. They had a strong disaster group community and traditional bond. These kinds of assets do not lead to a conflict and dispute between the villages including Oanmane Village, but rather, they became a new standing power on supporting their resiliency system. The result of the assessment of human resilience in the West Malaka Subdistrict is spatialized into visual data (i.e., Scoring Map of Human Resilience Adjustment) as seen in Fig. 9.3.

9.2.2 *Natural Resilience Adjustment (NRA): Assessment of Participative Talk on the Building Community Resilience*

Natural assets are a part of a resilience assessment adjustment. Natural assets of the community can be assessed through discussions with the community. The assessment of natural assets can be executed when the carrying capacity of the human assets is contributed to the utilization of natural assets. This can be seen in Fig. 9.4.

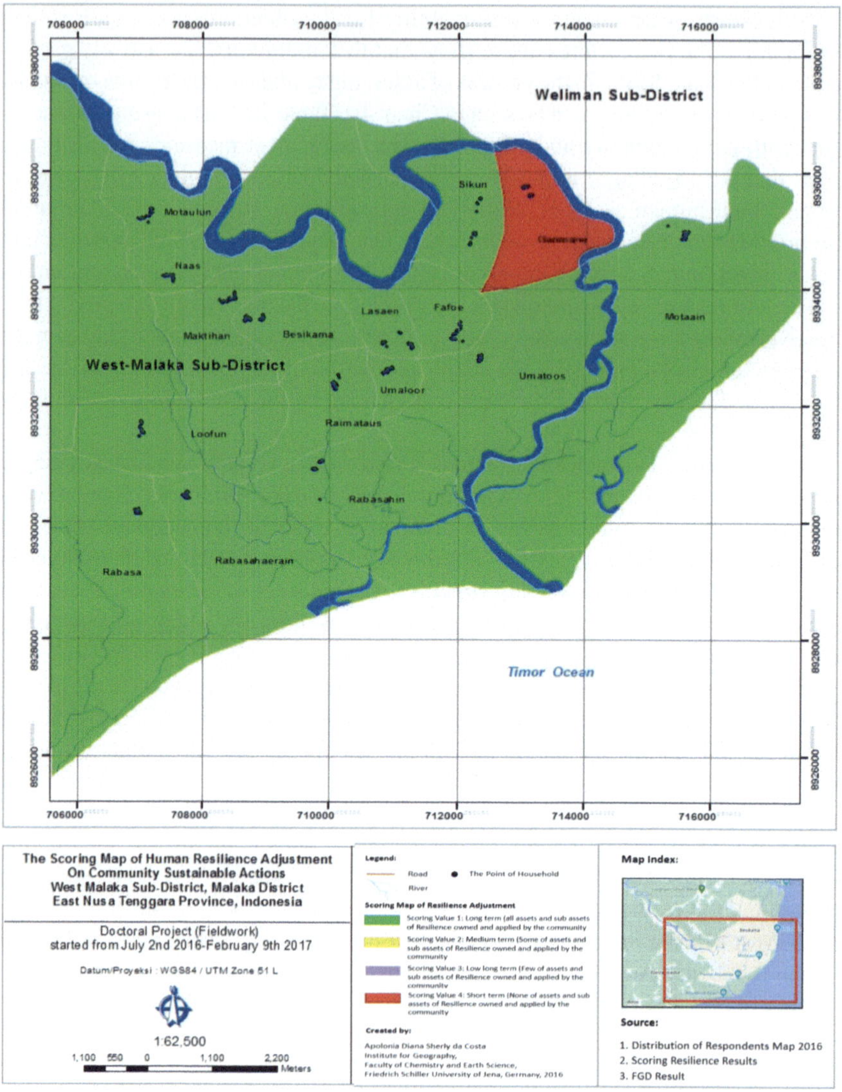

Fig. 9.3 Scoring map of Human Resilience Adjustment (HRA) on community sustainable actions in the West Malaka Subdistrict, Malaka District, East Nusa Tenggara Province, Indonesia. *Source* Author's Research Fieldwork on FGD and SMEs, November 19, 2016

The Scoring Level of the Natural Asset of Community Resilience Adjustment Based on FGD and SMEs can be seen in Chap. 10, Table 10.1, which shows participatory assessment of natural asset and its sustainability for the flood disaster of a resilient community through community resilience assessment. The result of FGD and SMEs with the community showed that Umatoos Village has class or level four

Fig. 9.4 Focus Group Discussion (FGD) and Stakeholder Meetings (SMEs) on natural resilience adjustment in the head of West Malaka Subdistrict's Office, Besikama Village, West Malaka Subdistrict, East Nusa Tenggara Province, Indonesia. *Source* Author's Research Fieldwork on FGD and SMEs, November 19, 2016

of human asset as indicated by a short term of utilizing the natural assets. While Fafoe Village and Oanmane Village have a class or level one, the long term in the utilization of natural assets. Subsequently, the other 13 villages belong to a class or level three, low long term (i.e., nearly moderate term). These rates can be indicated that the weakness of natural assets owned by Umatoos Village due to damage of physical assets such as supporting facilities of fishing, aquaculture, and fish and shrimp ponds. Umatoos Village which is geographically situated near the estuary of Benanain River is a resilient community that is supported through a wealthy of their natural assets.

It could leverage their natural assets. However, the geographical position of being adjacent to the estuary of the Benanain River has resulted in destructive impacts of their natural assets after a flood occurs. The community in Umatoos Village has a good carrying capacity through good cooperation with partners of other villages including Fafoe Village and Oanmane Village as their neighboring villages. They took advantage of natural assets such as coordination and corporation. Such coordination and corporation among villages has been applied to empower their system of resilience in terms of utilization of natural assets. However, it still does not eliminate the bureaucratic functions and local authorities from Umatoos Village, Fafoe village, and Oanmane Village to support the provision and reconciliation of using the physical facilities for the natural assets in Umatoos Village that was damaged by flooding.

Coordination of the recovery process took a long time. This is due to the fact that the recovery phase should be appropriated by the phase of facilities reconciliation of physical assets in Umatoos Village. Increased capacity was run by an agreement between Oanmane Village and Fafoe Village who have good adaptive phases in order to deal with their recovery process of their natural assets. Umatoos Village keeps on improving their coordination and corporations in terms of facilitating the physical resources to support the utilization of natural assets. It can be seen in the real condition of Umatoos Village that there stills to be needs a long time for their recovery process onto repair their physical assets. Recovery of these physical assets will be successful based upon the review process and the routine and regular learning of the community through their strength of social capital that adequately supports their natural and physical assets. A long-term period for the recovery phase of physical assets was experienced by Umatoos Village as has been calculated during a full effort in a year. In general, a full-applicative-adaptive phase could be successful if there is implementation of coordination and corporation between Umatoos Village, Fafoe Village, and Oanmane Village. The result of analyses with the community on FGD and SMEs has been displayed spatially (i.e., Visual Map that can be seen in Fig. 9.5).

9.2.3 Physical Resilience Adjustment (PRA): Assessment of Participative Talk on the Building Community Resilience

Physical assets can be understood as a driving asset that drives the other assets of resilience, namely human assets, natural assets, social assets, and economic assets. Physical assets bring a creative human resource to efficiently utilize the existing physical assets (Fig. 9.6).

The results of discussion with the community can be seen in Chap. 10, Table 10.1. Being a village, which is near the estuary of Benanain River, Umatoos Village has real physical assets that support their natural assets. Their natural assets are supporting them in the process of community resilience and the long-term recovery process. However, Oanmane Village has more disadvantages of their physical assets due to flood disaster than their natural assets that Umatoos Village and Loofun Village had, such as damage of public facilities. Umatoos Village and Loofun Village have a class or level one, long term. However, the other 13 villages are included in class or level two, moderate-term action. Picture of the result analysis with the community on FGD and SMEs can be seen in Fig. 9.7.

9.2.4 Social Resilience Adjustment (SRA): Assessment of Participative Talk on the Building Community Resilience

Social assets can be understood as a strength of social community in supporting other assets in *nine livelihood resources* that a community had and implemented for flood disaster response and to improve their community resilience. Binding, bridging, and connecting social capital of each individual are the social assets of resilience that could strengthen their basis scale of social power (i.e., community groups) which have a major role in flood response steps to respond to the flood disaster and the community recovery process (Fig. 9.8).

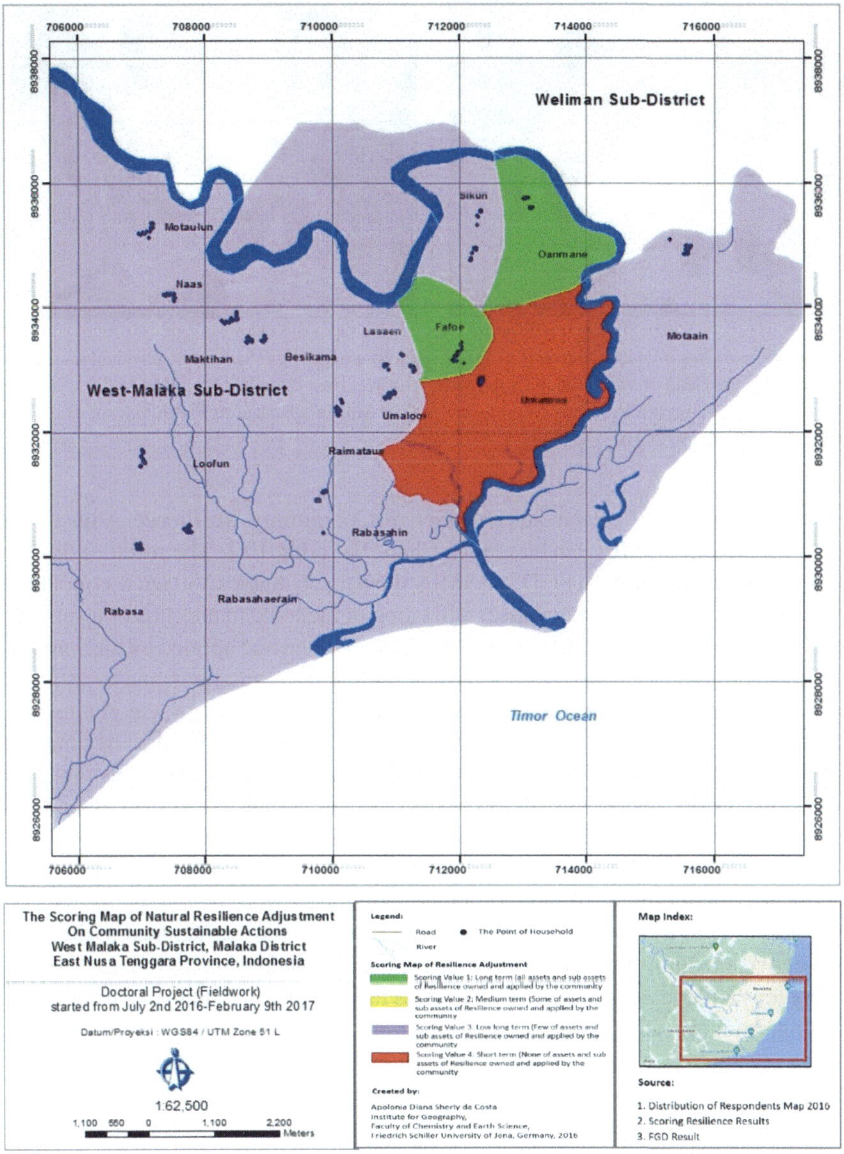

Fig. 9.5 Scoring map of Natural Resilience Adjustment (NRA) on community sustainable actions in the West Malaka Subdistrict, Malaka District, East Nusa Tenggara Province, Indonesia. *Source* Author's Research Fieldwork on FGD and SMEs, November 19, 2016

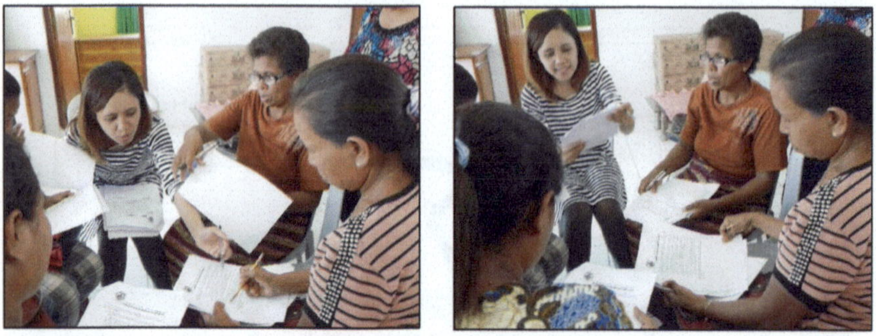

Fig. 9.6 Focus Group Discussion (FGD) and Stakeholder Meetings (SMEs) on physical resilience adjustment in the Head of West Malaka Subdistrict's Office, Besikama Village, West Malaka Subdistrict, East Nusa Tenggara Province, Indonesia. *Source* Author's Research Fieldwork on FGD and SMEs, November 19, 2016

The Scoring Level of the Social Asset of Community Resilience Adjustment Based on FGD and SMEs can be seen in Chap. 10, Table 10.1. The result of discussions with the community in FGD and SMEs show that Umatoos Village was included in the vulnerable group of social asset utilization or included in the class or level four. Level four explains that the subassets of resilience owned and applied by the community in Umatoos Village are unproductive or ineffective, inefficient, and insufficient for a long-term action. The reason for this that was explained by the community was regarding the vehicle. There was no vehicle or container established in Umatoos Village for handling the flood disaster management. However, during this time, the community in the Umatoos Village is still properly connected to the Community Group Disaster as well as Lofoun Village and Rabasahaen Village.

Lofoun Village and Rabasahaen Village have a class or level three, low long term (i.e., nearly moderate term), and the other 13 villages were belonging to class or level one, long term. Meanwhile, none of the 16 villages were included in the class or level two, medium-long term. However, the community that has a yield-agriculture in Umatoos Village does not have well-netted opportunities in their social capital of bonding, bridging, and linking.

The community recognizes that they have the ability to form and to implement the three elements of social capital, namely bonding, bridging, and linking. Even though the community has been experiencing challenges in facing the flood disaster in its application, they are really aware about their social capital that can assist them to increase their strategies of resilience. The community has a good cultural unity. There are some social measures such as *hakawak* or *gotong royong* or mutual assistance which have been a unifying system for community in the West Malaka Subdistrict. However, there was still a main obstacle in community's social strength in Umatoos Village; every individual of the society in Umatoos Village has a good social capacity, but not evenly across its social aspect. Thus, it triggers the community's social capacity and leads to a weakness of social trust possessed by the community in Umatoos Village. It becomes an inseparable part of the community unity and

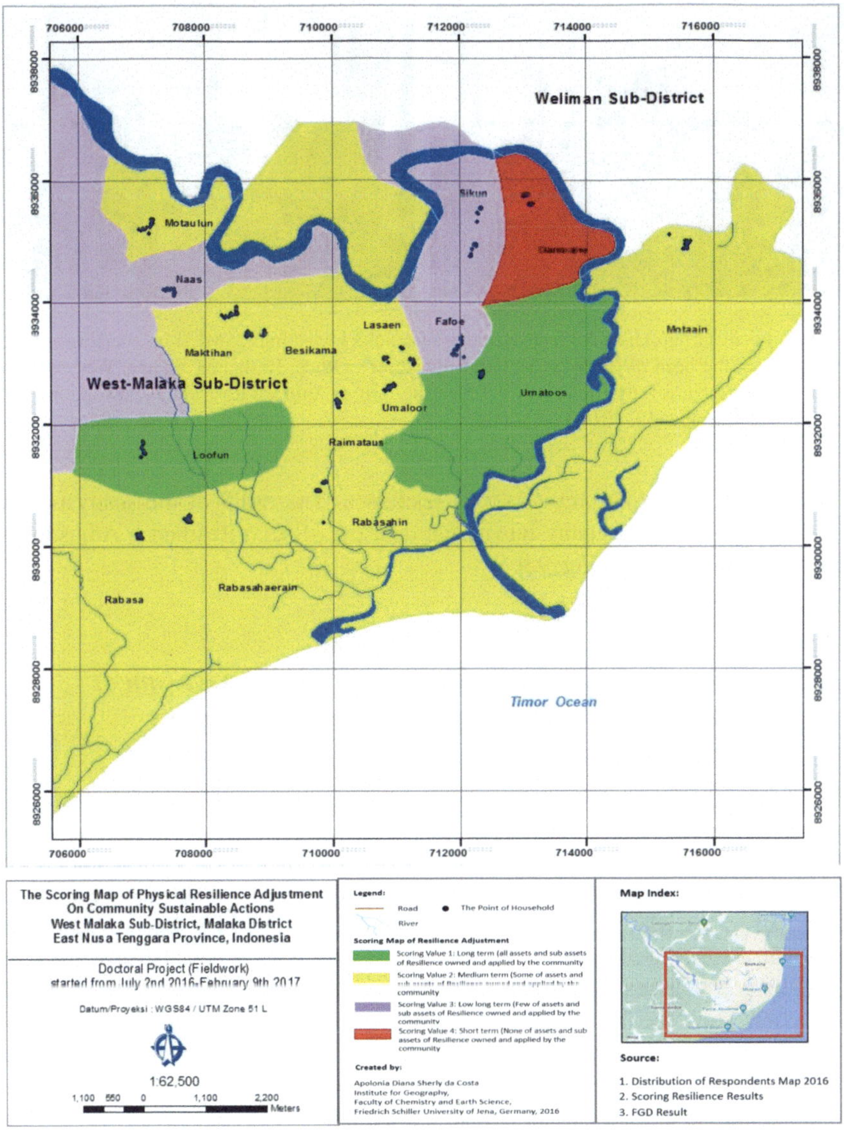

Fig. 9.7 Scoring map of Physical Resilience Adjustment (PRA) on community sustainable actions in the West Malaka Subdistrict, Malaka District, East Nusa Tenggara Province, Indonesia. *Source* Author's Research Fieldwork on FGD and SMEs, November 19, 2016

Fig. 9.8 Focus Group Discussion (FGD) and Stakeholder Meetings (SMEs) on social resilience adjustment in the head of the West Malaka Subdistrict's Office, Besikama Village, West Malaka Subdistrict, East Nusa Tenggara Province, Indonesia. *Source* Author's Research Fieldwork on FGD and SMEs, November 19, 2016

a strength of the community assets of resilience. The result of this analysis was spatialized by the community into visual data (i.e., Social Resilience Adjustment Map) that can be seen in Fig. 9.9.

9.2.5 Economic Resilience Adjustment (ERA): Assessment of Participative Talk on the Building Community Resilience

Economic assets are one of the determinant indicators of community welfare in utilizing the assets of resilience to handle flood disaster management in an integrated response. The community in the West Malaka Subdistrict has assumed that economic assets are assets which are vital to their survival for flooding. Because, without economic assets, the community cannot take any advantage of social assets, physical assets, natural assets except that a strong cultural can strengthen the social and cultural recoveries. Thus, the need of a real implementation of the Indonesian national and local government's policy (in Malaka Regency, East Nusa Tenggara Province, and Senayan, Jakarta) was important for the survival and welfare of the community in the West Malaka Subdistrict. As can be seen in Fig. 9.10, "FGD and SMEs on Economic Resilience" is presented.

The result of the discussion showed that Umatoos Village has class or level 1, the long term as a potential indicator of the ability of nature to be utilized for the purposes of profit and economic assets. The fact also showed that even though the natural assets could also be a natural disaster, the community remained optimistic in the recovery phase of utilizing their natural assets, physical assets, and social assets.

Although the availability of physical assets either from Fafoe Village and Oanmane Village can support recovery of natural assets owned by Umatoos Village, Fafoe Village, and Oanmane Village have been experiencing economic constraints in

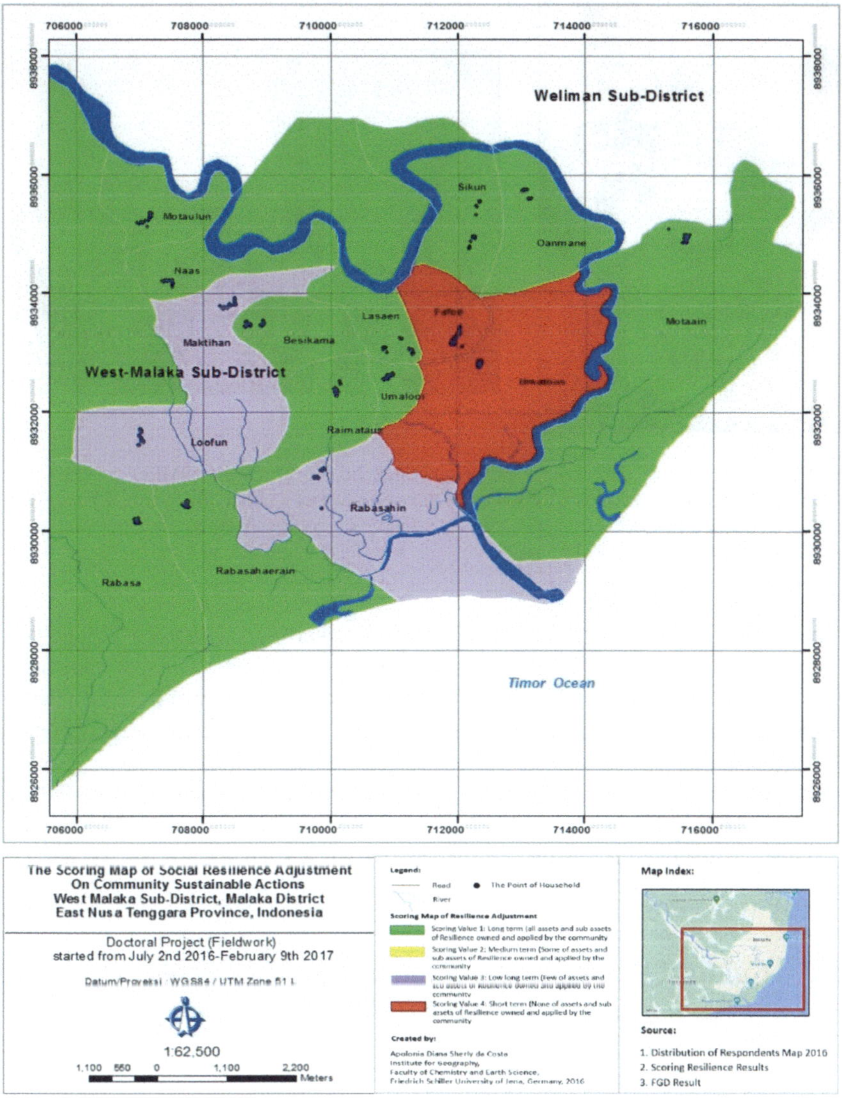

Fig. 9.9 Scoring map of Social Resilience Adjustment (SRA) on community sustainable actions in the West Malaka Subdistrict, Malaka District, East Nusa Tenggara Province, Indonesia. *Source* Author's Research Fieldwork on FGD and SMEs, November 19, 2016

expediting the physical resources at their disposal as manager of natural resources to support a worth economy for them. Thus, Oanmane Village and Fafoe Village require more economic capital due to the damage caused by flooding which has destroyed their economic assets. Fafoe Village and Oanmane Village are included in class or level 4, short term. While Sikun Village is the only village that has class or level

Fig. 9.10 Focus Group Discussion (FGD) and Stakeholder Meetings (SMEs) on economic resilience adjustment in the head of the West Malaka Subdistrict's Office, Besikama Village, West Malaka Subdistrict, East Nusa Tenggara Province, Indonesia. *Source* Author's Research Fieldwork on FGD and SMEs, November 19, 2016

3, low-long term (i.e., nearly moderate-term). Furthermore, Umatoos Village was occupied class or level 1, the long term because of the rich natural assets' indicators can be seen from the geographical position that nearby the estuary of Benanain River. While, class or level 3, the low long-term (i.e., nearly moderate term) were occupied by 12 other villages.

For more detail, the spatial distribution of the scoring map adjustment of economic resilience can be seen in Fig. 9.11. The Scoring Level of the Assets of Community Resilience Adjustment Based on FGD and SMEs can be seen in Chap. 10, Table 10.1.

9.2.6 Cultural Resilience Adjustment (CRA): Assessment of Participative Talk on the Building Community Resilience

Cultural asset is one vital asset of community resilience which drives the strength of the community recovery process in addition to the role of religion (i.e., the church) where the cultural assets are owned by the actor of resilience (i.e., community and institutions and the Indonesian government). Community, institutions, and the Indonesian government are able to change the adaptive pattern of a weak capacity, or short-term assets possessed by the community into them being strong and of long-term capacity. Cultural assets have become the strength of the community in the process of improving the quality of flood disaster management and improving the quality of long-term actions of a resilient community (Fig. 9.12).

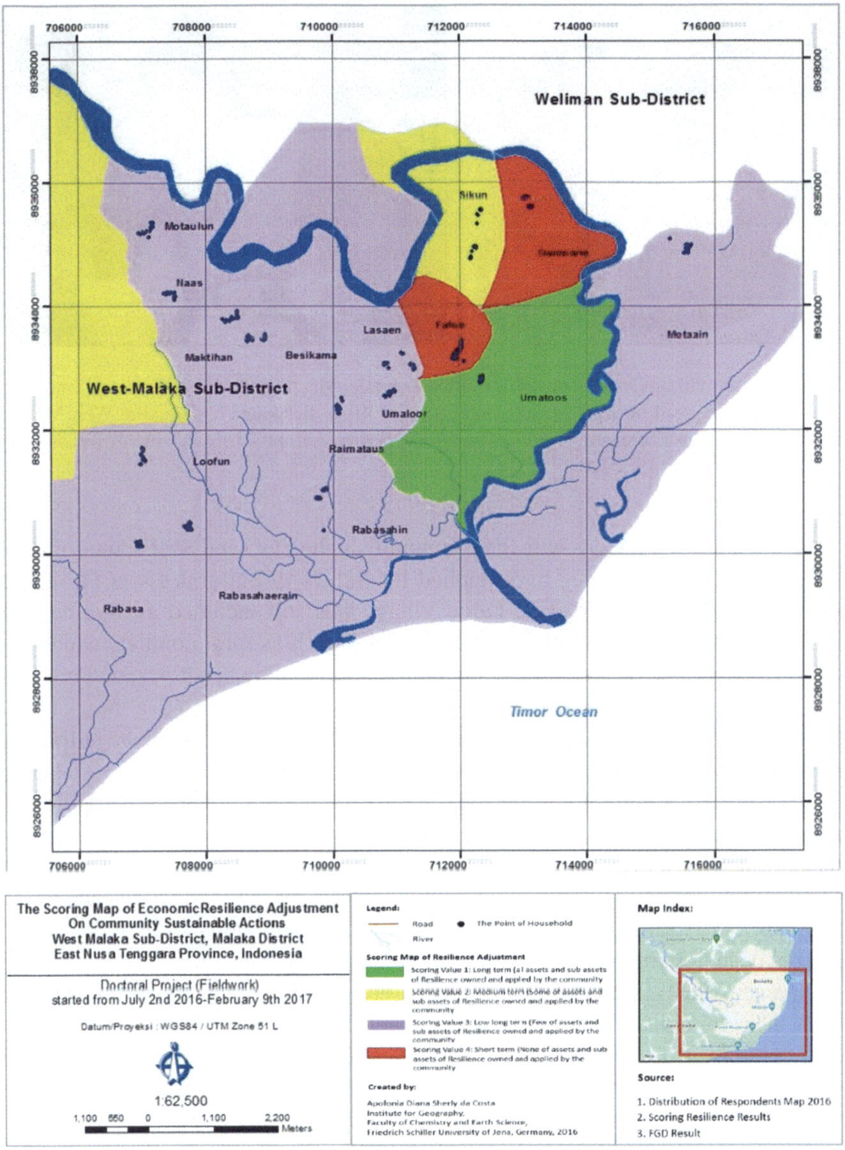

Fig. 9.11 Scoring map of Economic Resilience Adjustment (ERA) on community sustainable actions in the West Malaka Subdistrict, Malaka District, East Nusa Tenggara Province, Indonesia. *Source* Author's Research Fieldwork on FGD and SMEs, November 19, 2016

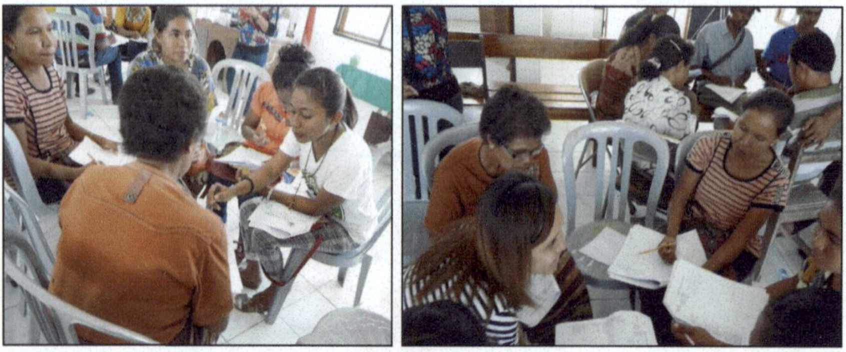

Fig. 9.12 Focus Group Discussion (FGD) and Stakeholder Meetings (SMEs) on cultural resilience adjustment in the head of West Malaka Subdistrict's Office, Besikama Village, the West Malaka Subdistrict, East Nusa Tenggara Province, Indonesia. *Source* Author's Research Fieldwork on FGD and SMEs, November 19, 2016

Based on the discussions with the community in the FGD and SMEs, the community said that although they have applied the values of cultural assets through a guidance of the cultural leaders, Fafoe Village was still included as a vulnerable village due to the negative effects of social cases such as rural conflicts which are still to be overcome between the community about land ownership, sanctions, and other customary agreements. It is known as a cornerstone of negative things in the social structures in which social order is impacting the sustainability of the utilization of other assets owned by the community.

However, the community in the West Malaka Subdistrict tends to be optimistic rather than pessimistic in facing socio-cultural issues. The community adheres to the rules that have been inherited. They consider that cultural rules and guidance of traditional leaders can enhance the positive effects for optimizing the capacity of communities to deal with flooding. In the process of community resilience, cultural assets play an important role in the review of the damaged assets as well as a source or container solution for damages to other assets owned by the public. In implementation, the community in the West Malaka Subdistrict had fully been authorized their socio-cultural environment by the traditional leaders as official or primary mover and spiritual facilitator for them as to alleviate the negative effects of floods that were threatening their lives and livelihood.

The spiritual power of the ancestors through the role of traditional leaders in the West Malaka Subdistrict has been spreading to every individual to surrender their prayers and belief in the goodness that produces a strong faith of community to the ancestors. It was through guidance by the traditional leaders. Impression and moral messages that support a harmonious life based on the values of a noble culture in protecting and preserving the natural environment, social environment, and the physical environment for the community in the West Malaka Subdistrict became part of the role of traditional leaders and the community in keeping the cultural values. Cultural values are represented in the value of the cultural assets that have been executed and implemented by traditional leaders to the community in the West Malaka Subdistrict. The community has practiced this in their everyday life as a custom value and habit. A diversity of conflict is part of the community challenges in strengthening their cultural assets. Thus, although there is one village, namely Fafoe Village that has a conflict of social culture, there was a very negative interference with the role of the community and traditional leaders' roles in dealing with flooding and increasing community resilience. The result of discussion was spatially created in visual data; the adjustment map of cultural resilience assessment can be seen in Fig. 9.13.

9.2.7 Religion Resilience Adjustment (RRA): Assessment of Participative Talk on the Building Community Resilience

Religion is a means of spiritual service provider of both intangible and tangible means that can strengthen the capacity of community resilience in the face of the problems, in this case, flood disaster. The community in the West Malaka Subdistrict appreciates the role of religion (i.e., the Catholic Church and its religious services) in contributing aid for the community as an applicative-human-service of faith and ecumenical. Although Maktihan Village was included in class or level 4, for the short-term asset of religion, the Catholic Church's equalization aid is for the whole community in the West Malaka Subdistrict. Intangible services deal with the ministry of prayer and worshiping together provided by the priests and sisters before and after the flood. Meanwhile, tangible services deal with direct ministry or service that was provided by the priests, sisters in terms of assistantship before, during, and after the flood has occurred such as clean water, shelter and tents for the displaced, treatment and care, prayer facilities and also as a facilitator for connecting and informing all of the structural levels of the World Caritas Institutions which more assistance for the affected people of flood disaster in the West Malaka Subdistrict can be given. The role of religion has been playing a vital role in supporting the community recovery process in the West Malaka Subdistrict. Figure 9.14, "FGD and SMEs on Religion Resilience Adjustment with community and Fig. 9.15. Scoring Map of Religion Resilience Adjustment are presented."

9.2.8 Law or Legal Resilience Adjustment (LRA): Assessment of Participative Talk on the Building Community Resilience

Regulation tools can be channeled through the establishment of the rules that are well-accommodated by the regulators—namely legislative council' rules which drive the

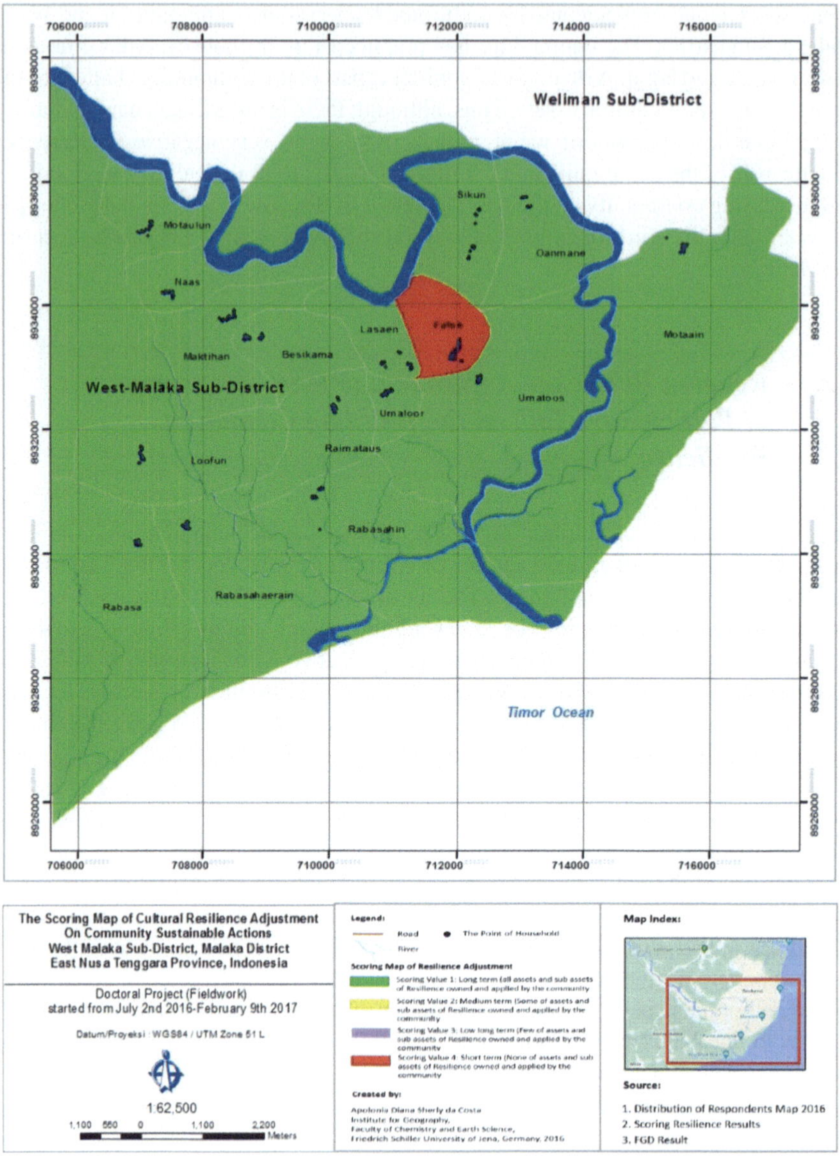

Fig. 9.13 Scoring map of Cultural Resilience Adjustment (CRA) on community sustainable actions in the West Malaka Subdistrict, Malaka District, East Nusa Tenggara Province, Indonesia. *Source* Author's Research Fieldwork on FGD and SMEs, November 19, 2016

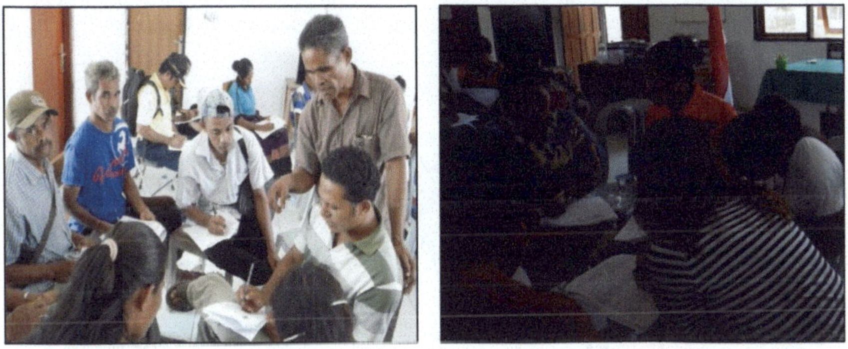

Fig. 9.14 Focus Group Discussion (FGD) and Stakeholder Meetings (SMEs) on Religion Resilience Adjustment in the Head of West Malaka Subdistrict's Office, Besikama Village, the West Malaka Subdistrict, East Nusa Tenggara Province, Indonesia. *Source* Author's Research Fieldwork on FGD and SMEs November 19, 2016

rules as well as the application of the rules implemented by the executive council. However, the result of assessment about the implemented regulation for the people of West Malaka Subdistrict was not yet implemented. The people assessed that law or legal asset is not yet implemented by the executive council even in the absence of the legal aspects of the protection of protected areas and protection of prone areas to river floods on communities in West Malaka Subdistrict. Law or legal adjustment resilience can be understood as a means of implemented the Indonesian government policy through the establishment of rules that could be applied and be able to change the welfare aspects of marginalized communities, affected communities due to flood disaster and society in war or conflict. Community discussion in FGD and SMEs produces a sharp and a critical appraisal to the Indonesian government to settle their obligations with the right target and prioritizing on the welfare of the people in West Malaka Subdistrict. Figure 9.16 Focus Group Discussion (FGD) and Stakeholder Meetings (SMEs) on Law or Legal Resilience Adjustment and Fig. 9.17 Scoring Map of Law or Legal Resilience Adjustment are presented.

9.2.9 Political or Policy Resilience Adjustment (PolRA): Assessment of Participative Talk on the Building Community Resilience

Political or policy resilience adjustment can be understood as the implemented idea based on the manuscript' regulations. The manuscript' regulations contained the legal points or ideas that should be emergently implemented into a political or policy resilience for the affected community in West Malaka Subdistrict (Fig. 9.18).

The Scoring Level of the Political Asset of Community Resilience Adjustment Based on FGD and SMEs can be seen in Chap. 10, Table 10.1. On the political

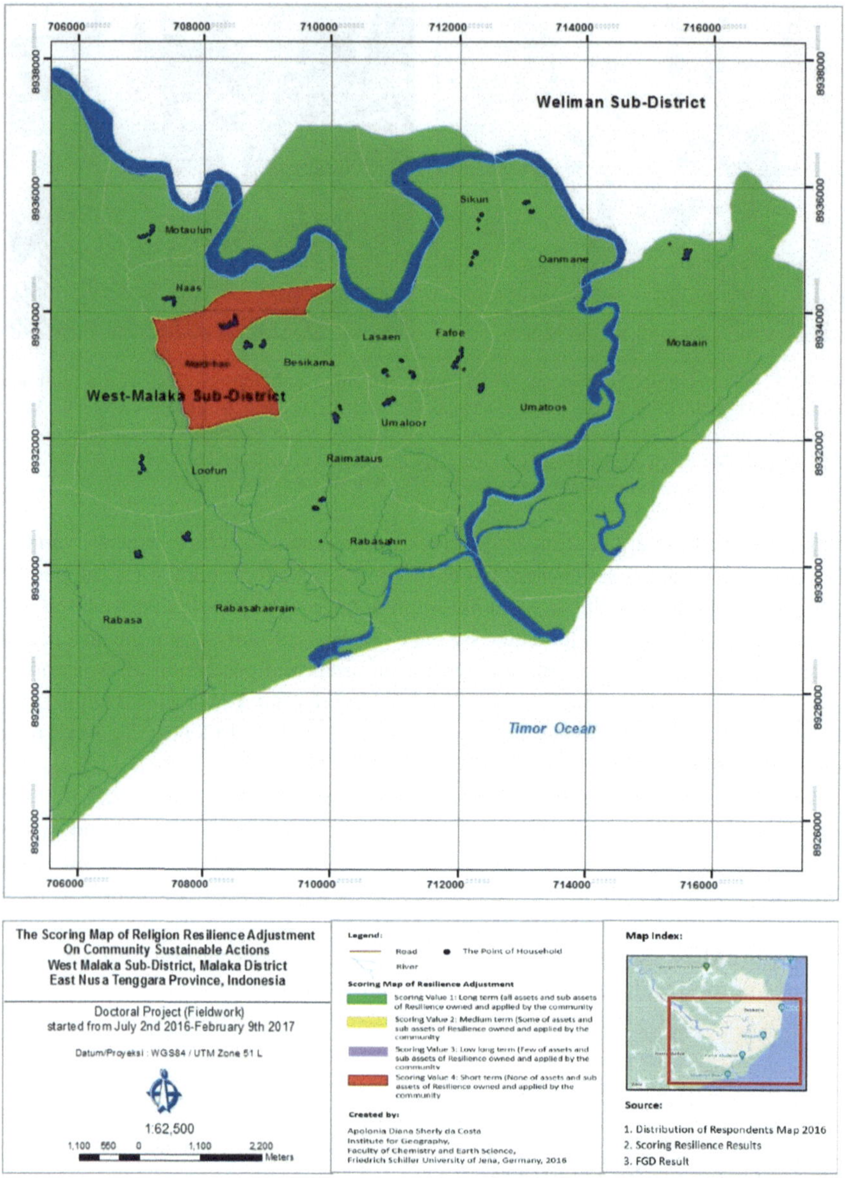

Fig. 9.15 Scoring map of Religion Resilience Adjustment (RRA) on community sustainable actions in the West Malaka Subdistrict, Malaka District, East Nusa Tenggara Province, Indonesia. *Source* Author's Research Fieldwork on FGD and SMEs, November 19, 2016

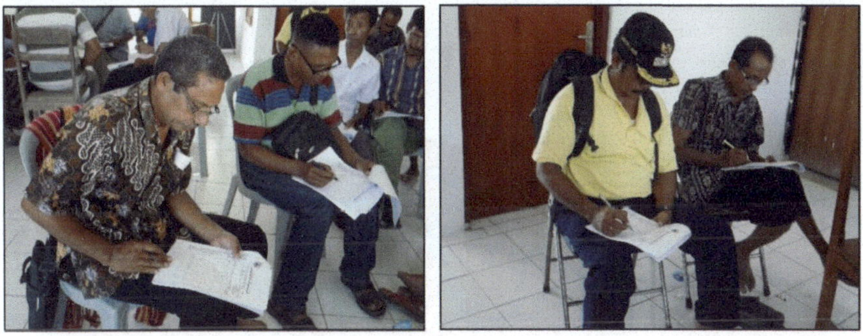

Fig. 9.16 Focus Group Discussion (FGD) and Stakeholder Meetings (SMEs) on Law or Legal Resilience Adjustment in Head of West Malaka Subdistrict's Office, Besikama Village, West Malaka Subdistrict, East Nusa Tenggara Province, Indonesia. *Source* Author's Research Fieldwork on FGD and SMEs, November 19, 2016

assessment policy or assets, the community recognizes that there is no realization of the Indonesian government in handling the flood disaster and community resilience with a sufficient and an integrated solution. The result of assessment according to community' assessment through FGD and SMEs was spatially created into visual data or Political Map of the Political or Policy Resilience adjustment that can be seen in Fig. 9.19.

9.3 Discussion and Summary

The discussion and summary are in two parts for building community resilience adjustment for community sustainable development through FGD and SMEs. *First*, this part discusses and concludes an assessment of the essential theory "adjustment" as supposed by Walker et al. (2004) and community risk perception. *Second*, it discusses and concludes a measurement of the practice of the community resilience adjustment. Discussion and conclusion are presented below.

First, the target solution through an assessment using the scheme's motions and analysis (phases of participative talk) from the FGD and SMEs in discussing community resilience adjustments was successfully achieved in this research project of this book. The use of risk perception in the FGD and SMEs is an effective, motivational, deep, and significant medium for drawing information from the community. Risk perception is a practical and methodical community-based participatory approach using the theory of resilience provided by Walker et al. (2004). Walker et al. state that risk perception embraces resilience as an adaptive manner, which means as an adaptive capacity or adjustment, rather than as an adaptation. An adjustment upholds absorptive capacity and collective actions, which have been assessed in this study and show that communities have a solid understanding of their resilience assets in addressing flood disaster. Through risk perception, the community delivers its target

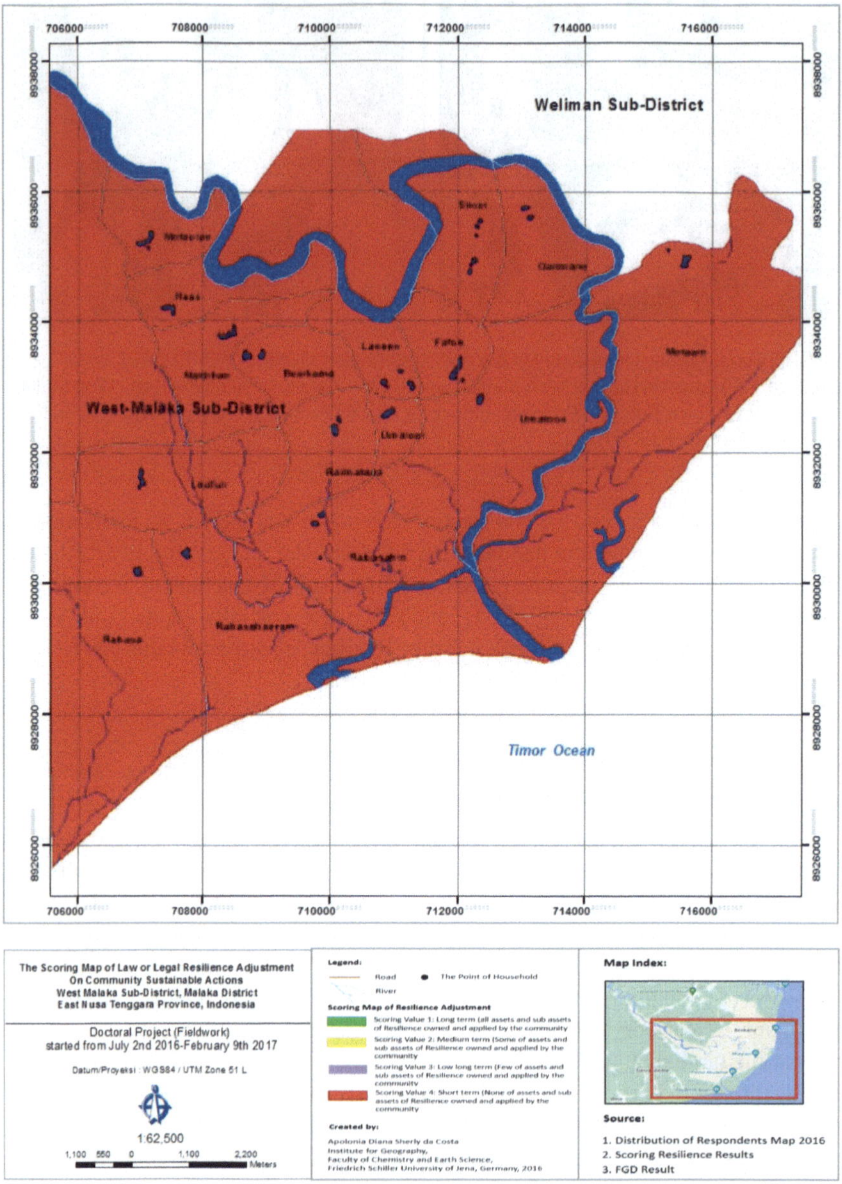

Fig. 9.17 Scoring map of law or Legal Resilience Adjustment (LRA) on community sustainable actions in West Malaka Subdistrict, Malaka District, East Nusa Tenggara Province, Indonesia. *Source* Author's Research Fieldwork on FGD and SMEs, November 19, 2016

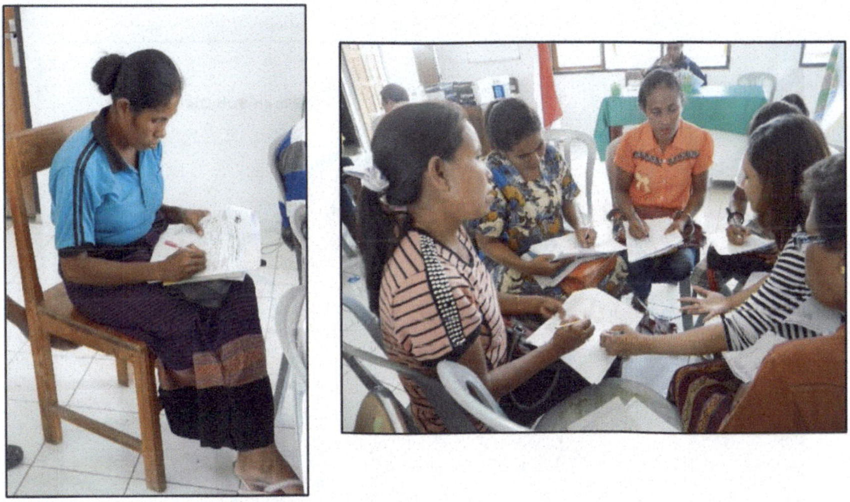

Fig. 9.18 Focus Group Discussion (FGD) and Stakeholder Meetings (SMEs) on political or policy resilience adjustment in Head of West Malaka Subdistrict's Office, Besikama Village, West Malaka Subdistrict, East Nusa Tenggara Province, Indonesia. *Source* Author's Research Fieldwork on FGD and SMEs, November 19, 2016

solutions, which reflect its appreciation of its livelihood resource benefits (resilience asset benefits) and how it utilizes its livelihood resource benefits to support its efforts to cope with flood disaster.

Previous authors and researchers have rarely invented and discussed resilience adjustment for natural disasters/hazards. Most of them have only discussed a single analysis of adaptation combined with an analysis of coping, or resilience for managing natural disasters. This book tries to forefront its use of the term "adjustment." Then, transformation of the term of "adaptation" is switching to "adjustment," as the term adjustment explicitly described by Walker et al. as human actions practice or absorb more invaluable efforts. In practice, an adaptation is invaluable if it contains absorptive capacity and supportive capacity from human abilities, which means a good understanding combined with an effective absorptive capacity and supportive capacity is termed as an adjustment. Given that human abilities in understanding a flood disaster while having an absorptive capacity and supportive capacity, in this particular project of this book, the community in Malaka has both. The community sought various ways to adjust to flood disasters, in terms of creating, building, and maintaining their assets of resilience. Such ways were nurtured and grown by the process of encountering floods.

Second, this chapter explains that the community delivers its risk perception in the form of aspirations, solutions, and critiques. These risk perceptions have been obtained through community-based participative open-platform discussion and assessment with the main theme: "Building community resilience for flood disaster." The community creates its own measurement system (level or score) of building

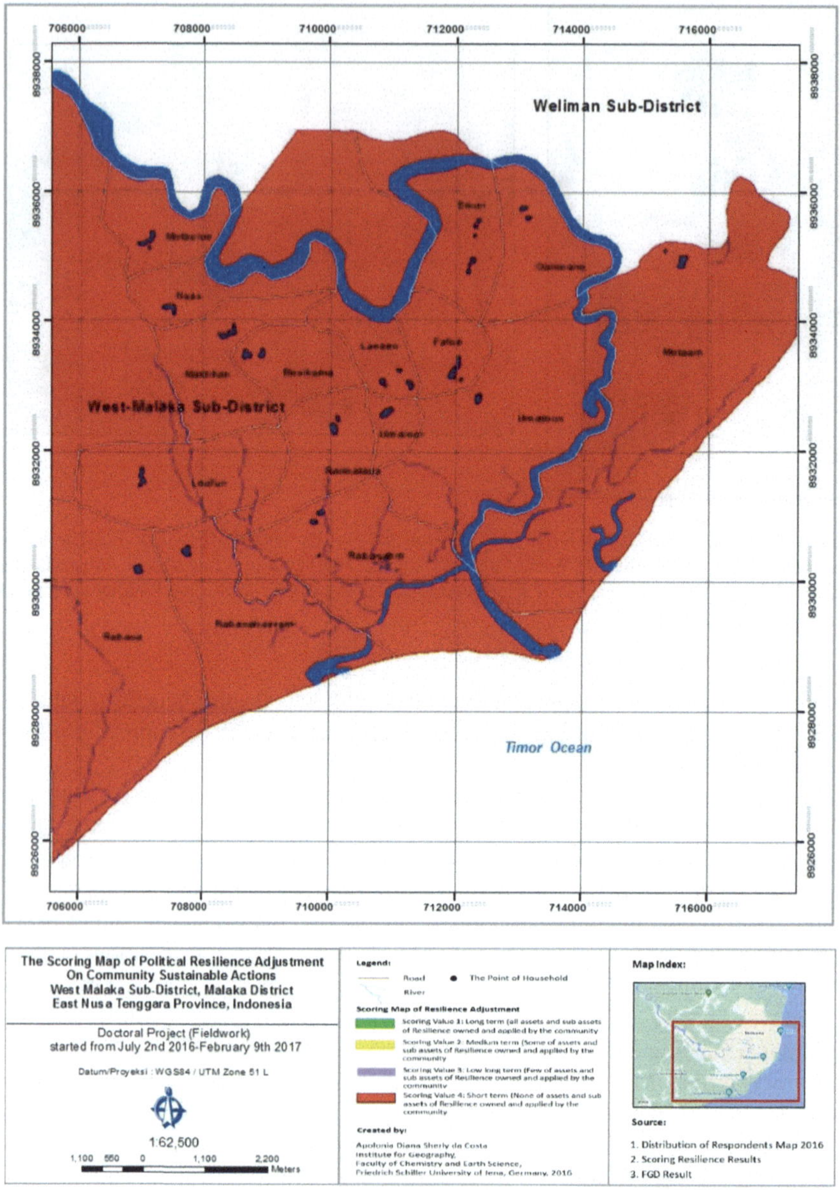

Fig. 9.19 Scoring map of Political or Policy Resilience Adjustment (PolRA) on community sustainable actions in West Malaka Subdistrict, Malaka District, East Nusa Tenggara Province, Indonesia. *Source* Author's Research Fieldwork on FGD and SMEs, November 19, 2016

resilience for flood disaster. The community conveyed information about the four classes or scoring values of its nine resilience adjustments or the nine livelihood resources; these are human resilience adjustment (HRA), natural resilience adjustment (NRA), physical resilience adjustment (PRA), social resilience adjustment (SRA), economic resilience adjustment (ERA), cultural resilience adjustment (CRA), religion resilience adjustment (RRA), law or legal resilience adjustment (LRA), and political resilience adjustment (PolRA). The *nine livelihood resources* are discussed and concluded below.

(a) HRA

The community scrutinized HRA to establish whether they were skillful, knowledgeable, and owned and implemented by them (15 villages except Oanmane Village). Oanmane Village's HRA was positioned in level four (scoring value four), which means that its resilience adjustment is unstable and not productive, effective, efficient, or sufficient for long-term action. The community in Oanmane Village faces problems in recovering damaged human assets (i.e., health) and the economic factor, and physical recovery factor takes a longer period.

(b) NRA

NRA was not implemented long term by Umatoos Village, as witnessed by its level four in the assessment, which means it faces the overwhelming risk of physical damage, such as lack of vehicles after a flood. Its geographical position adjacent to the estuary of the Benanain River has caused Umatoos Village to experience the destructive impact of flood disaster on its natural assets.

In contrast, Fafoe Village and Oanmane Village have adopted a positive cooperation and coordination in an effort to provide some assurance relief for their natural assets. Long-term NRA was achieved by Fafoe Village and Oanmane Village, which were rated at level one, indicating sufficient NRA in utilizing natural assets. The other 13 villages were rated level three (i.e., nearly moderate term).

(c) PRA

Although Umaoos Village scored four (short term) for NRA, they do not have any constraints in accessing physical assets as their location is near the estuary of the Benanain River. Umatoos Village scored one (long term) but is not effective in terms of operating PRA because their NRA is at risk; this indicates that once flood strikes the villages, this village has prepared all the equipment for fishing fish and shrimp ponds in the Benanain River, considering their location is closest to estuary of Benanain River. Lofoun Village scores one (long term) for PRA and helps Umatoos Village escape and temporarily store Umatoos village's equipment. Lofoun Village is 5–7 km from Umatoos Village and is safe from floods (moderate vulnerable) when they strike. While Oanmane Village scores four (short term) of their PRA, the other 13 villages score two (moderate or medium term) for PRA. However, Umatoos Village and Lofoun Village assisted the PRA of Oanmane Village and the other 13 villages, showing applied coordination and cooperation. The impact of the flood disaster made people strengthen their cooperation and coordination in accessing and operating PRA.

(d) SRA

In SRA, Oanmane Village and the other twelve villages score one (long term) and none of the villages score two (medium long-term). A score of three (nearly medium-term) was achieved by Lofoun Village and Rabasahaen Village. Umatoos Village scores four, indicating that their village was not as effective in dealing with other assets (PRA as earlier discussed) and needed access to SRA from Oanmane Village, Lofoun Village and the other twelve villages. Umatoos' vehicle or containers located in their village are at risk, even though they had sufficient PRA in terms of fishing equipment. The village is also connected to a networking group that the villages established themselves and engages key leaders and volunteers of the 16 villages, namely the Community Group Disaster. This group networking was also used by Lofoun Village and Rabasahaen Village, and the other 13 villages.

(e) ERA

In obtaining ERA, Umatoos Village shows ERA actions through increasing their income. Umatoos Village scores one, even though they have constraints in accessing NRA and SRA. An optimistic statement is that: "*Natural assets could also be a natural disaster and vice versa as told by the three hundred and twenty two respondents at the sixteen villages in the study area which confirms that natural disaster is not a bane but a boon*" based on the discussion on this section and previous sections of the study. The closer to the estuary of the Benanain River, the easier it is to collect more NRA for ERA and continue to collect more ERA for NRA. Furthermore, the other 12 villages score three, and Sikun Village is the sole village from the other 16 villages which scores two. This indicates that the village is in a transition between accessing either SRA, NRA, or ERA, which is caused by their village being geographically situated at the point of a broken dike. Even though the wall has been reconstructed by the Indonesian government, villages have worried since 2011 when flood struck the district and destroyed the dike and left them in a transition of creating their ERA. They are effectively carrying out their activities in the making of NRA (utilizing natural assets for ERA through valuing their SRA; dividing household chores) but their ERA is not sufficient in the long run. Fafoe Village and Oanmane Village score four, which reinforce their efforts to stabilize their ERA, which is affected by the actual flood.

(f) CRA

The community acknowledges that its CRA is owned and implemented as part of its daily life. One village has an obstacle in addressing floods using its CRA, namely Fafoe Village. The obstacle is that there are negative rural conflicts due to social issues such as land ownership, sanctions, and other customary agreements. These issues were not fully exposed by the community during discussions, as these are sensitive or sacred issues known only to a limited group. However, the customary leaders are in the process of finding a solution. The community agreed upon the role of customary leaders representing one of the CRA to help and guide them in their daily lives and in the phase of dealing with flood disaster. The community understands and obeys the cultural rules taught by the customary leaders.

(g) RRA

Like CRA, the community adheres to RRA, as represented by the role of the religious actors who provide humanitarian relief (priests, sisters, and brothers in the Catholic Church) and who have been supporting them in the phase of coping with flood disaster. The support is both intangible and tangible in form.

Tangible support includes distributing stocks of foods and drinks after floods occur and also stocks of medicines, while sisters treat injured people who are affected by flood disaster, *(after flood disaster occurs or ex-post)*. The religious actors also coordinate with the Community Group Disaster in providing tents or shelter, through guidance from the customary leaders *during flood disaster*, and coordinate with customary leaders and Community Group Disaster to deliver messages of the alerting system in the church *before flood disaster occurs or ex-ante*. Intangible support deals with unseen or invisible power (i.e., prayers, worship, and singing together as an expression of gratitude to the Lord) to strengthen community faith to be more optimistic and positive in addressing the difficulties of life including flood disaster. Intangible support also includes helping the community to absorb positive stimulus to be emotionally stronger to deal with flood. Even though the role of religious actors is not fully as a provider or facilitator for solving the community's problems, the community admitted that its RRA is effective and quite sufficient; however, further assistance is needed from the Indonesian government and other institutions. Despite the positive aid and the manifestation of RRA, the community score Maktihan Village at four for the short-term asset of religion because of its location and because RRA through the role of priests, sisters, and brothers was not that effective and sufficient due to transportation (PRA); the village's position is a little far away from the central village in Besikama and Umatoos Village.

(h) LRA

On the surface, regulation is well designed and established by the legislative councils, but the problem with LRA is that regulation was not implemented in a way that would guarantee a prosperous life for the people who live near the Benanain River watershed, in a rural and remote area. The community concludes that law-makers should realize that regulation requires cooperation with the executive council which is the actor in implementing the regulation. The community delivers its criticism, saying that it is "an old issue that might be always an old issue to be solved." This statement affirms their criticism about their social welfare (see discussion on PolRA) and concerns their resilience adjustment to reach a sustainable path to each of their resilience assets. Furthermore, they also emphasized the issue of protected areas in the upstream area and the issue of disaster risk management and its protection in the downstream area (their place) for their life and their resilience assets. They recognize that there is an Indonesian government regulation or *peraturan pemerintah* (PP) (number 37: point a, b, and c from 2012) concerning *river watershed resource management and protection and environmental protection.* This regulation is, on one hand, helpful for guiding the community considering their rural area is dominated by the role of female leader who acts as a customary leader, but on the other hand the main problems (social welfare and resilience for flood disaster and environmental issues) seem to be ignored by the Indonesian government.

The community questioned the role of government which must be in charge of solving its problems and building a sustainable path through a holistic solution.

(i) PolRA

The community is very critical of the role of the Indonesian government in helping and increasing their assets of resilience associated with living in a rural, remote area that is prone to flood hazard. In addition, the community pinpoints the issue of socialization, where concern from the government is lacking. In the assessment and measurement phases, the community admitted that none of its PolRA has ever been implemented by the government, with the exception of a few examples of assistance which were effective but are not yet complete. The community comes to a solution of PolRA that the need of the implementation of its resilience adjustment (i.e., HRA, NRA, PRA, SRA, ERA, CRA, RRA, and realization of the regulation (LRA)). It is a crucial issue for the community to achieve its social welfare, and therefore carrying capacity would be realized not only by the community but also by the government. Furthermore, the community highlights the issue of socialization as important for its skills and knowledge (HRA) to run the other resilience adjustments. Health issues are also important in running their daily activities and optimizing their resilience assets. The coping mechanism of saving food stocks and eating one or two times per day should be a temporary mechanism and should not be practiced in the long run, as the community needs balanced nutrition, a point that should be noticed by the government. At the end of the PolRA discussion, the community also highlighted child and spatial issues. They tend to give inputs on the issue of their children and their resiliency to cope with flood disasters. They are aware that advances in technology are important, especially for an alerting system and easing daily activities, ensuring resilience mobilities and growth and also communication. However, on the other hand, technology may cause their cultural adherence (local wisdom) to be slowly ignored by future generations or disappear. The cultural adherence that they have been maintaining for many years taught them how essential it is to run life with local wisdom. They teach their children in this modern society about cultural values, and their children recognize these values, but the community is still worried about whether attaining new coping strategies with practical traditional ways will be practiced and maintained by their children in the future.

They also worry whether the government will address this issue. This discussion touchingly and emotionally affects the sense of consciousness, positivity, and adaptivity that they own and implement, which will be different in the future spatial development. Thus, resilience is important to achieve a sustainable livelihood development. The community's path of sustainable livelihood development in its space and its action (current situation) should be thought about from now on, not only by the community itself but also by the local and national Indonesian governments and other institutions' missions.

References

Peraturan Pemerintah (PP) (2012) Indonesian Government Regulation number 37 year 2012: Pengelolaan Daerah Aliran Sungai. President Republic of Indonesia's Regulation Achieves. Jakarta, Indonesia

Walker BH, Gunderson LH, Kinzig AP, Folke C, Carpenter SR, Schultz L (2006) A handful of heuristics and some propositions for understanding resilience in social-ecological systems. Ecol Soc 11(1):13. [online] URL: http://www.consecol.org/vol11/iss1/art13/

Walker B, Holling CS, Carpenter SR, Kinzig A (2004) Resilience, adaptability and transformability in social–ecological systems. Ecol Soc J 9(2)

Chapter 10
Final Discussion, Conclusion of the Book, and Recommendation for Further Authors

This final discussion of this book, I used surveys and interviews to gather information on community risk perception regarding building community resilience for flood disaster. Surveys and interview data were obtained from the 320 households in the research location, Malaka-Timor, Indonesia. Using surveys and interviews, a qualitative research method is intended to answer (i.e., assessing and measuring) my research objectives two and research questions two in this book. The third research question was answered using two field participation methods with 37 key informants and leaders, namely FGD and SMEs, Malaka-Timor, Indonesia (See Table D.1 in Appendix D: "List of Participants in FGD and SMEs"), combined with results of resilience adjustment maps highlighting spatial issues with resilience adjustment and disaster management (See Chap. 9 "The Assessment of Building Community Resilience Adjustment for Community Sustainable Development Through FGD and SMEs"). As field participation methods, FGD and SMEs are relevant, significant, inventive, and effective in achieving the best results and solutions for building community resilience for flood disaster.

The two research objectives and research questions involving interviews with 320 households in Malaka-Timor have been introduced and discussed in Chaps. 5, 6, and 7. Chapter 5 details the components of households' socio-economic data, which include gender, age, education, job, and religion. This data is the baseline of reference for the discussion in Chaps. 6 and 7. The community's social economic characteristics is shown by preliminary data for assessing community resilience assets as discussed in the Sustainable Livelihoods Framework (SLF) defined by FAO and ILO (2009: 11) and Mayunga (2007) qtd. in Bahadur et al. 2010: 5 in Chap. 6.

Chapter 6 describes each asset of resilience that has been seen as the determinant of resilience assessment in this study and also outlines an integrated system of resilience, which consists of five resources or so-called *pentagon resources*, as mentioned by FAO and ILO (2009: 11) and Mayunga (2007), qtd. in Bahadur et al. (2010: 5). These resources are human capital, natural capital, physical capital, social capital,

A. D. S. da Costa, *Building Resilience for Flood Disaster in Malaka-Timor, Indonesia*, Springer Natural Hazards, https://doi.org/10.1007/978-3-031-63272-3_10

and economic capital. This study discovers nine resources or so-called *nine resources* that were owned and implemented by the community in Malaka-Timor, Indonesia.

The community's *nine resources* are four resources in addition to their *pentagon resources*. The four additional resources owned and applied by the community are: cultural resources, religion resources, law, or legal resources, and political or policy resources. These resources have been discussed in Chap. 6. Each community determines its own socio-cultural endurance by applying the *nine resources* in the local context. Social-cultural endurance depends on the community's cultural capital of resilience which can also be understood as their cultural values. These values distinguish them from other communities in the context of building community resilience for flood disasters. Hereinafter, these values are the contribution of the female leader (i.e., gender role) in the research location that provides a positive stimulus in leading and guiding the community and also cooperates closely with religious institutions in response to flood disaster (Chap. 5). The role of the female leader as the customary leader in the research location reveals that the community's local cultural values may enrich our knowledge about the context of building community resilience for flood disaster. Furthermore, it is also officially confessed by the national and international enactment's regulations. Indonesian government regulation (number 76, year 2000, article 2) states that "*the role of the organization of the Indonesian government and rural development should embody an active role of the community, so that people have and will continue to take part in being responsible for the development of life together as the fellow of villagers*" (The Watershed Management Bureau of Benanain-Noelmina 2011: 29–30).

The discussion in Chap. 7 addresses "Human risk perception in relation to flood characteristics" and is followed by a view on the international regulation, *The United Nations International Strategy for Disaster Reduction (UN-ISDR), the Hyogo Framework for Action (HFA) 2005–2015.*

Human risk perception is useful to contribute to rural development. This study uses human risk perception that embraces a comprehensive analysis in regard to sustainable livelihood resources (i.e., *nine resilience assets*). Such comprehensive analysis aims to achieve sustainable livelihood resources. The results with regard to the *nine resources* or *nine resilience assets* in this study will be a useful reference for subsequent studies.

Community's risk perception enables them to gain knowledge about disaster risk management. Communities convey their understanding of risk management in relation to particular flood characteristics (i.e., periodic flooding, flood depth, flood duration, flood distribution, and flood casualties). This discussion answers the first and the second questions of this study. It is discussed in Chaps. 6 and 7.

Chapter 7 discusses *an understanding of the flood characteristics in risk management* that a community has perceived—its sensitivity and adaptivity to a natural environment that is vulnerable to flood disaster. People are living near to a river and estuary (i.e., Benanain River), and therefore, they have a good understanding of their situation in terms of flood characteristics. The flood characteristics perceived by the community are *periodic flooding, duration of flood, flood depth, flood distribution (river buffer), and community assets of resilience for flood risk management within each village's profile prone to flood disaster and maps* in the research location.

In addition, the community's *nine resources* in regard to flood casualties (i.e., flood risk) are also discussed in Chap. 7 in the context of "the distribution of human assets of resilience impacted by the flood risk in the research location". The discussion of this content highlights an understanding of community risk perception that deals with resilience adjustment and risk management to handle each asset of resilience.

Chapter 8 discusses the relationship between building community resilience and flooding risk factors in the sustainable development model. It sets out an integrated ecological point of view, building on previous researchers' work (i.e., *Venn's sustainable livelihood development model*) and adds a new model that integrates building community resilience and the risk factor of flooding with the idea of human risk perception in order to achieve effective, efficient, and sufficient coping strategies for the community to maintain a sustainable livelihood development. This study puts forward an integrated ecological solution (i.e., the carrying capacity of human ability or community resilience) to deal with flood disasters. It substantiates the idea of "*carrying capacity*" which should be balanced for an integrated ecosystem capacity. The argument complements the idea of Venn that a community's "*carrying capacity*" addresses its understanding of its coping capacity for implementing resilience in response to flood disaster.

The idea of sustainable livelihood development leads to an understanding of the community resilience system in coping with flood disasters. This discussion is presented in Chap. 9, which discusses *"The Assessment of Community Assets for Sustainable Development and Measuring Community Resilience Adjustment on Sustainable Actions through FGD and SMEs"*. Chapter 9 sets out final findings and discussion on the community's resilience adjustment, as supposed by the theory of Walker et al. (2000). Walker et al. concluded that the term "*adaptive capacity*" or "*adjustment*" is more appropriate for the use by practitioners or researchers in case studies rather than the word "*adaptation*". Walker et al. implicitly transformed the term "adaptation" into the term "adjustment" which means "*absorptive capacity*". The term "absorptive capacity" describes the community's resilience adjustment (*nine resources*) that was collected by the community to withstand flood disaster. It is discussed in the findings and discussions in Chap. 9.

In each section of the assessment and discussion (Chap. 9), the results of community engagement through FGD and SMEs and spatial distribution maps (i.e., scoring maps) of each asset of resilience are described. There are nine resources or *nine resources* of building community resilience: human resilience adjustment (HRA), natural resilience adjustment (NRA), physical resilience adjustment (PRA), social resilience adjustment (SRA), economic resilience adjustment (ERA), cultural resilience adjustment (CRA), religion resilience adjustment (RRA), law or legal resilience adjustment (LRA), and political resilience adjustment (PolRA). Chapter 9 also discusses the results of assessment and measurement of community resilience in coping with flood disasters using FGD and SMEs as field approaches. These approaches were set into two motions and involve five-submotions in answering the third research questions of the second objective in this study. Meanwhile, this chapter discusses the conclusion and recommendation of the study as presented below.

10.1 Conclusion of the Book: Findings

In order to understand the findings and discussion based on the research questions and research objectives of this book, a concise presentation (i.e., Scoring Level of Each Asset of Community Resilience Adjustment based on FGD and SMEs) is provided in Table 10.1.

Based on Table 10.1, discussion of each scoring level of each asset of community resilience adjustment is presented in Sect. 10.1.1.

10.1.1 Findings of the Book: The Nine Assets or the Livelihood Resources on Building Community Resilience

1. Human assets on building community resilience: the community has given their objective assessment that Oanmane Village is categorized into class (scoring value four) meaning that none of the assets and subassets of resilience owned and applied by the community are productive, effective, efficient, or sufficient for long-term actions. Whereas, the other 15 villages belong to class or level one, long-term actions.
2. Natural assets on building community resilience: Umatoos Village that has the level or class one of human assets, long term, has actually the level or class four of natural assets, short term. Meanwhile, Fafoe Village and Oanmane Village have the class or level one, long term in utilizing their natural assets. Whereas, the other 13 villages belong to class or level three, low long term (i.e., nearly moderate term).
3. Physical assets on building community resilience: Oanmane Village does not have constraints in the physical assets that support their natural resources, but rather, the constraints are in the physical assets (i.e., public facilities) which were damaged by floods. As compared with Umatoos Village and Lofoun Village which have class or level one, long term, the other 13 villages have class or level two, medium or moderate long term.
4. Social assets on building community rebelonging: Umatoos Village belongs to a vulnerable group in the utilization of their social assets or belongs to a class or level four. Class or level four explains that subassets of resilience of the social assets that are owned and applied by the community in Umatoos Village are unproductive, ineffective, inefficient, and insufficient for long-term action. The reason for this assessment was explained by the community in that there is not a

Table 10.1 The scoring level of the assets of community resilience adjustment based on FGD and SMEs

No.	The 16 affected villages experienced flood disaster in the West Malaka Subdistrict, Malaka District, in the West Timor Island, East Nusa Tenggara Province, Indonesia	*The scoring level of the assets of community resilience adjustment*								
		Scoring level (Class: 1–4) *Note* Class 1: Long-term (effective, efficient, and sufficient) Class 2: Medium/moderate-term or nearly long-term (effective, efficient, and low sufficiency) class 3: nearly moderate-term or low long-term (effective, inefficient, and insufficient) Class 4: Short-term (ineffective, inefficient, and insufficient)								
		Human Resilience Adjustment (HRA)	Natural Resilience Adjustment (NRA)	Physical Resilience Adjustment (PRA)	Social Resilience Adjustment (SRA)	Economic Resilience Adjustment (ERA)	Cultural Resilience Adjustment (CRA)	Religion Resilience Adjustment (RRA)	Law or Legal Resilience Adjustment (LRA)	Political Resilience Adjustment (PoLRA)
1	*Lasaen*	1	3	2	1	2	1	1	4	4
2	*Umatoos*	1	4	1	4	1	1	1	4	4
3	*Sikun*	1	3	2	1	3	1	1	4	4
4	*Oanmane*	4	1	1	1	4	1	1	4	4
5	*Besikama*	1	3	2	1	2	1	1	4	4
6	*Rabasa*	1	3	2	1	2	1	1	4	4
7	*Rabasahain*	1	3	2	3	2	1	1	4	4
8	*Rabasahaerain*	1	3	2	1	2	1	1	4	4
9	*Umaloor*	1	3	2	1	2	1	1	4	4
10	*Fafoe*	1	1	2	1	4	1	1	4	4
11	*Raimataus*	1	3	2	1	2	1	1	4	4
12	*Loofoun*	1	3	1	3	2	1	1	4	4
13	*Motaain*	1	3	2	1	2	1	1	4	4

(continued)

Table 10.1 (continued)

No.	The 16 affected villages experienced flood disaster in the West Malaka Subdistrict, Malaka District, in the West Timor Island, East Nusa Tenggara Province, Indonesia	*The scoring level of the assets of community resilience adjustment*								
		Scoring level (Class: 1–4) *Note* Class 1: Long-term (effective, efficient, and sufficient) Class 2: Medium/moderate-term or nearly long-term (effective, efficient, and low sufficiency) class 3: nearly moderate-term or low long-term (effective, inefficient, and insufficient) Class 4: Short-term (ineffective, inefficient, and insufficient)								
		Human Resilience Adjustment (HRA)	Natural Resilience Adjustment (NRA)	Physical Resilience Adjustment (PRA)	Social Resilience Adjustment (SRA)	Economic Resilience Adjustment (ERA)	Cultural Resilience Adjustment (CRA)	Religion Resilience Adjustment (RRA)	Law or Legal Resilience Adjustment (LRA)	Political Resilience Adjustment (PoLRA)
14	*Motaulun*	1	3	2	1	2	1	1	4	4
15	*Maktihan*	1	3	2	1	2	1	4	4	4
16	*Naas*	1	3	2	1	2	1	1	4	4

Source Author's Research Fieldwork: FGD and SMEs, July 9 2016 to February 9 2017 strong social facility, or a container established in Umatoos Village in terms of flood disaster management. During this time, the Umatoos Village community is still well-connected with disaster community groups in Lofoun Village and Rabasahaen Village which have class or level of assessment three, low long term (i.e., nearly moderate term). Whereas the other 13 villages belong to a class or a level one, long term. Meanwhile, no village is included in a class or a level one, moderate or medium long term. However, the community in Umatoos Village, where their source of livelihood is in the field of agriculture, has a netted a chance to properly implement their social capital elements (i.e., bonding, bridging, and linking).

5. Economic assets on building community resilience: Fafoe Village and Oanmane Village are categorized into class or level four, short term. Whereas Sikun Village is the sole village that belongs to class or level three, low long term (i.e., nearly moderate term). Furthermore, Umatoos Village is in class or level one, long term due to an indicator of a wealth of natural assets from the geographical position of being near the estuary of the Benanain River. Meanwhile, class or level two, low long term (i.e., moderate term) is occupied by the other 12 villages in the West Malaka Subdistrict.
6. Cultural assets on building community resilience: in the process of community resilience, cultural asset plays an important role in the review of damaged assets due to flood disaster and as a source or container of solutions for other assets that were damaged and experienced by the community. In its implementation, a community puts full authority to the customary leaders as the main guide, mobilizer, and spiritual generator for the community which was dimmed due to the negative or threatening effects of the flood disaster. Cultural values become the value of cultural assets that have been practiced and applied by the customary leaders to the community in the West Malaka Subdistrict. Also, cultural values are thought of as a customary value (i.e., a custom that is already embedded in their hearts and practiced in their everyday life). The diversity of conflicts are a part of the community's challenge in strengthening their cultural assets. Thus, although there is one village (i.e., Fafoe Village) that has a conflict concerning social culture, it does not negatively interfere with the role of the community and customary leaders in handling the problem of flooding or increasing the recovering power of the community.
7. Religion assets on building community resilience: religion is a means of intangible and tangible spiritual service providers that can help to strengthen the capacity of community resilience in facing problems, in this case flood disaster. Although Maktihan Village belongs to the class or level four, short term of religion asset, the distribution of church aid is implemented in all levels of the West Malaka community. Intangible service means prayer services and worshiping together with the priests, sisters, brothers, and communities before and after flood disaster. Meanwhile, the tangible service in question is the direct service from priests, sisters, and brothers in helping before and after the flood disaster has occurred such as clean water, shelters and tents for evacuation, health treatment and care, prayer facilities, and facilitating, connecting, and informing other

institutions of charity in the worldwide level to have support of intangible and tangible services.

8. Law or legal assets on building community resilience: Law or legal resilience adjustment can be understood as a means of the Indonesian government policy through the establishment of applicable rules and the ability to change aspects of the welfare of marginalized communities, affected communities of floods, and communities in war or conflict and prone areas of natural disaster cases. The result of the community discussions in FGDs and SMEs has brought a sharp and critical assessment of the Indonesian government to address their responsibilities with the right target and prioritizing the welfare of the community in the West Malaka Subdistrict.
9. Political or policy assets on building community resilience: in the assessment of political or policy assets, the community recognizes that there has been no realization from the Indonesian government in handling an integrated and an adequate flood disaster management for building community resilience.

10.1.2 Findings of the Book's Part 2: The Community Risk Perception About the Risk Factors of Flooding, i.e., Flood Characteristics (River-Buffer Class Distribution, Flood Frequency, Flood Duration, Flood Depth, and Damage of Flooding as Are Perceived by the Community in Accessing the Nine Assets or So-Called the Nine Livelihood Resources for Building Community Resilience)

1. **River-Buffer Class Distribution**

   (a) The village which is not prone to flood disaster is Besikama Village with a radius of 100–250 m from the Benanain River mouth to the community settlements.

   (b) The village which belongs to the prone area of flooding are Maktihan Village and Umaloor Village with a radius > 25–100 m from the danger of flooding from the Benanain River mouth to the community house.

   (c) The other 13 villages in the West Malaka Subdistrict, namely *Lasaen, Fafoe, Umatoos, Raimataus, Rabasa, Rabasahain, Rabasahaerain, Lofoun, Motaulun, Motaain, Naas, Oanmane, and Sikun* belong to "vulnerable" and "more vulnerable" categories concerning flood hazard with each village having only a radius of 0–25 m; therefore, it is an estimated highest risk that would be experienced by these villages.

2. **Flood Frequency (Period of Flooding)**

All 16 villages in the West Malaka Subdistrict have a similar duration of flood disaster which was beginning from 1939, 1959, 1975, 1979, and 1999. It was a 20 cycle of flood disaster. However, it has turned out to be more dangerous and actively occurring after the last in 1999. The reason why floods are more dangerous to the resilience system of affected communities living along the Benanain River and the Timor Sea is the excess of human activity (many of the East Timorese refugees from the post-referendum civic political war in mid-1999) in contributing negative actions to the ecological balance of nature through illegal tree felling along the upper reaches of the Benanain River, tidal flooding from the Timor Sea, extreme climate change from storms and non-stop rains for months in the rainy season, the unavailability of sturdy concrete embankments provided by the Indonesian central and local governments, but instead local independent efforts provided by local communities, stakeholders, and local traditional leaders to make embankments from non-concrete materials. I added here that, morphologically, the Benanain River is a meandering river with meandering curves that trigger high sediment deposition due to weak medium erosion. If the rainy season arrives with a high level of rainfall intensity, 100–200 mm/per day, and flooding occurs, the appearance or impact on erosion is smaller on the vertical side of the river, and greater on the horizontal side of the river. The Benenain river, and the type of land (plains where people live) are fine alluvial because alluvium deposits are more than 80% which is a soil with a dominance or concentration of low permeability levels with a slope of 15%, relatively flat, so that the flow of water tends to slow down, and the level of flood vulnerability increases every rainy season in the Malaka-Timor region. Also, See Fig. 7.8 on Chap. 7 "River buffer class map", which from this map we can see the appearance of a meandering river where local people (respondents/households) settled in Malaka, building their houses, and carrying out daily activities close to the Benanain River. It was explained that the villages are prone to flood disasters/hazards, e.g., Lasaen, Fafoe, Umatoos, Raimataus, Rabasa, Rabasahain, Rabasahaerain, Lofoun, Motaulun, Motaain, Naas, Oanmane and Sikun are categorized as "vulnerable" and "more vulnerable" to flood hazard with a radius of 0–25 m, the distance between the respondents' houses and the Benenain River; Maktihan Village and Umaloor Village are moderate prone to flood hazards where each with a radius distance of > 25–100 m, and Besikama Village is in the low risk of flood disaster due to a distance of the respondents' houses to the river is with a radius of 100–250 m. A flood disaster has occurred annually at the second round from 2000 until 2016. The year 2016 was the date of conducting this research. Meanwhile, the total year of cycle of the flood occurrence was 98 and the total number of floods experienced in the West Malaka Subdistrict was 43.

3. **Flood Duration**

Based on the interview with the 320 respondents in the West Malaka Subdistrict, the longest duration of flood that entered the respondents' homes was 25–30 days (a month) with 31.3 % from a 100 respondents reporting this. These 100 respondents are from 16 villages in the West Malaka Subdistrict. The community tried to solve the risk of flood by building an 80 cm of their wall door. Different flood durations that were experienced by the community have resulted in determining the level or size in building the wall door. While moderate or medium flood duration was 10–15 days occupying 3.1% homes from the respondents compared to people who have a small percentage of 7.8% (5–10 days); these people implemented a mitigation strategy by establishing or raising their door wall approximately 30–50 cm with materials or wood materials to prevent the water from entering their home.

4. **Flood Depth**

Flood inundation depth determines the length of the duration of the flood. The depth of the flood can be said to be a matter of disturbing public activities in the West Malaka Subdistrict. However, applied adaptive society that is hereditary in practice can be easier for them to cope with the risk of flooding. In general, people have relatively similar capabilities in handling flood disaster risk.

The deepest flood depth that occupied a 100 or 31.3% respondents was up to the range of 251–300 cm. However, no one has ever experienced the risk of flooding their homes more than 300 cm. Meanwhile, the moderate or medium flood depth that occupied the homes of 60 respondents or 18.7%, was 151–200 cm. The lowest depth of flood that occupied ten or 3.1% respondents' homes reached 0–50 cm. The community in the West Malaka Subdistrict has applied an adaptive strategy by establishing or raising the wall door (i.e., stage houses where the stage house in their local language, *Tetum Therik Malaka*, is called *Uma-Fehan*) which is part of physical resilience represented in their local culture. Community has similar experiences by building stage houses ranging from 65 to 250 cm. This effort was applied by the community to deal with flood inundation from the deepest flood depth, moderate or medium flood depth, and less flood depth.

5. **Damage of the Nine Assets of Resilience in the Community After Flood Risk**

(a) **Human Assets**

There are four subassessments of human assets that were impacted by flood risk (i.e., health, consuming nutritional food, drinking water quality, and education (knowledge)) from human assets of resilience that were impacted by the flood risk in the West Malaka Subdistrict. The community in the West Malaka Subdistrict has encountered obstacles in maintaining their health due to the danger of flooding that occurred. There were 25 respondents who thought that they experienced health problems after the flood disaster in every flood incident that has thus occurred. This is indicated by the indication of the presence of several diseases such as flu, fever, diarrhea, and typhus. The impact of the flood disaster seen from the human indicator has degraded the performance of the human ability to take preventive action before the flood disaster, during the flood disaster, and after the flood disaster.

(b) **Natural Assets**

The West Malaka Subdistrict community has good adaptive resilience based on their experience in overcoming flood risk to their natural assets. The same opinion is explained by all 320 respondents by utilizing natural assets such as the river as a source of food consumption, soil as a source of food consumption, fisheries as a source of income, and soil as a source of gardening. These natural assets are helpful in the recovery process.

(c) **Physical Assets**

From the ten physical assets damaged by the floods in the community, there are five physical assets in which all respondents experienced a negative impact. The five physical assets that all 320 respondents experienced damaged to are the public internet kiosk, the public telephone kiosk, the traditional market, the community health center, and homes. There are two kinds of physical assets that were impacted by the flood risk (i.e., damaged to transportation (motorcycle) and damage to irrigation structures and canals).

The community also complained that flood risk caused damage to electrical power, causing outages. As can be seen in the data, there were 150 respondents who experienced this kind of damage to their physical assets. Meanwhile, there are 120 respondents who are noted damage to the public school and the public road, preventing access to the school.

(d) **Social Assets**

The community experienced barriers to their social assets that they have due to the risk of flood disaster. Entirely, the community has applied the system of *hakawak or gotong royong* or mutual assistance. It can be proven by the 320 respondents who said that this kind of social asset can lead them to unifying their social power to manage the flood disaster. However, their mutual assistance is cooperatively manifested after the flood disaster or ex-post such as cleaning the mud from the flood from other's homes. This system of mutual assistance has also brought a positive impact to the process of early warning systems in that information related to hazards was run by individual connection (there were 180 respondents), household or family connections (there were 200 respondents), and disaster community group connection (there were 220 respondents). These kinds of social assets are seen as community social adaptation that has the power to be able to solve the case of the flood disaster in the period before the flood disaster has happened or ex-ante and after the flood disaster has happened or ex-post. While no activities during the flood disaster happened as the community strove to save their lives through moving to a safe shelter or place. This social adaptation is working to deal with some cases of social issues such as human rights and discrimination of each household (there were five respondents), fear and anxiety (there were 16 respondents) of the social conflict and the lack of implementation of the Indonesian government assistance toward the case of flood disaster.

(e) **Economic Assets**

The community has economic risks caused by the flood disaster (i.e., low income). There were 320 respondents that explained their income is decreasing as caused by the flood disaster as they need to cope with and to recover from. The community has an adaptation of economic resilience. However, their adaptation was destroyed due to the flood disaster. There were five kinds of economic consequences that were experienced by the 160 respondents in dealing with the flood risk, namely access to loan or credit from the local bank, *Bank Rakyat Indonesia (BRI) in* Betun City, the capital city of Malaka District, obstacles in transfer system via colleagues, animal diseases in the community's livestock, and limited saving. These assets are needed to fulfill their daily necessities and to handle the flood risks that cause several damages to other assets of resilience such as physical assets and social assets. The community also has been connected to insurance from a local insurance company in Betun City, the capital City of the Malaka District. They require a diversity of financial solutions in order to solve several damages caused by the flood disaster; the damages were to two kinds of their livelihood resources (i.e., damaged to fisheries and shrimp ponds (there were fifty respondents) and damaged to farming equipment motilities (there were 66 respondents)).

(f) **Cultural Assets**

People in the West Malaka Subdistrict have a strong system of cultural resilience in the practice of everyday life and in adaptive practice of flood disaster. This can be analyzed based upon the risk of floods that have an impact on the cultural resilience of the community. Although the community owns and applies every cultural asset they possess, they must continue to practice their cultural endurance. All respondents gave their opinion on the five types of assets of cultural resilience that experienced obstacles due to flood disaster. These five kinds of cultural assets were obstacles of applied-routine traditional rituals for preventing systems (before the flood disaster happened or ex-ante), obstacles of applied-routine traditional rituals for mitigation systems (before the flood disaster happened or ex-ante, during the flood disaster, and after the flood disaster happened or ex-post), obstacles of applied-routine traditional rituals for resiliency systems (before the flood disaster happened or ex-ante, during the flood disaster, and after the flood disaster happened or ex-post), preventing and protecting the flood depth as so that it could not reach the customary leader's home (before the flood disaster happened or ex-ante, during the flood disaster, and after the flood disaster or ex-post), and limited accessibility and management with the religious leaders in the church and other institutions to handle the case of flood disaster.

(g) **Religion Assets**

There were nine kinds of religion assets that were impacted by the flood risk, namely

1. Less of facilitator of disaster community charity.
2. Increasing the early warning information.
3. Less advocation for the marginalized community.
4. They needed more pastoral hospitality to assist the community.
5. Actions were postponed for reparation and reconciliation of the infrastructure but focusing as a shaper and facilitator on strengthening the faith of the community.
6. Financial resources assistance such as water and food for the community during the flood disaster happened.

7. As responder working quickly for the interest of the community in onset crises cases including flood disaster.
8. Providing pastoral services while considering the flood disaster has brought to a limited of resilience corporation with the customer leaders in the prone area of food disaster; and
9. Religious leaders manifested a few of their abilities to assist with the role of the customary leaders in the case of the flood disaster.

The case of flood disaster did not totally disturb all the operational systems of both the religious leaders' hospitality and the customary leaders' services, but it caused a temporal limited assistance for the community. However, all nine kinds of the religion assets that were impacted by the flood disaster are still actively handled by the religious leaders. There was only one kind of religion asset (i.e., actions for basic training and support) which was actively implemented by the religious leaders without being interrupted by the flood.

As the task of religious leaders who are really concerned about ecumenical and charity actions, basic training and support has been always the agenda of assisting the community before the flood disaster has happened or ex-ante.

(h) **Law or Legal Assets**

There was one kind of the law or legal asset (i.e., the broken public road) that was impacted by the flood risk, and therefore, assistance from the Indonesian government was necessary. The public road is seen as the main mode of access for community activities. Once the community activities are disturbed by the flood disaster, more resilience asset activities should be implemented effectively, efficiently, and sufficiently or adequately. Every asset of resilience that the community had explains more of the actions of the Indonesian government to each of them. The data has shown that there was no intervention on behalf of the Indonesian government both for the case of an integrative flood disaster management (i.e., early warning system), prevention, mitigation, or an integration of resilience management (i.e., improving human, natural, physical, social, economic, cultural, and religion assets.

(i) **Political or Policy Assets**

The implementation of policy actions driven by the Indonesian government was still slow in actions for assisting the community in the West Malaka Subdistrict. There were sixty respondents who explained that the implementation of building resilience with institutions was still in progress or in other words it is seen as being in a 'rehearsal period' and has a long way to reach being a "solution-based community approach with the community who experienced the flood disaster".

The active institution is the Non-Government Organization (NGO), named Community Association for Disaster Management (CADM), which is always welcome and ready for discussing the resilience themes for the flood disaster with every institution both in Indonesia and foreign countries. As a single actor of resilience, the community seeks their own solutions to cope with flood disaster. Their social asset of resilience (i.e., bridging, bonding, and linking) that they had and had implemented (i.e., disaster group platform) for a depth-discussion and sharing of experiencing the flood disaster into coping mechanisms (i.e., the ability to cope with flood disaster) for a strong resiliency system.

10.2 Recommendation for Further Authors

1. Since this research of this book has critically assessed an integrated assessment of the nine assets of community resilience or so-called the nine livelihood resources that the community had and implemented to handle the flood disaster, the further or upcoming research should be conducted on each specific asset of community resilience related to human or social geography perspectives in dealing with flood disaster in the West Malaka Subdistrict, East Nusa Tenggara Province, Indonesia.
2. This book is the first book that examines the relationship between human risk perception of building resilience of the affected communities of flood disaster and flood disaster management in Malaka-Timor, i.e., the West Malaka Subdistrict, East Nusa Tenggara Province, Indonesia. Therefore, further empirical research is strongly required to do an adequate assessment through conducting fieldwork-research onsite in Malaka-Timor for each of the human risk perceptions separately either to build community resilience for the flood disaster without considering all integrated aspects of risk factors of flood disaster or flood disaster analyses and flood scenarios in more specific rescue engineering ways and Artificial Intelligence for Flood's Early Warning System, Flood Affected Regions Prone to Casualties, and Flood Predictive Analytics for Disaster Recovery to inform the affected communities and the readers about the hazard level in the flood hazardous area in Malaka-Timor, Indonesia.
3. Since this research of this book was conducted in the downstream area (i.e., the most prone area to flooding), it is therefore required for further authors to also conduct research projects in the upstream area concerning a critical assessment on integrated watershed management and community resilience where the upstream area is known as the border area between Atambua City, Belu District, East Nusa Tenggara Province Indonesia, and Kovalima District of Timor Leste.
4. The low assets of resilience are law or legal and political or policy where these assets of resilience are the vital assets supporting community assets to deal with flooding. Further authors are expected to conduct the domain of law or legal assets and political or policy-related access to political power and assets to have a final critical assessment and solution regarding the integrated building resilience in Malaka-Timor, Indonesia
5. Even though I have assessed all of the assets of resilience that the community had, there is a need to review the assessment of building community resilience, not only limited to human risk perspective but a wider analysis on another perspective of geography studies toward flood disaster.
6. This research project of this book that was conducted by me was successful as the assessment lies on integrated research using nine parameters of analyses or so-called the nine livelihood resources. However, there is a need for spatial planning

and policy research and assessments. Hence, I call for more research and books to be completed by further authors from other institutions throughout the world in order to assess the environmental humanities aspect thorough the research in the West Malaka Subdistrict, East Nusa Tenggara Province, Indonesia.

7. The strong cultural values through the role of the customary leaders in the West Malaka Subdistrict of Malaka District have led the research in unique ways. Upcoming authors are expected to put the cultural assets on community resilience assessment to cope with flooding.
8. For further book's project, it is required to do a comparison concerning urban flooding with the use of the nine assets of community resilience assessment to deal with flood disaster as to examine the existence and malleable values of each asset of resilience in different geographic areas in the world. Since this research project of this book was conducted in a rural-riverine flooding in downstream villages in Malaka-Timor, Malaka Regency, Indonesia, the upcoming research project should focus on characteristics of the people in urban areas in middle-stream and upstream cities/villages of North Central Timor Regency and South Central Regency, Indonesia, and other Indonesia's urban areas and the society who are facing the urban-flood disaster.

10.3 Summary

I assessed the indicator of community sustainable livelihood development (i.e., building community resilience) for flood disaster using community risk perception (i.e., 320 affected communities who experienced flood risk in 16-flooded villages and 37 stakeholders). Community risk perception information was derived through surveys and interviews with 320 households, using field participation and discussion of the 37 stakeholders' outcomes in the FGD and SMEs as well as spatial distribution of each resilience asset, i.e., resilience adjustment maps that the community understood and perceived as dealing with flood disaster (See Chap. 9 "The Assessment of Building Community Resilience Adjustment for Community Sustainable Development Through FGD and SMEs"). For the overall conclusion of my empirical research, in this book is that I invented nine resilience adjustment assets for the assessment based on actual data of the affected-communities and stakeholders in Malaka-Timor, Indonesia, from conducting the research on-site.

References

Bahadur AV, Ibrahim M, Tanner T (2010) The resilience renaissance? Unpacking resilience for tackling climate change and disasters. Strengthening climate resilience discussion Paper 1. First published by the Institute of Development Studies of University of Sussex, The United Kingdom

Food and Agriculture Organization and International Labor Organization, FAO and ILO (2009) The livelihood assessment tool-kit. Analyzing and responding to the impact of disasters on the livelihood of people. Policy and Support Branch Communication Division, Geneva and Rome

The Watershed Management Bureau of Benain Noelmina (2011) Rencana Pengelolaan Daerah Aliran Sungai Terpadu di DAS Noelmina. Kementrian Kehutanan Direktorat Jenderal Bina Pengelolaan DAS dan Perhutanan Sosial. Kupang, Indonesia

Information Sheet

Information Sheet:

Apolonia Diana Sherly da Costa is the first and the only one researcher/scientist in Indonesia and Globally who conducted this empirical research regarding this topic onsite in Malaka-Timor, Indonesia. She is therefore the first and the only one author who writes scientific books on this topic based on empirical research data in Malaka-Timor, Indonesia, starting from her first-empirical research in 2012, and her second-empirical research in 2016-2017 (which has produced this book).

This empirical research was conducted on July 9 2016 until February 9 2017 (seven months) in all 16 flood-affected villages i.e., Lasaen, Umatoos, Fafoe, Besikama, Raimataus, Oanmane, Lofoun, Naas, Sikun, Rabasa, Rabasahain, Rabsahaerain, Motaain, Maktihan, Umaloor, and Motaulun (320 affected households as sampling who experienced flood hazards) in West Malaka Sub-district, Malaka District, West Timor Island, East Nusa Tenggara Province, Indonesia which borders from the south of the Benanain river watershed and West Malaka with the territory of Timor Leste which has resulted an unpublished dissertation thesis at the Friedrich Schiller University of Jena, Germany, 2018 titled "Building Community Resilience for Flood Disaster in West Malaka Subdistrict of Malaka District in the West Timor Island East Nusa Tenggara Province Indonesia", and further she neatly designed/modified, edited, and revised; added more discussion on Sustainable Livelihood Development Model on Chapter 8, Vulnerability of Flood's discussion on Chapter 10 and also some revisions on other Chapters, and completed, and published as this book titled "*Building Resilience for Flood Disaster in Malaka-Timor, Indonesia: Risk Perception from Human Geography*" published by the Publisher, Springer Nature A.G Switzerland and Berlin, Germany 2024.

She is also an author of the other scientific book published this year, 2024 by Regiospectra Verlag Berlin, Germany titled "*Flood Resilience Perceptions of Community-Based Participatory Research in Malaka-Timor, Indonesia*". This book discusses mainly on flood resilience-based geospatial participatory research study e.g., delineations of flood hazard map with stakeholders onsite, and focused on the 60 affected households as sampling in the most flood-affected villages in south-downstream area, closest to Abudenok estuary of Benanain river i.e., Lasaen, Umatoos, and Fafoe in Malaka-Timor Indonesia (West Malaka Sub-district, Malaka District, West Timor Island, East Nusa Tenggara Province). The use of empirical data, method, technical of geography study, and social science approaches and literatures were different. This book was based on her first-participatory empirical research onsite exploring the three flooded-villages in August 2012.

This empirical research for this book published by the Publisher, Springer Nature A.G Switzerland and Berlin, Germany 2024, titled "*Building Resilience for Flood Disaster in Malaka-Timor, Indonesia: Risk Perception from Human Geography*", that was conducted by researcher/scientist/author has received official permission from the Department of Investment Service and One Stop/Gate Integrated Services, East Nusa Tenggara Province, Indonesia in July 2016, and has also received official permission from the Regional Government of Malaka District in Betun City of Central Malaka Sub-district, and formal-structural levels of permits for field research in West Malaka Sub-district and village heads from the 16 flooded villages. For further information regarding this matter, please contact the author via email.

The author declares that this empirical research for this book published by the Publisher Springer Nature A.G Switzerland and Berlin, Germany 2024 was conducted independently by herself onsite in Malaka-Timor, Indonesia on July 9 2016 until February 9 2017, and there are no fundamental issues that need to be conflated regarding the permission of conducting the research, the origin of data, the ownership of data, the use of study literatures, and the research methods and research tools in use before, during, and after the research.

Researcher/Scientist (Author):

Apolonia Diana Sherly da Costa

Münster, 2024

A. D. S. da Costa, *Building Resilience for Flood Disaster in Malaka-Timor, Indonesia*, Springer Natural Hazards, https://doi.org/10.1007/978-3-031-63272-3

Appendices

Appendix A1

DRAFT QUESTIONNAIRE FOR EMPIRICAL RESEARCH (ENGLISH VERSION)

Compiled by Researcher/Scientist (Author):

Apolonia Diana Sherly da Costa

A. D. S. da Costa, *Building Resilience for Flood Disaster in Malaka-Timor, Indonesia*, Springer Natural Hazards, https://doi.org/10.1007/978-3-031-63272-3

Data Questionnaire for the Interviewee (Household) in West Malaka Region, East Nusa Tenggara Province, Indonesia

Questionnaire Number :

Interviewee :

Date :

Time :

House Number :

GPS Number :

District/Sub district/Village :

Family Information

Name	Sex (M/F)	Age in 2016	Last Education	Job / Source of Income	Position in Community

I. Flooding Disaster based on Community Perception and Community Decision Making

1. Human Process Recovery (Human Capital)
   a. Did flood threat your daily activities and your occupation ? (If you answer either yes / no, Please give detail description !)

      Yes : ....
      ....
      ....

      No : ....
      ....
      ....

   b. How is your perception about flood ?
      - No Problem : ....
      - Scary : ....
      - Hazardous/Disastrous : ....

   c. What do you think about the dike of river?
      - Avoiding you from Flood's Strike : ....
      - Very Helpful and Optimal : ....
      - Useless : ....
      - Others : ....

   d. How did you deal with the flood?
      - Asking assistance from:
        ✓ Family/ Relatives : ....
        ✓ Friends : ....
        ✓ Community Group if any : ....
        ✓ Community Leaders if any : ....
        ✓ Another Villages if any : ....
        ✓ Tycoons if any : ....
      - Church (Individual Assistance from Monks and Priests): ....
        ....
      - Other Charities: ....
        ....

2. Social Process Information (Social Capital)

a. Social Aspect

Social Network	Family	Neighborhood	Government	Other
Social Status	Homogeny	Heterogenic	Combination	Other
Social Relationship	Strong	Weak	Moderate	Other
Social Norms				
Social Cost				

b. Social Status

- Since did you stay in this area?

☐ 0-5 years ☐ 5-10 years ☐ 10-20 years ☐ > 20 years

- What is your social position in this area ?

☐ Villager/officer ☐ Head of Village ☐ Youth Organization Leader

☐ Organization Leader ☐ Religion Leader ☐ Village Elder

☐ Community Member

- Did you ever get financial assistance from your family living outside this area when Flood?

....

....

- How are the local people facing and overcome the situation when Flood swept their home (Measurement process of escaping the flood) ?

....

....

3. How is social strategies that has been applied by you (affected community) in facing flood event (Before, During and After Flooding) ?

....

....

4. Recognize the affected community's agricultural land ownership

Land Type	Value	Commodity	Harvest	Remark
Wet Rice/Paddy	ha			
Dry Land	ha			
Homestead	ha			

5. Are you aware that your area is a flood prone area?

....

....

6. Why do you still live in this flood prone area even though you knew it ?

....

....

7. What are you doing to face this condition ?

....

....

8. What are you going to do to reduce / minimize the effect of flood ?

....

....

9. Why do you decide to stay in this prone area even if you knew about it ? Please give the reason!

....

....

3. Physical Recovery (Physical Capital)

a. How do you deal with flooding in your agricultural land ?

....

....

c. What's your action to deal with flooding for your house's building ?

....

....

4. Economic Recovery (Economic Capital and Political Capital)

a. When was the flood event occurred in this area, did you have any saving to cope with ?

....

....

b. Did you use your saving for financing the repair as the impact of flooding ?

....

....

c. Did you ever sold your assets for financing the repair cost as the impact of flooding ?

....

....

d. Did you use all of the proceeds to finance the repair cost ?

....

....

e. Did you get any help / aid from your family / your colleagues / your neighbor / Charities Organization / Church for the repairing ?

....

....

f. Did you borrow any funds for the repair cost ?

....

....

g. Did you get any supporting fund from Government / Non- Government in the repairing ?

....

....

h. Source of Income:

Job / Occupation	Primary (P)	Secondary (S)	Remark
Agriculture (Own the Land)			
Agriculture Labor			
Entrepreneur			
Government Employment			
Private Employment			
Trader			
Other, Please Specify on Remark Box			

i. Properties Ownership

Properties	Value	Remark
Car		
Motorcycle		
Cycle		
Bicycle		
Boat		
Cow		
Sheep		
Goat		
Horse		
Dog		
Bird		
Buffalo		
Chicken		
Duck		
Furniture		
Electronic	TV Radio	Refrigerator
	Other	
Other		

j. Crop Calendar

Activity	Jan	Feb	Mar	Apr	May	June	July	Aug	Sep	Oct	Nov	Dec
Paddy Plant -Land Cultivation -Paddy Planting -Maintenance -Harvesting Palawija (Maize & groundnut, etc) -Land Cultivation -Paddy Planting -Maintenance -Harvesting												

4. Social Recovery and Social Mechanism (Social Capital)

1. How do communities do relate to this area that has flood threat ?

....

....

2. Do you have social network / aid (relative, government) to help when flood strikes ?

....

....

5. Cultural Recovery (Capital)

a. Is there any certain religious rituals or any thus alike that conducted whether by individual or elders when post-flood event sweeps and strikes your house, crops, and livestock? Please mention and give description !

....

....

b. What are social norms of this area that formed/built the unity of community ? Please mention and give description !

....

....

c. What are cultural activities that you think can strengthen social unity to be in a good corporation in handling a case / a threat situation (flood event) ? Please mention and give description !

....

....

d. How's big is the role of Church on this community (especially of giving financial assistance / aids, socialrecovery, and individual's empowerment ? Please mention the Church's role and give a description of it!

....

....

e. Could you describe the role both Church (Whether Church's Financial assistance or individual's financial assistance from Priest, Deacon, and Monk ?

....

....

f. How's the positive outlook or influence of your religion leaders and your community leaders in decision making for the adversity (i.e., Flooding) ?, Do their role or influence enhances your resilience to face theflooding ?

....

....

5. Psychology recovery Process

a. Do you still feel trauma with Flooding?

If yes, give your description ....

If no, give your description ....

b. How much time did you need to relieve the trauma?

....

....

c. Was there any assistance from Government / NGO to reduce your traumatic ?

....

....

If yes, mention the program and give description !

....

....

If not, give description !

....

....

d. How did you recover from your trauma ?

....

....

6. Institutional Respond (Social and Political Capitals)

a. Was there any action from the Government related to mitigation ?

....

....

If yes, mention the action ? Give Description

....

....

If no, give description :

....

....

b. Was there any rule related to disaster risk reduction from the local government that emerged after the occurrences of the flood event ? ....

If yes, mention the rule ? Give description !

....

....

If no, give description

....

....

c. What's your opinion about the Government's policy to handle flood case ?

....

....

d. Do you think there should be a new well-implemented policy to overcome the case of flooding in your area ?

....

....

7. Law (Legal) capital

a. Do you think the law products have been succeed in implementing them into a real action of minimizing and reducing the flood event in your area ?

....

....

b. Do you know one or some law products / regulation (s) ? Have these been satisfied / suitable to beimplemented ? Please mention if you know, then give description !

....

....

c. What's your opinion about a good management deal with watershed management of Benanain River? and What is the task or responsibility of Government to overcome the flood case ?

....

....

d. Regarding, one of causes of flooding, what's your opinion about the lack of land cover in upstream area? Is the source problem due to human intervention / human activities that has been led to flooding occurred in downstream area (your area) ?, Do these cases should take serious action by the Government for sustaining a Benanain Watershed management in these Districts; South Centre of Timor, North Centre of Timor, Belu and one neighbored-Country (Republic Democratia de Timor Leste) ? Please give description !

....

....

e. What do you think about the communication and coordination among three Districts and one neighbored-Country (Republic Democratia de Timor Leste) dealing with Flood's mitigation measures ?

....

....

II. Flood Impact for the Community Resilience

1. What kind of damage has happened to your house's building ? Please mention them and explain!

   ....

   ....

2. What kind of damage has happened to your agricultural land or communal agricultural land if you had it? Please give description !

   ....

   ....

3. What kind of damage has happened to your building's content and livestock?

   ....

   ....

4. What kind of efforts or your strategies that you applied to face and minimize the floodcase?

   ....

   ....

5. How do you (the community) do when they face flood event ?

   ....

   ....

6. How is your (the community) health condition after flood event ?

   ....

   ....

7. How could you be built your resiliency ?

   ....

   ....

8. Do your existing strategies within your capitals or assets on resilience could optimal be implemented in facing the flood event in this periodical's flooding ? How do you think ?

   ....

   ....

9. Mention the impact of flooding which resulted to public health was compromised? Please give meyour description !

   ....

   ....

10. How was your individual resilience, group resilience and community resilience working on, to withstand and adapt with flooding ? Are these helpful and take a long optimal action further for the next return flooding periodically ?

    ....

    ....

III. Other Factor

1. Vulnerable Phase Factor which applied by the community to face the Risk of Flooding

Activity	Before	During	After

2. Risk of Economic Matters

"":note: give mark on the table ! (Rp / Rupiah= Indonesian Currency)

Income Per Month :

< Rp. 200.000,- = < 15 Euro : (1) ....

Rp. 200.000, to Rp. 500.000,- =
15 Euro to 45 Euro : (2) ....

Rp. 500.000,- to Rp 750.000,-
= 45 Euro to 70 Euro : (3) ....

>Rp. 1.000.000,- = 85 Euro : (4) ....

3. Risk of Physical' vulnerability

Building Type	Settlement	Shop	Traditional Market, Church, School, Restaurant	Other
Floor Material	(1) Cement	(2) Wood	(3) Ground	Other
Building Material	(1) Cement	(2) Wood	(3) Bamboo	Other
Building ground level	(1) 1st floor	(2) 2nd floor	(3) 3rd floor	(4) More than 3rd floor if any
Rooftop material	(1) Tile	(2) Zinc	(3) Roofed	Other, concrete roof
Building Age	Year			
Spacious Yard	m^2			
Building Area	m^2			
The Length of the Road	m^2			

Thank you for your participation and corporation
God bless you

Appendix A2

Data Pertanyaan untuk masyarakat Malaka Barat (Kepala Keluarga) di Kecamatan Malaka Barat , Kabupaten Malaka, Provinsi Nusa Tenggara Timur, Indonesia

Nomor Pertanyaan :

Kepala Keluarga yang diwawancarai :

Tanggal :

Waktu :

Nomor Rumah :

Nomor GPS :

Dusun/Desa :

Informasi Keluarga

Nama	Jenis Kelamin (Pria/Wanita)	Usia Pada 2016	Pendidikan Terakhir	Pekerjaan (Sumber Pendapatan)	Posisi di masyarakat

I. Bencana Banjir berdasarkan Persepsi Masyarakat dan Pengambilan Keputusan Masyarakat

1. Proses Pemulihan Modal Manusia (Modal Manusia)

a. Apakah banjir mengganggu aktivitas sehari-hari anda dan pekerjaan anda? (Jika anda menjawab Ya / Tidak, Tolong berikan penjelasan yang detail!

Ya: ....
....
....

Tidak: ....
....
....

b. Bagimana persepsi anda terhadap Banjir ?
- Bukan Sebuah Masalah : ....
- Hal yang menakutkan : ....
- Bahaya/Bencana : ....

c. Apa pendapt anda tentang Tanggul Banjir?
- Menghindari anda dari bahaya Banjir : ....
- Sangat Membantu dan Sudah Optimal : ....
- Tidak berguna : ....
- Lain-Lain : ....

d. Bagaimana anda menghadapi Banjir?
- Meminta Bantuan dari:
  - ✓ Keluarga/ Relasi terdekat : ....
  - ✓ Teman-Teman : ....
  - ✓ Kelompok Komunitas Jika ada : ....
  - ✓ Pemimpin Komunitas Jika ada : ....
  - ✓ Desa-Desa Lainnya Jika ada : ....
  - ✓ Dermawan/Dermawati Jika ada : ....
- Gereja (Bantuan Perorangan/Individu dari Biarawan dan Biarawati):
  ....
  ....
- Bantuan Karitas lainnya jika ada:
  ....
  ....

2. Proses Informasi Modal Sosial (Modal Sosial)

a. Aspek Sosial

Jaringan Sosial	Keluarga	Rumpun Tetangga	Pemerintah	Lainnya
Status Sosial	Homogenik	Heterogenik	Combination	Lainnya
Hubungan Sosial	Kuat	Lemah	Sedang	Lainnya
Norma Sosial				
Dana Sosial				

b. Status Sosial

- Sejak kapan anda menetap di Wilayah ini (berapa lama nya)?

  ☐ 0-5 tahun ☐ 5-10 tahun ☐ 10-20 tahun ☐ > 20 tahun

- Apakah posisi sosial anda di wilayah ini ?

  ☐ Penduduk/Pekeraj Kantoran ☐ Kepala Desa ☐ Pemimpin Organisasi Pemuda-Pemudi

  ☐ Pemimpin Organisasi Desa/Petuah Desa ☐ Pemimpin Agama ☐ Pemimpin Adat

  ☐ Anggota Masyarakat

- Apakah anda pernah mendapat bantuan keuangan dari keluarga anda yang menetap di luar Desa Bencana ini?

  ....

  ....

- Bagaimana masyarakat lokal menghadapi dan menangani situasi bahaya banjir dan ketika banjir terjadi menyapu rumah masyarakat local (Proses Pengukuran dari Penanganan Banjir) ?

  ....

  ....

- Bagaimana strategi sosial yang anda sudah laksanakan khususnya dan Masyarakat terdampak bencana pada umumnya dalam menghadapi banjir (Sebelum, Sedang dan Sesudah Banjir)?

  ....

....

- Mengapa anda masih menetap di area bencana banjir ini walaupun anda mengetahuinya ?

....

....

- Apa yang anda lakukan dalam menghadapi kondisi ini?

....

....

- Apa yang anda segera lakukan untuk menurunkan/meminimalisirkan efek dari banjir?

....

....

- Apakah anda pernah meminimalisir efek banjir dengan strategi baru anda sendiri dan atau dengan bantuan tetangga, kelompok masyarakat atau organisasi tertentu? Apakah strategi anda pernah berhasil atau justru gagal?

....

....

3. Mengenal Kepemilikan Lahan Pertanian dari Masyarakat Terdampak pada modal alam (Modal Alam)

Tipe Lahan	Luas	Komoditas	Panen	Keterangan
Tanah Basah/Beras	ha			
Tanah Kering	ha			
Penginapan	ha			

- Apa sajakah modal alam yang dapat digunakan untuk meningkatkan daya pulih atau ketahanan anda dala menghadapi bencana banjir?

....

....

- Jika anda sadar bahwa Desa anda ialah rawan bencana banjir, dan anda tetap menetap di Desa anda, bagaimana modal alam yang anda miliki dan upayakan mampu mengatasi dan beradaptasi dengan persoalan najir?

....

....

- Apakah anda memiliki strategi baru dalam mendayagunakan modal alam yang anda miliki?

4. Pemulihan Fisik (Modal Fisik)

a. Bagaimana anda mengatasi banjir pada peralatan pertanian, kegiatan pertanian, irigasi dan lahan pertanian anda?

....

....

b.Apa sajakah tindakan pencegahan anda dalam menghadapi banjir yang mengancam bangunan rumah anda?

....

....

5. Pemulihan Ekonomi dan Kebijakannya (Modal Ekonomi dan Modal Politik)

a. Kapan kejadian banjir terjadi di area ini, apakah anda mempunyai tabungan untuk mengatasinya?

....

....

b. Apakah anda menggunakan tabungan anda untuk menanggulangi perbaikan rumah, perabot rumah, Ternak, harta, barang-barang anda sebagai akibat dari dampak banjir?

....

....

c. Apakah anda pernah menjual asset anda seperti barang-barang dan harta anda untuk dana penanggulangan sebagai akibat dari dampak banjir ?

....

....

d. Apakah anda sudah menggunakan seluruh proses jalan keluar untuk dana penanggulangan banjir ?

....

....

e. Apakah anda mendapat bantuan dari keluarga anda / kolega anda/ tetangga anda/Organisasi Karitas /Gereja dalam reparasi banjir?

....

....

f. Apakah anda meminjam dana untuk dana penanggulangan banjir?

....

....

g. Apakah anda mendapat dukungan dana dari Pemerintah/Non-Pemerintah dalam reparasi banjir?

....

....

h. Sumber Penghasilan

Pekerjaan	Primer (P)	Sekunder (S)	Keterangan
Pertanian (Pemilik Lahan)			
Pekerja Pertanian (Petani)			
Wirausahawan/wati			
Pekerja Pemerintah (PNS)			
Pekerjaan Swasta			
Pedagang			
Lainnya, Tolong Spesifikan (dituliskan pada Kotak Keterangan jika ada)			

i. Kepemilikan Properti

Properti	Nilai/Harga	Keterangan
Mobil		
Motor		
Sepeda		
Traktor		
Perahu		
Sapi		
Domba		
Kambing		
Kuda		
Anjing		
Burung		
Babi		
Ayam		
Bebek		
Perabot Rumah Tangga		
Elektronik	TV Radio Kulkas Lainnya	
Lainnya		

J. Kalender Musiman

Aktifitas	Jan	Feb	Mar	Apr	Mei	Juni	Juli	Aug	Sep	Oct	Nov	Des
1. Tanaman Padi: – Lahan Ladang – Penanaman Padi – Proses Penyiraman dll – Hasil 2. Palawija (Jagung & Kacang, dll) – Lahan Ladang – Penanaman Padi – Proses Penyiraman dll – Hasil												

6. Pemulihan Sosial and Mekanisme Sosial (Modal Sosial)

1. Bagaimana menurut anda Tindakan Masyarakat terkait terhadap wilayah bencana yang terancam banjir setiap saat ?

....

....

2. Apakah anda memiliki jaringan sosial / bantuan (relasi dan Pemerintah) untuk membantu anda ketika banjir merusaki aset / barang anda?

....

....

7. Pemulihan Budaya dan Agama (Modal Budaya dan Modal Agama)

a. Apakah ada ritual agama tertentu atau sejenis yang pernah dilakukan baik dari individu atau petuah adat ketika bencana banjir telah terjadi dan merusaki rumah anda, hasil panen anda dan ternak anda? Tolong sebutkan dan beri penjelasan!

....

....

b. Apa saja norma-norma sosial di area ini yang membentuk/membangun kesatuan dan kekuatan dari masyarakat ? Tolong sebutkan dan beri penejelasan!

....

....

c. Apa saja aktifitas-aktifitas budaya yang menurut anda dapat memperkuat kesatuan masyarakat menjadi sebuah koorporasi yang baik dalam menangani sebuah masalah / sebuah situasi ancaman (kejadian Banjir)? Tolong sebutkan dan beri penejelasan!

....

....

d. Bagaimana besar nya peranan dari Gereja pada masyarakat (khususnya dalam memberikan bantuan keuangan, pemulihan sosial dan pemberdayaan individu? Tolong sebutkan dan beri penejelasan!

....

....

e. Dapatkah anda jelaskan peranan Gereja (antara Bantuan Keuangan Gereja atau bantuan keuangan dan bantuan bantuan lain secara pribadi dari Biarawan dan Biarawati, Jika ada?

....

....

f. Bagaimana gambaran positif atau pengaruh positif dari pemimpin agama anda dan pemimpin masyarakat anda yang diberikan dalam pengambilan keputusan untuk sebuah masalah (dalam hal ini Bencana Banjir)? Apakah peranan mereka atau pengaruh mereka meningkatkan ketahanan / daya pulih untuk menghadapi banjir?

....

....

8. Proses Pemulihan Psikologi

a. Apakah anda masih merasa trauma dengan Banjir?

Jika Ya, tolong berikan penjelasan ....

Jika Tidak, tolong berikan penjelasan ....

b. Berapa kali atau lamanya anda membutuhkan waktu untuk mengatasi trauma anda?

....

....

c. Apakah pernah ada beberapa bantuan dari Pemerintah Daerat setempat / Non-Pemerintah seperti NGO, dll untuk memulihkan atau meminimalisir trauma anda?

....

....

Jika Ya, tolong berikan penjelasan

....

....

Jika Tidak, tolong berikan penjelasan

....

....

d. Bagaimana tindakan anda yang dilakukan untuk memulihkam trauma anda?

....

....

6. Respon dari Institusi (Modal Sosial and Politik)

a. Apakah ada langkah dan tindakan mitigasi dari Pemerintah Daerah Setempat?

....

....

Jika Ya, tolong berikan penjelasan

....

....

Jika Tidak, tolong berikan penjelasan

....

....

b. Apakah ada beberapa aturan yang ditetapkan terkait pengurangan resiko bencana dari Pemerintah Daerah setelah banjir terjadi ? ....

Jika Ya, tolong berikan penjelasan

....

....

Jika Tidak, tolong berikan penjelasan

....

....

c. Apakah pendapat anda tentang kebijakan Pemerintah untuk menangani masalah banjir?

....

....

d. Apakah menurut anda seharusnya ada sebuah implementasi kebijakan yang baru untuk menyelesaikan masalah banjir di desa anda?

....

....

9. Modal Hukum

a. Apakah menurut anda produk-produk hukum atau aturan-aturan hukum yang and ketahui sudah sukses di dalam terapannya dan tindakannya yang nyata untuk pengurangan resiko bencana di wilayah anda?

....

....

b. Apakah anda mengetahui salah satu atau beberapa produk-produk hukum atau aturan-aturan hukum? Apakah produk-produk hukum atau aturan-aturan hukum ini memuaskan anda di dalam penerapan nya? Bagaimana pendapat anda?

....

....

c. Apakah sajakah pendapat anda tentang sebuah penanganan yang baik dalam mengatasi kerusakan Daerah Aliran Sungai (DAS) Sungai Benanain dari hulu ke hilir? Apakah anda mengetahui, sudahkah Tanggungjawab dari Pemerintahan menyelsaikan persoalan ini dalam pola penanganan masalah banjir?...

....

d. Mengenai, salah satu penyebab Banjir, Apa pendapat anda tentang kurangnya penutup lahan atau hutan di wilayah hilir Daerah Aliran Sungai (DAS) Sungai Benanain? Apakah sumber permasalahan tersebut dikarenakan oleh perilaku manusia/aktifitas manusia yang mengakibatkan banjir terjadi di wilayah hilir (wilayah yang anda tempati) dari Daerah Aliran Sungai (DAS) Sungai Benanain? Bagaimana pendapat anda, apakah seharusnya Pemerintah segera mengambil tindakan tegas, serius dan nyata untuk keberlanjutan lingkungan dan ekosistem dari Daerah Aliran Sungai (DAS) Sungai Benanain dan bagaimana situasi dan masyarakat di perbatasan dari Timor Leste terkait penanganan banjir ? Tolong berikan penjelasan!

....

....

e. Apa pendapat dan penilaian anda tentang komunikasi dan koordinasi dalam langkah dan tahapan mitigasi bencana banjir pada Daerah Aliran Sungai (DAS) Sungai Benanain antara 4 distrik (Timor Tengah Selatan/TTS, Timor Tengah Utara/TTU, Belu dan Malaka) dan satu negara tentangga kita, Timor Leste?

....

....

10. Dampak Banjir bagi Daya Pulih atau Ketahanan Masyarakat

1. Apa sajakah jenis-jenis kerusakan yang terjadi pada bangunan rumah anda? Tolong sebutkan kerusakan-kerusakan tersebut dan tolong jelaskan!

   ....

   ....

2. Apa sajakah jenis-jenis kerusakan yang terjadi pada lahan pertanian anda baik lahan pertanian pribadi atau lahan pertanian bersama masyarakat setempat lainnya didalam kelompok tani, jika ada ? Tolong berikan penjelasan!

   ....

   ....

3. Apa sajakah jenis-jenis kerusakan yang terjadi pada is dalam rumah anda dan ternak anda?

   ....

   ....

4. Apa sajakah jenis-jenis usaha dan strategi yang sudah anda laksanakan untuk menghadapi dan mengurangi resiko bencana banjir?

   ....

   ....

5. Apa yang anda dan masyarakat setempat lakukan ketika banjir mengancam anda dan rumah anda beserta seluruh harta anda, apakah ada tindakan gotong royong (bersama-sama saling membantu) dalam penanganan banjir?

   ....

   ....

6. Bagaimana menurut anda, kondisi kesehatan anda dan kondisi kesehatan Masyarakat setelah kejadian banjir? Bagaimana daya pulih kesehatan ada dan masyarakat setempat yang anda ketahui?

   ....

   ....

7. Bagaimana anda mampu membangun system daya pulih atau ketahanan anda dalam menghadapi banjir ?

   ....

   ....

8. Apakah strategi daya pulih atau ketahanan anda yang ada sekarang dengan modal-modal atau asset-aset dalam daya pulih anda mampu diterapkan dengan optimal dalam menghadapi kejadian banjir pada periode atau waktu banjir yang terjadi pada tahun ini atau tahun sebelumnya ? Bagaimana pendapat anda ?

   ....

   ....

9. Sebutkan dampak dari Banjir yang menyebabkan kesehatan masyarakat menjadi terganggu? Tolong berikan penjelasan !

   ....

   ....

10. Bagaimana ketahanan atau daya pulih anda secara individu, kelompok dan ketahanan masyarakat pada umumnya, untuk bertahan dan beradaptasi dengan banjir ? Apakah ketahanan atau daya pulih anda membantu anda mampu mengambil langkah inisiatif atau langkah lanjutan untuk mempersiapkan diri anda menghadapi ancaman banjir yang akan kembali terjadi di periode berikutnya?

    ....

    ....

II. Faktor Lainnya

1. Tahapan Faktor Kerentanan yang dilakukan olehMasyarakat untuk menghadapi Resiko dari Banjir.

Aktifitas	**Sebelum**	**Sedang**	**Sesudah**

2. Resiko Ekonomi

":Keterangan: Tolong berikan jawaban ya atau tidak pada kolum berikut!
(Rp
/ Rupiah=Mata Uang Indonesia)
Penghasilan Setiap Bulan :

Tipe Bangunan	Perumahan	Toko	Pasar Tradisional, Pasar Swalayan, Gereja, Sekolah, Restoran	Lainnya
Bahan Lantai	(1) Semen	(2) Kayu	(3) Fondasi	Lainnya
Bahan Bangunan	(1) Semen	(2) Kayu	(3) Bambu	Lainnya
Tingkatan Fondasi Bangunan	(1) Lt. 1	(2) Lt. 2	(3) Lt. 3	(4) Lebih dari Lt.3
Bahan Atap	(1) Genten g	(2) Seng	(3) Daun	Lainnya, Atap Konkrit
Usia Bangunan	Tahun			
Luas Teras	m^2			
Lokasi Bangunan	m^2			
Panjang dari Jalan Umum	m^2			

Lebih dari < Rp. 200.000,- = < 15 Euro : (1)................................
Pas Rp. 200.000, to Rp. 500.000,- =
15 Euro to 45 Euro : (2) ....
Pas Rp. 500.000,- to Rp 750.000,-
= 45 Euro to 70 Euro : (3) ....
Kurang dari >Rp. 1.000.000,- = 85 Euro : (4)................................

3. Kerentanan dari Resiko Fisik

Trimakasih atas partisipasi dan koorporasi anda
Tuhan Senantiasa Melindungi dan Memberkati

Appendix B: Data from Chapter V

Graph: Age of respondents (Households) in 14 villages of the West Malaka Subdistrict of Malaka District in the West Timor, East Nusa Tenggara Province, Indonesia

1. Besikama Village (Fig. B.1)

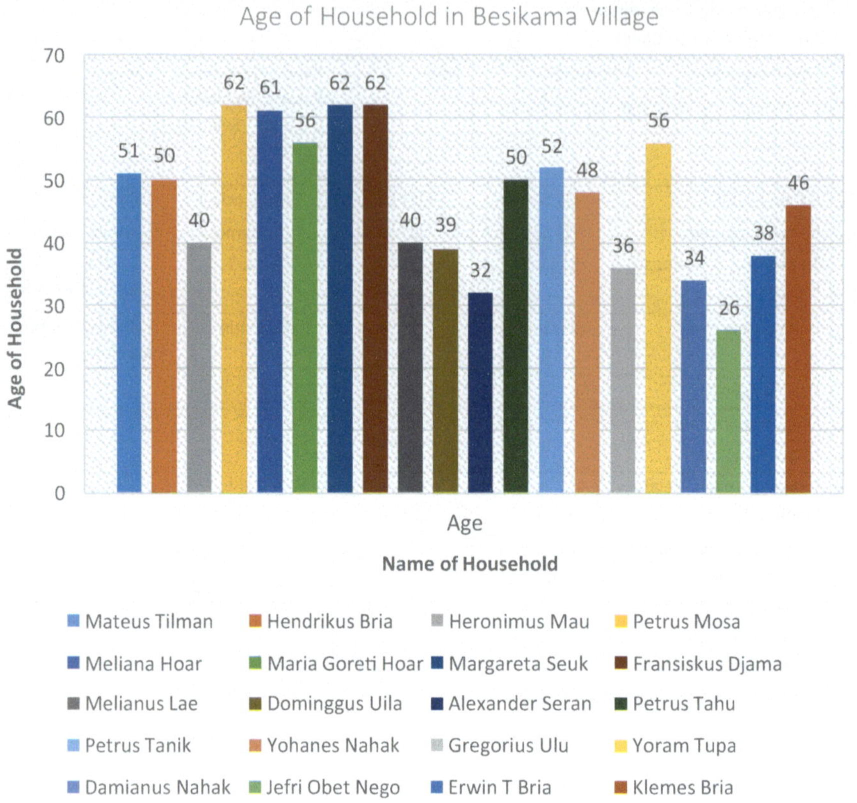

Fig. B.1 Graphic of age of respondents in Besikama Village. *Source* Author's Research Fieldwork, July 9, 2016, to February 9, 2017

2. Lasaen Village (Fig. B.2)

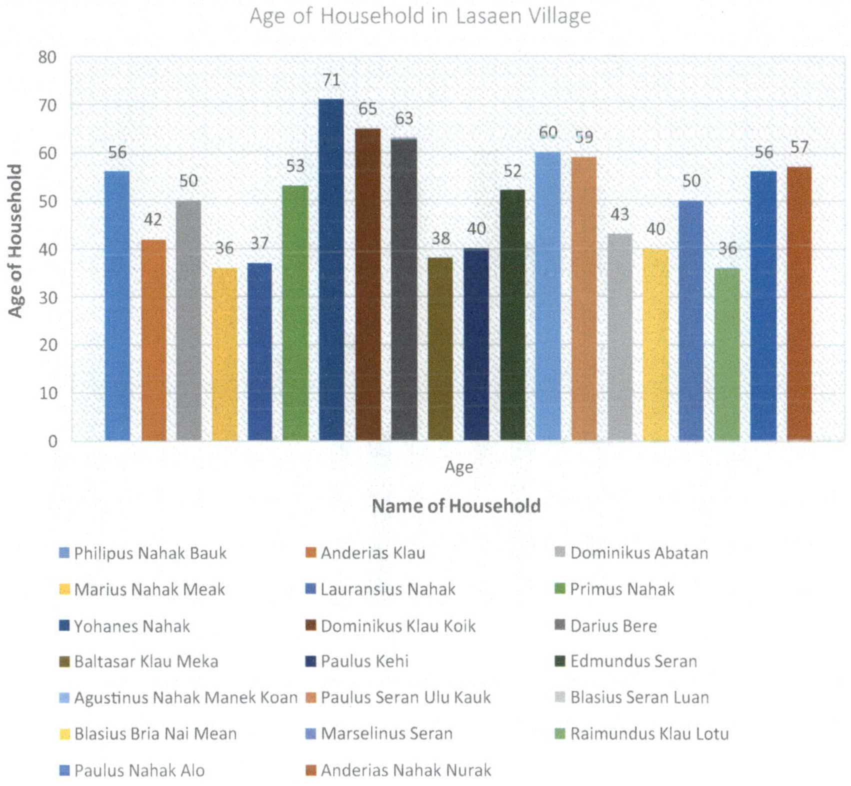

Fig. B.2 Graphic of age of respondents in Lasaen Village. *Source* Author's Research Fieldwork, July 9, 2016, to February 9, 2017

3. Umatoos Village (Fig. B.3)

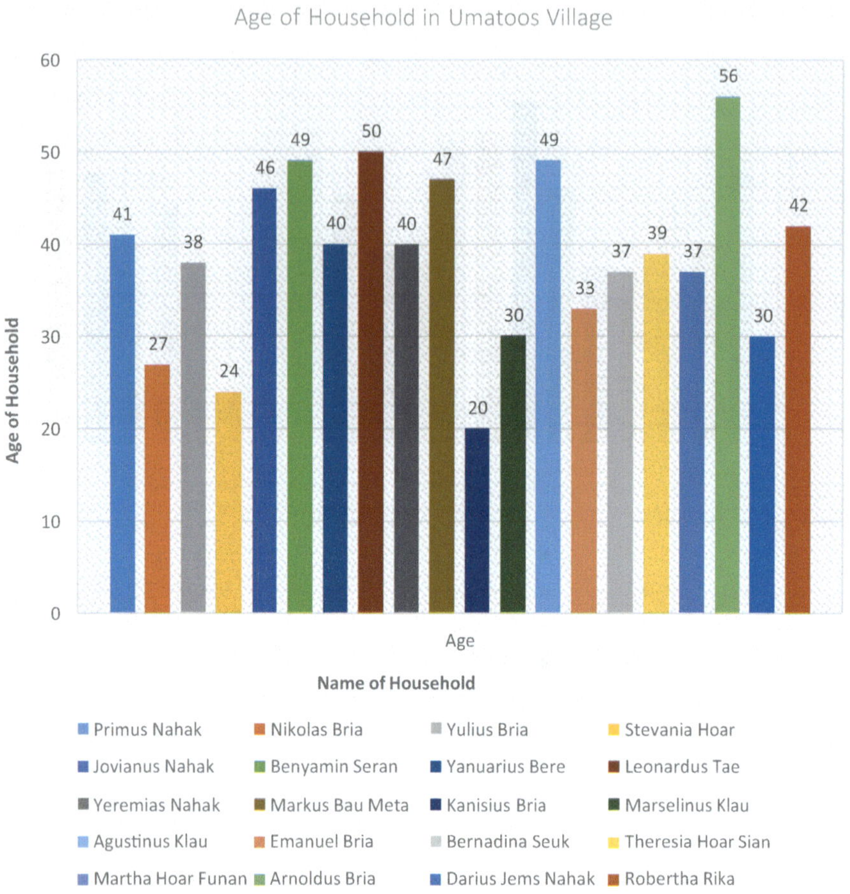

Fig. B.3 Graphic of age of respondents in Umatoos Village. *Source* Author's Research Fieldwork, July 9, 2016, to February 9, 2017

4. Fafoe Village (Fig. B.4)

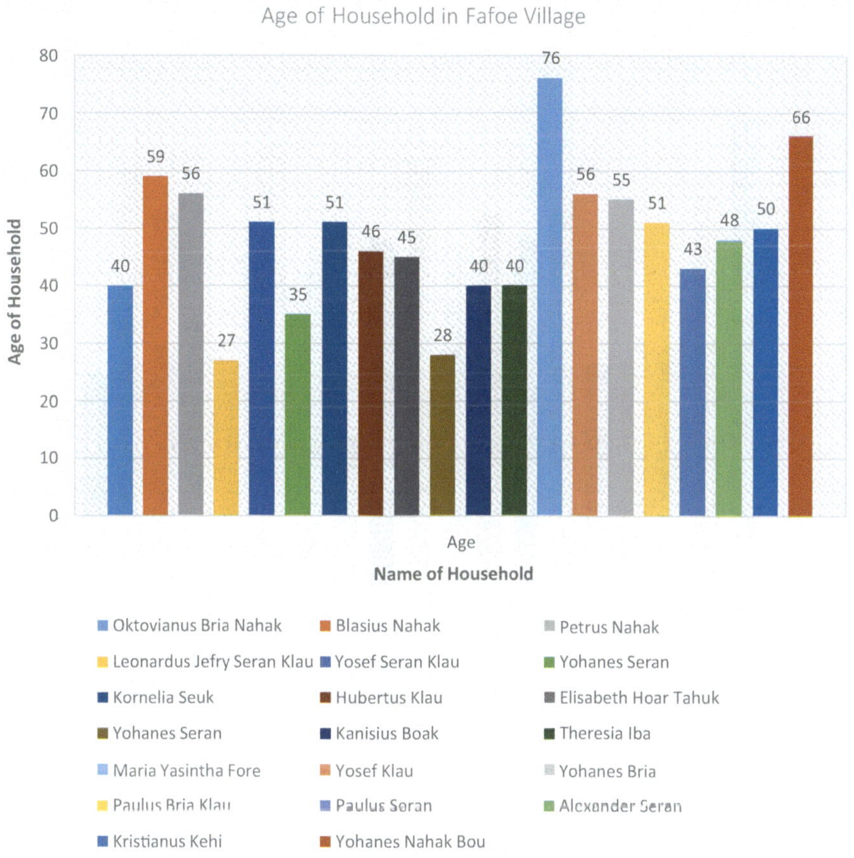

Fig. B.4 Graphic of age of respondents in Fafoe Village. *Source* Author's Research Fieldwork, July 9, 2016, to February 9, 2017

5. Raimataus Village (Fig. B.5)

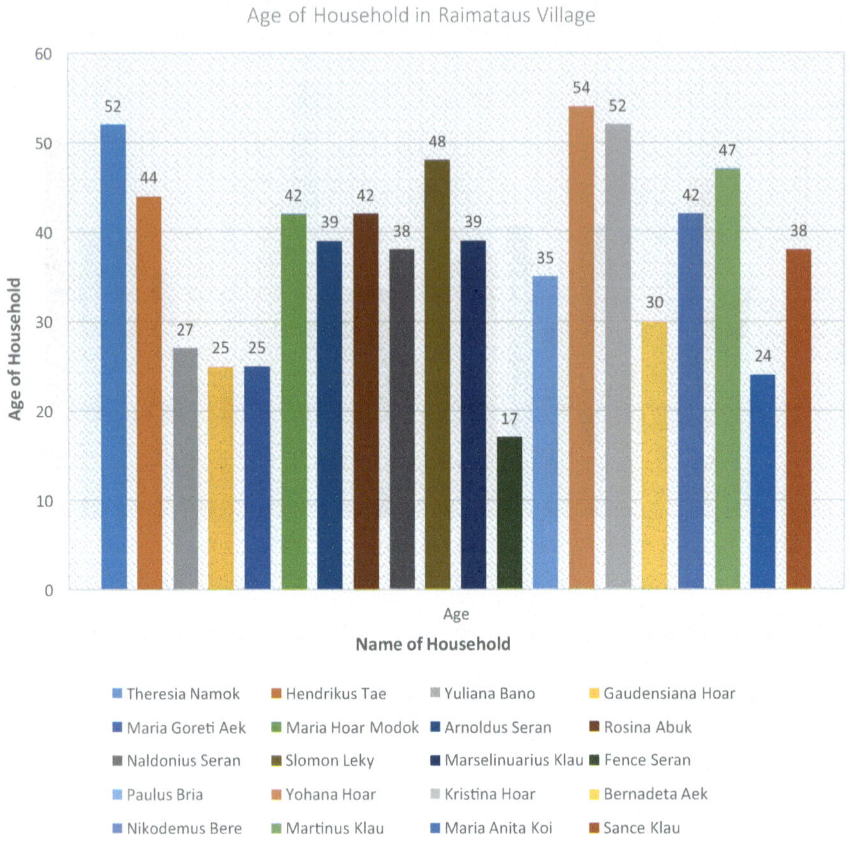

Fig. B.5 Graphic of age of respondents in Raimataus Village. *Source* Author's Research Fieldwork, July 9, 2016, to February 9, 2017

6. Naas Village (Fig. B.6)

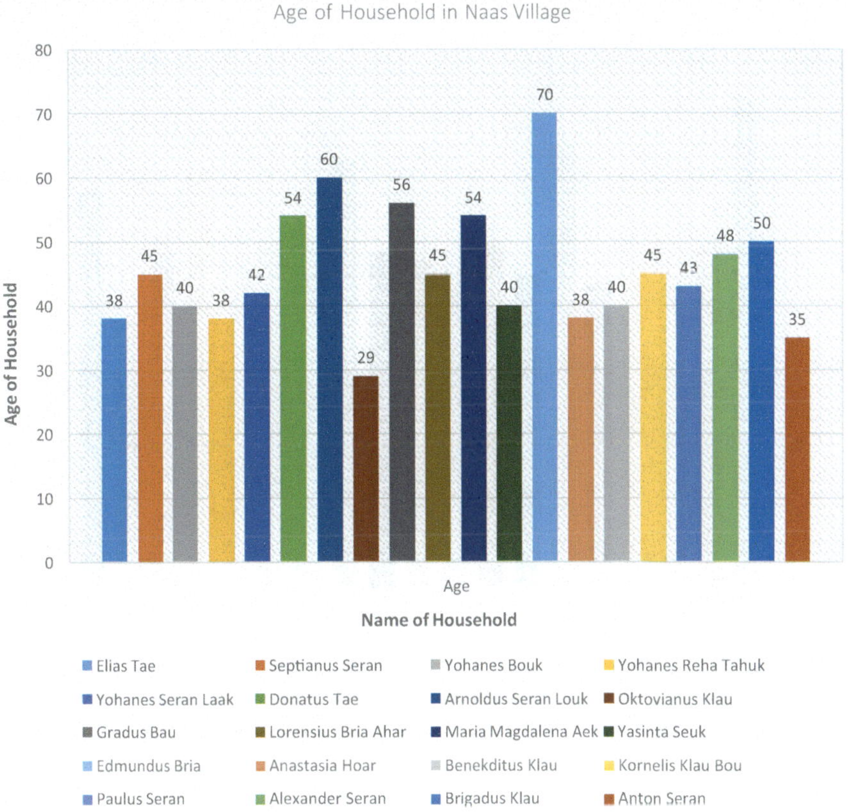

Fig. B.6 Graphic of age of respondents in Naas Village. *Source* Author's Research Fieldwork, July 9, 2016, to February 9, 2017

7. Lofoun Village (Fig. B.7)

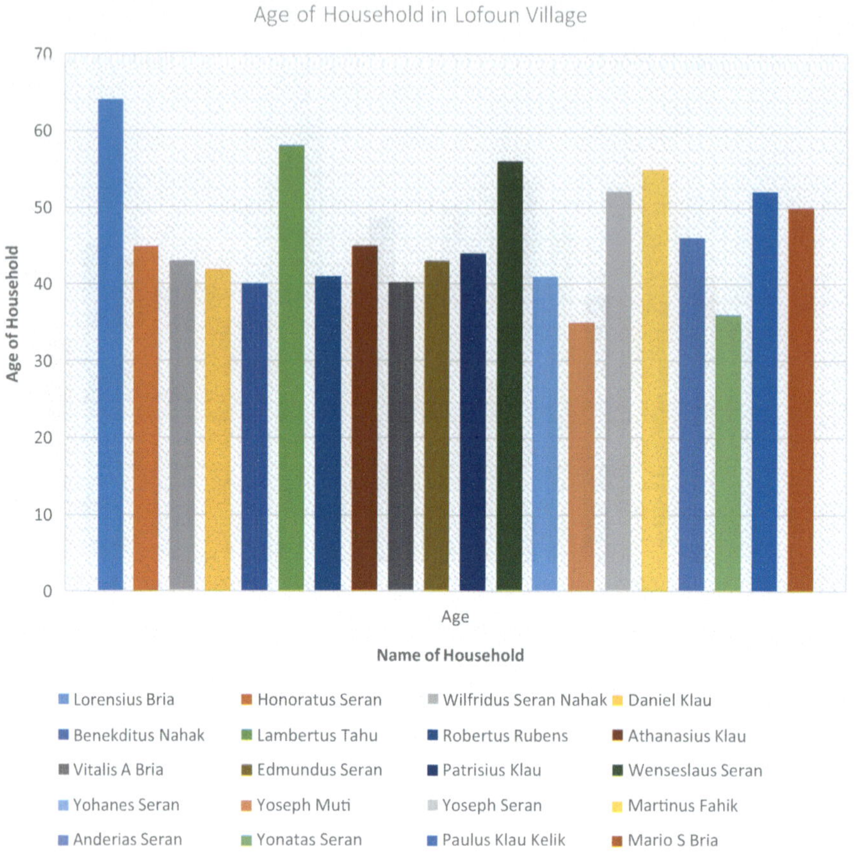

Fig. B.7 Graphic of age of respondents in Lofoun Village. *Source* Author's Research Fieldwork, July 9, 2016, to February 9, 2017

8. Maktihan Village (Fig. B.8)

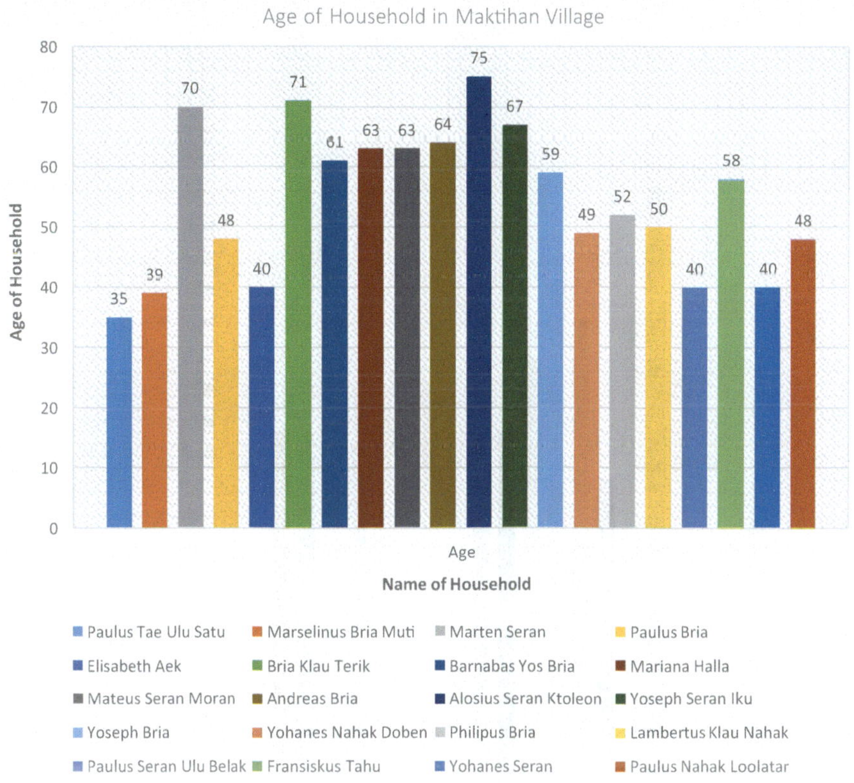

Fig. B.8 Graphic of age of respondents in Maktihan Village. *Source* Author's Research Fieldwork, July 9, 2016, to February 9, 2017

9. Motaain Village (Fig. B.9)

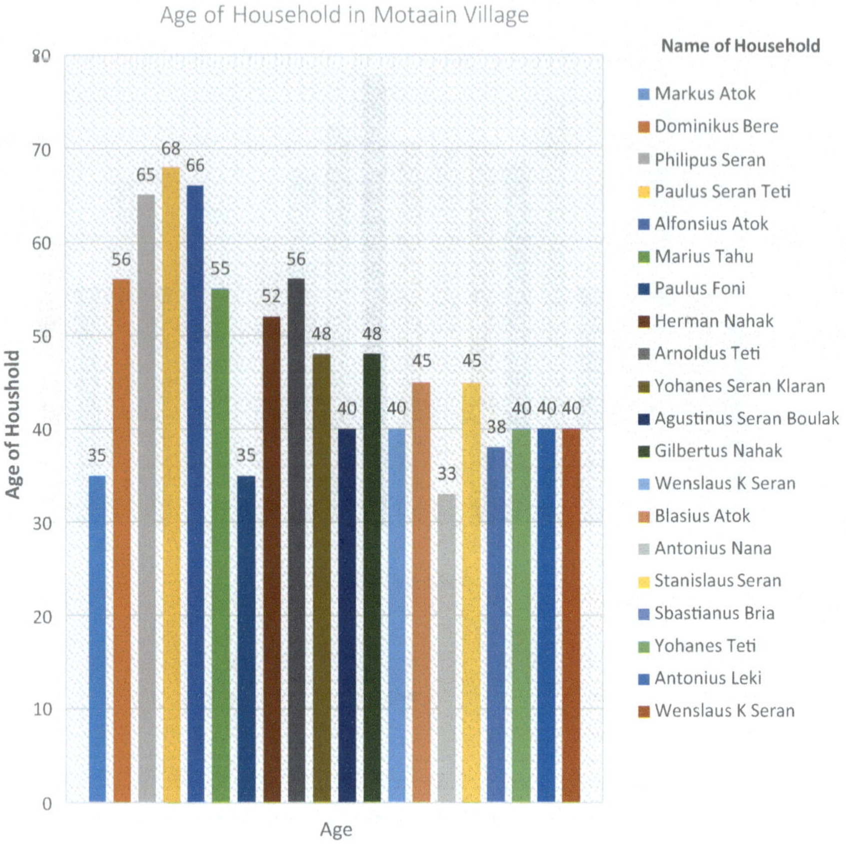

Fig. B.9 Graphic of age of respondents in Motaain Village. *Source* Author's Research Fieldwork, July 9, 2016, to February 9, 2017

10. Motaulun Village (Fig. B.10)

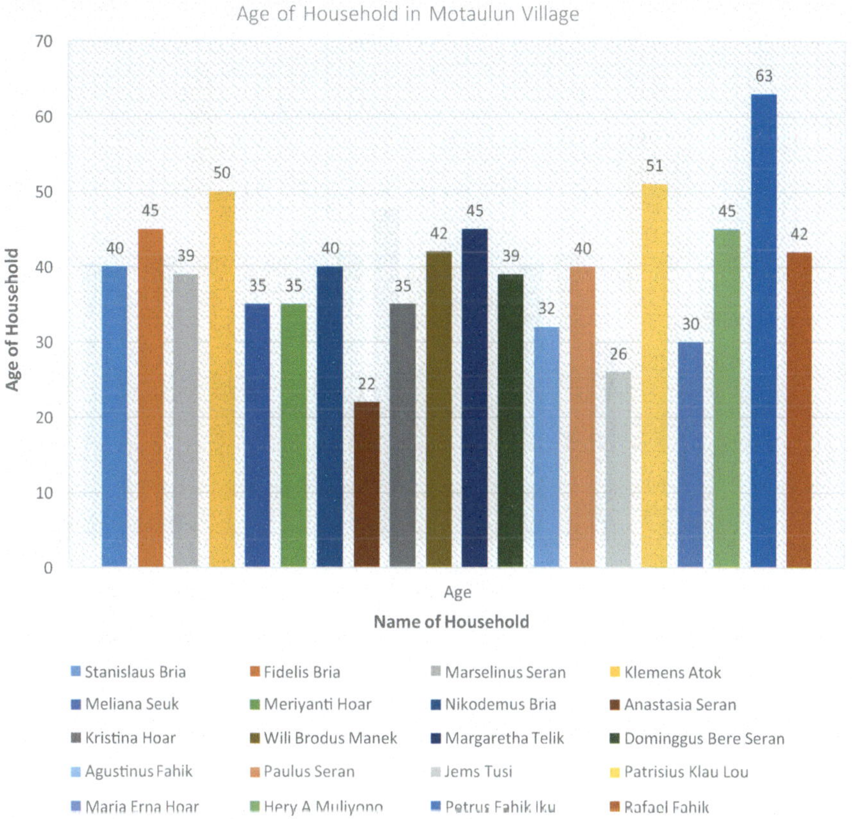

Fig. B.10 Graphic of age of respondents in Motaulun Village. *Source* Author's Research Fieldwork, July 9, 2016, to February 9, 2017

11. Oanmane Village (Fig. B.11)

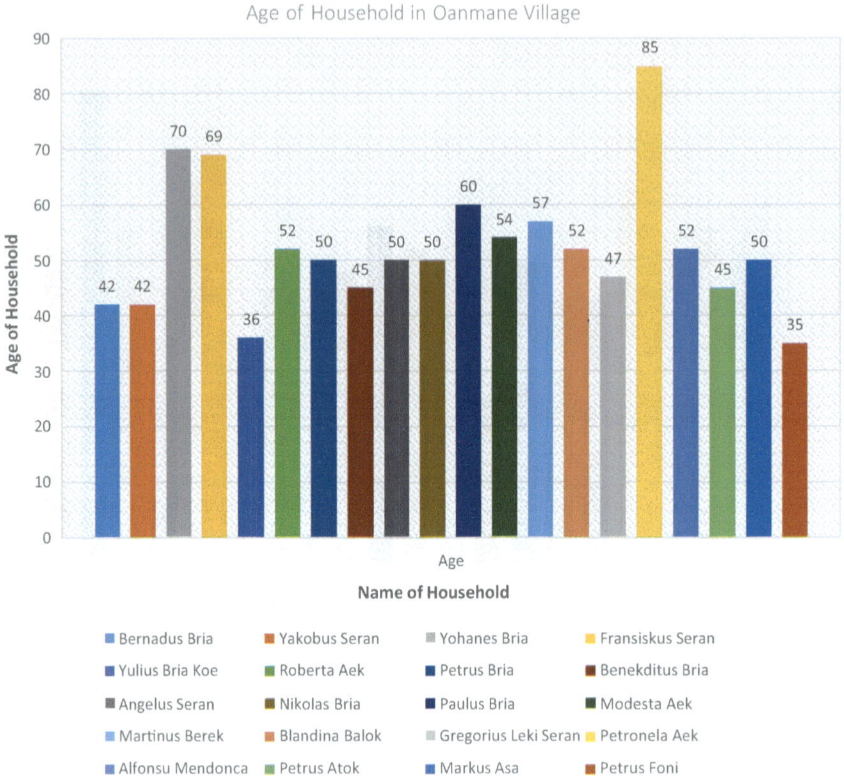

Fig. B.11 Graphic of age of respondents in Oanmane Village. *Source* Author's Research Fieldwork, July 9, 2016, to February 9, 2017

12. Rabasa Village (Fig. B.12)

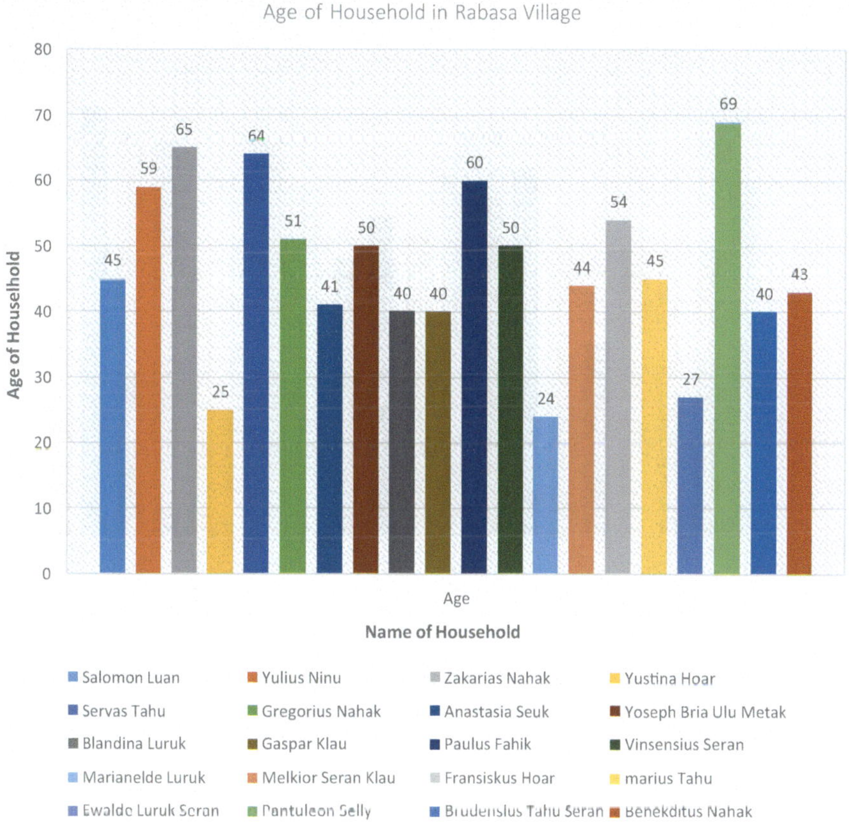

Fig. B.12 Graphic of age of respondents in Rabasa Village. *Source* Author's Research Fieldwork, July 9, 2016, to February 9, 2017

13. Rabasahain Village (Fig. B.13)

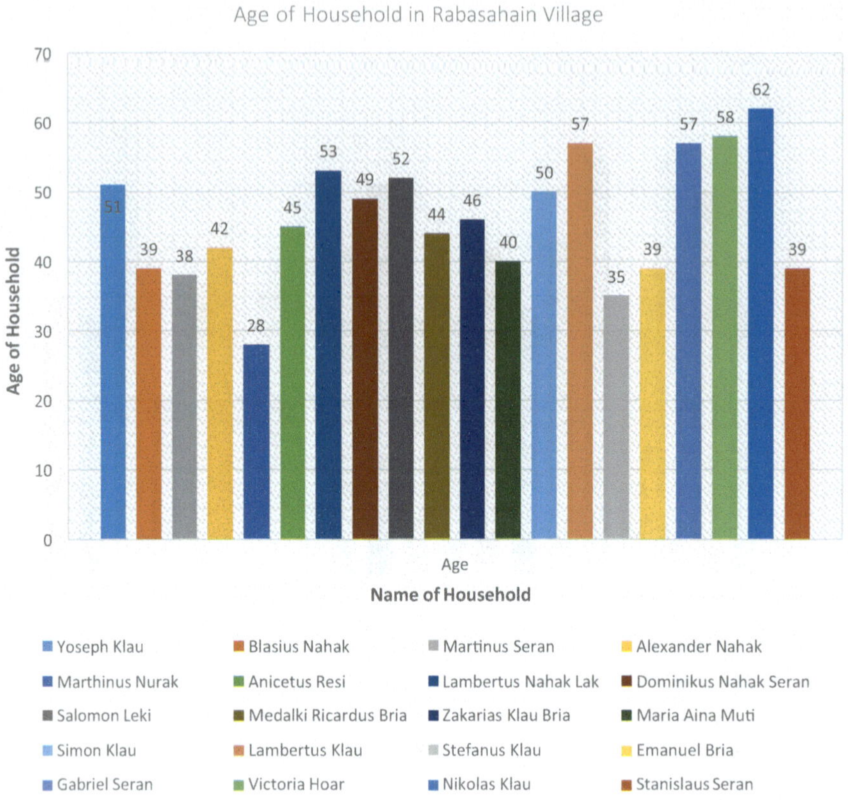

Fig. B.13 Graphic of age of respondents in Rabasahain Village. *Source* Author's Research Fieldwork, July 9, 2016, to February 9, 2017

14. Rabasahaerain Village (Fig. B.14)

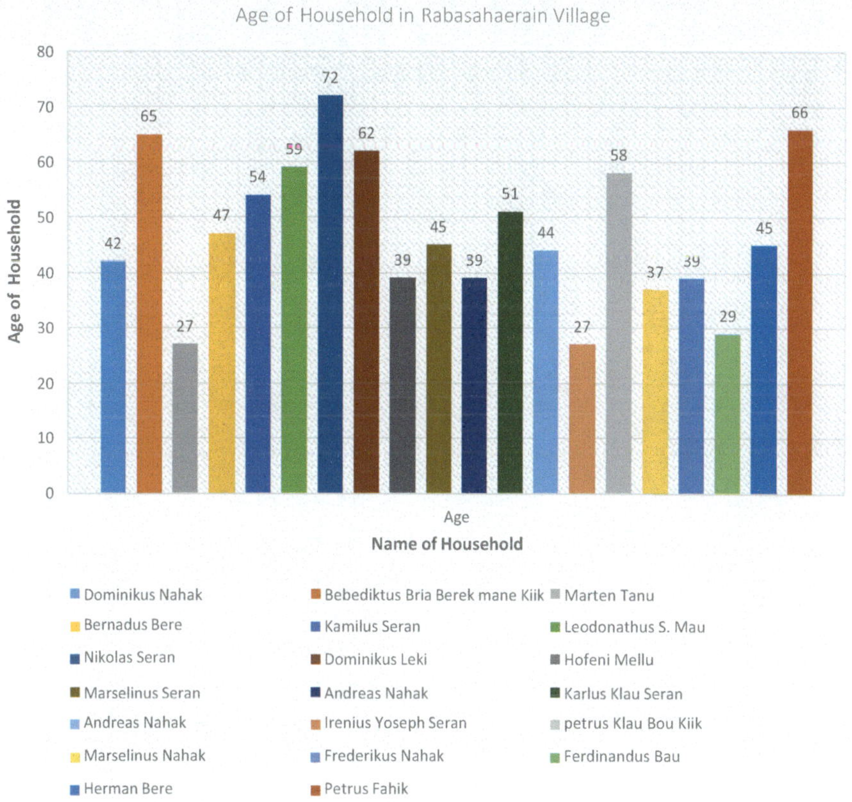

Fig. B.14 Graphic of age of respondents in Rabasahaerain Village. *Source* Author's Research Fieldwork, July 9, 2016, to February 9, 2017

Appendix C: Data from Chapter VII

Distribution map of respondents (Households) in 13 villages of the West Malaka Subdistrict of Malaka District in the West Timor, East Nusa Tenggara Province, Indonesia

1. Umatoos Village (Fig. C.1)

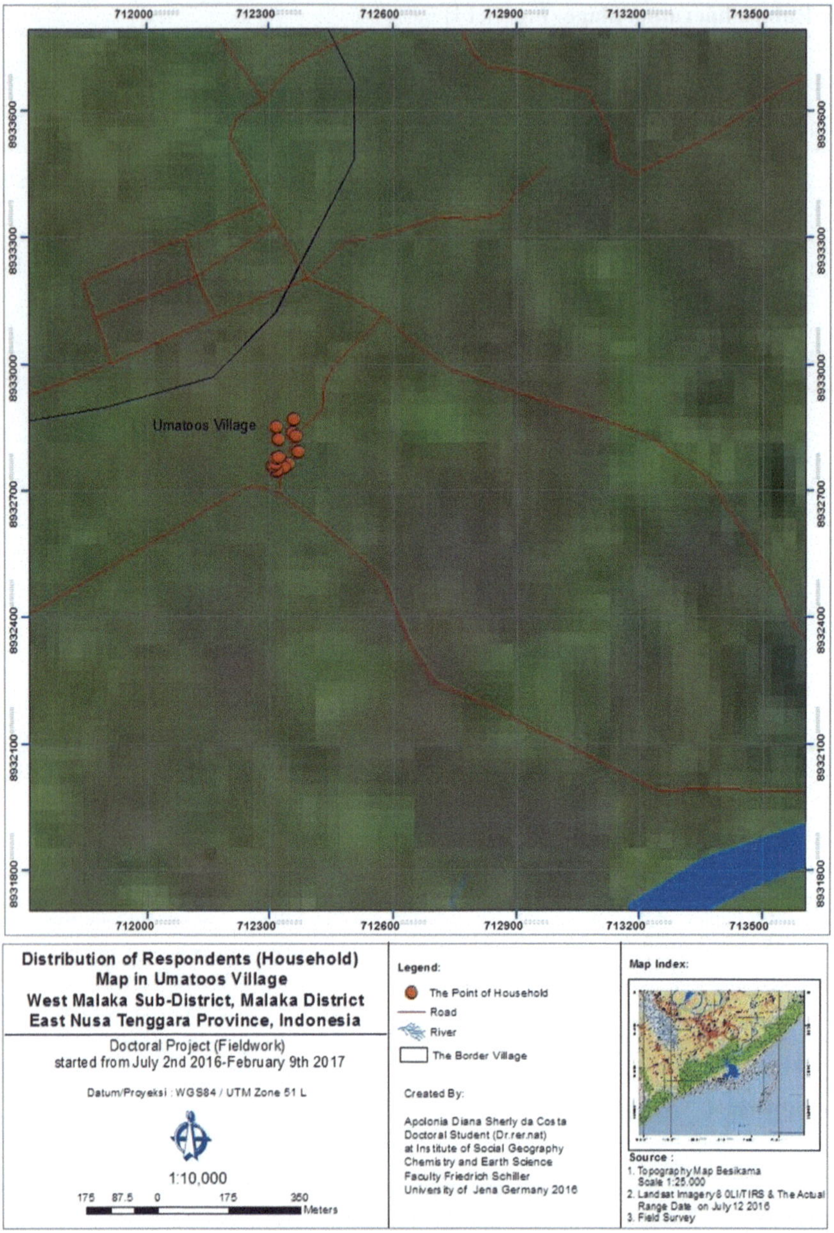

Fig. C.1 Distribution of respondents (Household) map in Umatoos Village, West Malaka Subdistrict, Malaka District, East Nusa Tenggara Province, Indonesia. *Source* Author's Research Fieldwork, July 9, 2016, to February 9, 2017

2. Fafoe Village (Fig. C.2)

3. Raimataus Village (Fig. C.3)

4. Motaulun Village (Fig. C.4)

5. Motaain Village (Fig. C.5)

6. Maktihan Village (Fig. C.6)

7. Rabasa Village (Fig. C.7)

8. Rabasahain Village (Fig. C.8)

9. Rabasahaerain Village (Fig. C.9)

10. Oanmane Village (Fig. C.10)

11. Lofoun Village (Fig. C.11)

12. Naas Village (Fig. C.12)

13. Sikun Village (Fig. C.13)

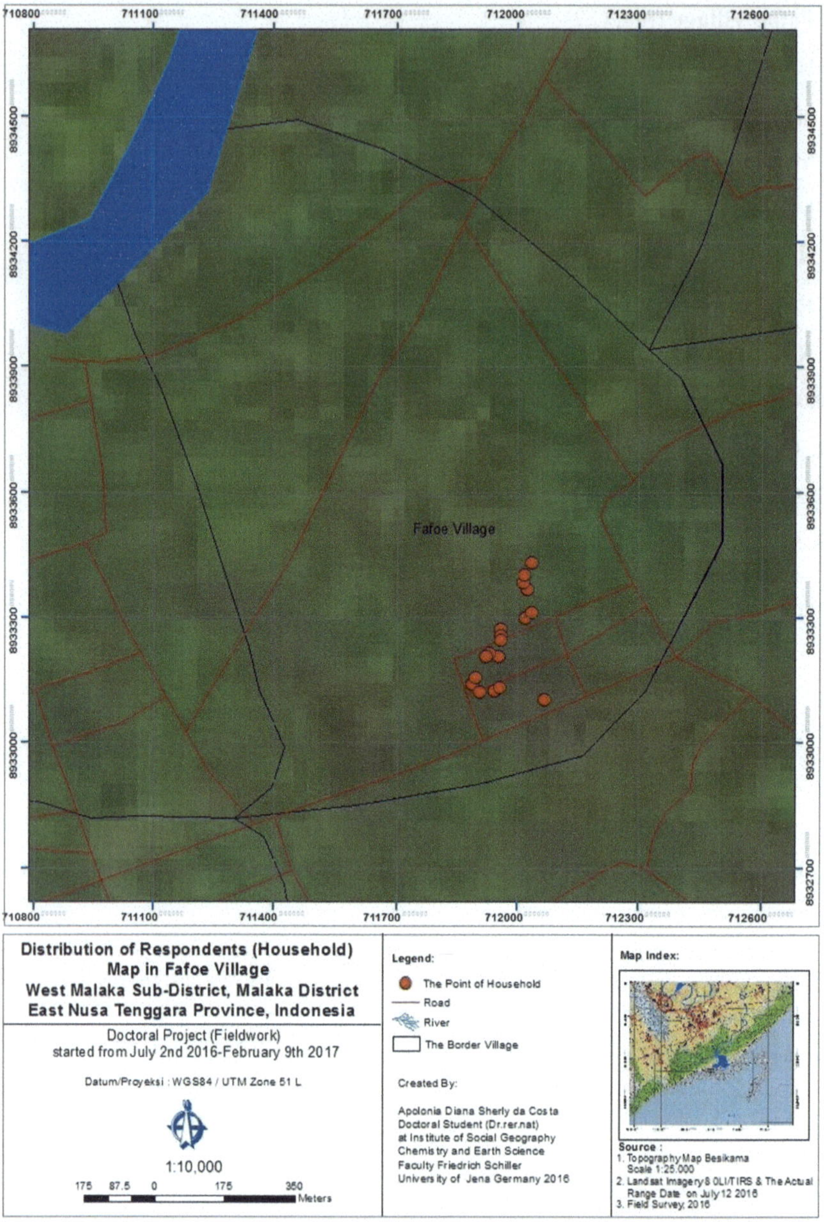

Fig. C.2 Distribution of respondents (Household) map in Fafoe Village, West Malaka Subdistrict, Malaka District, East Nusa Tenggara Province, Indonesia. *Source* Author's Research Fieldwork, July 9, 2016, to February 9, 2017

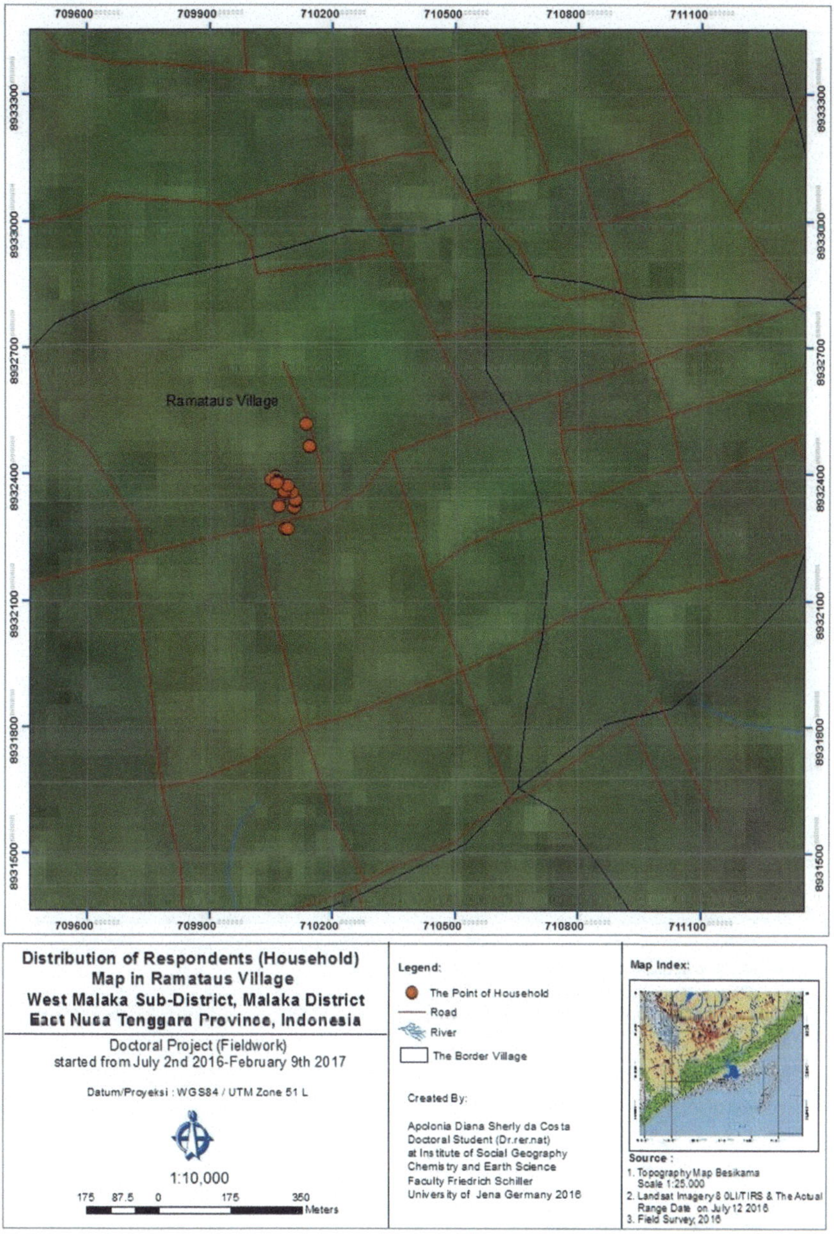

Fig. C.3 Distribution of respondents (Household) map in Raimataus Village, West Malaka Subdistrict, Malaka District, East Nusa Tenggara Province, Indonesia. *Source* Author's Research Fieldwork, July 9, 2016, to February 9, 2017

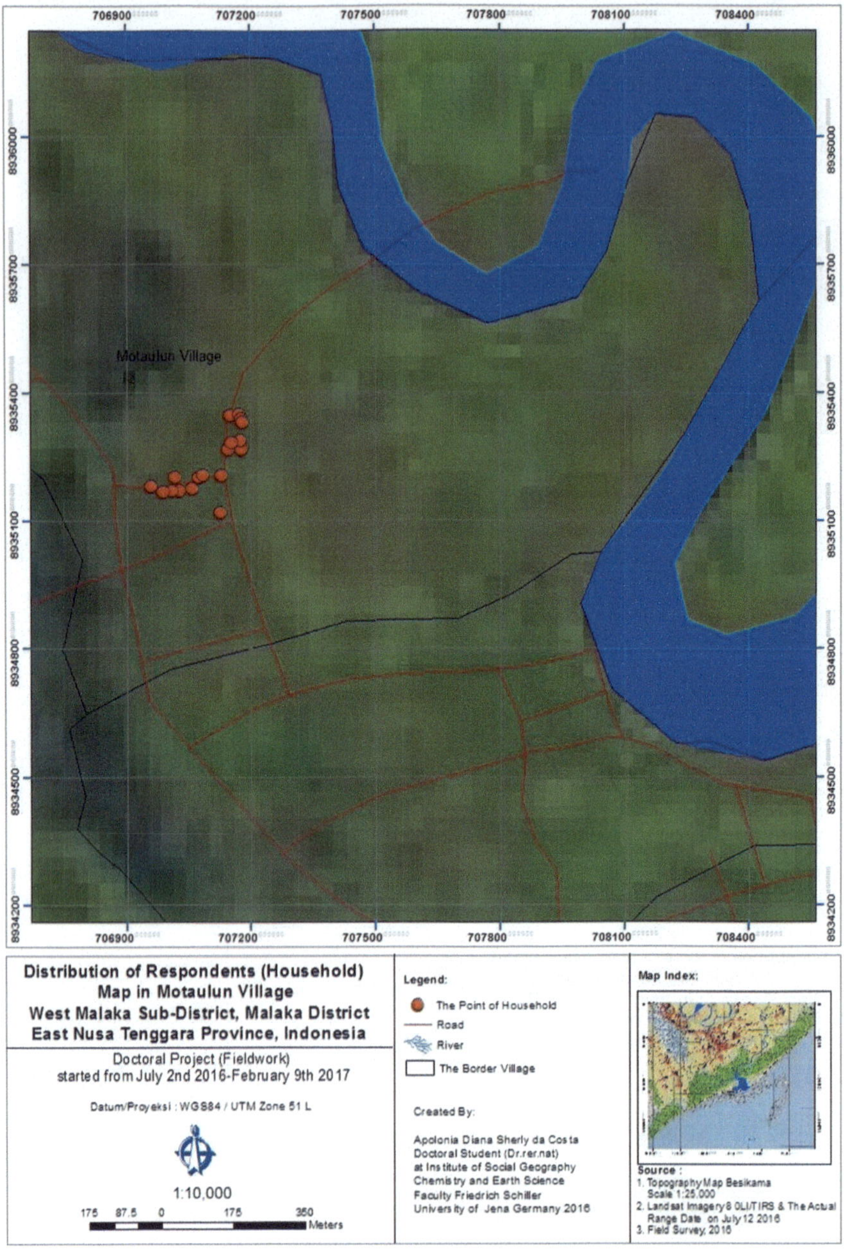

Fig. C.4 Distribution of respondents (Household) map in Motaulun Village, West Malaka Subdistrict, Malaka District, East Nusa Tenggara Province, Indonesia. *Source* Author's Research Fieldwork, July 9, 2016, to February 9, 2017

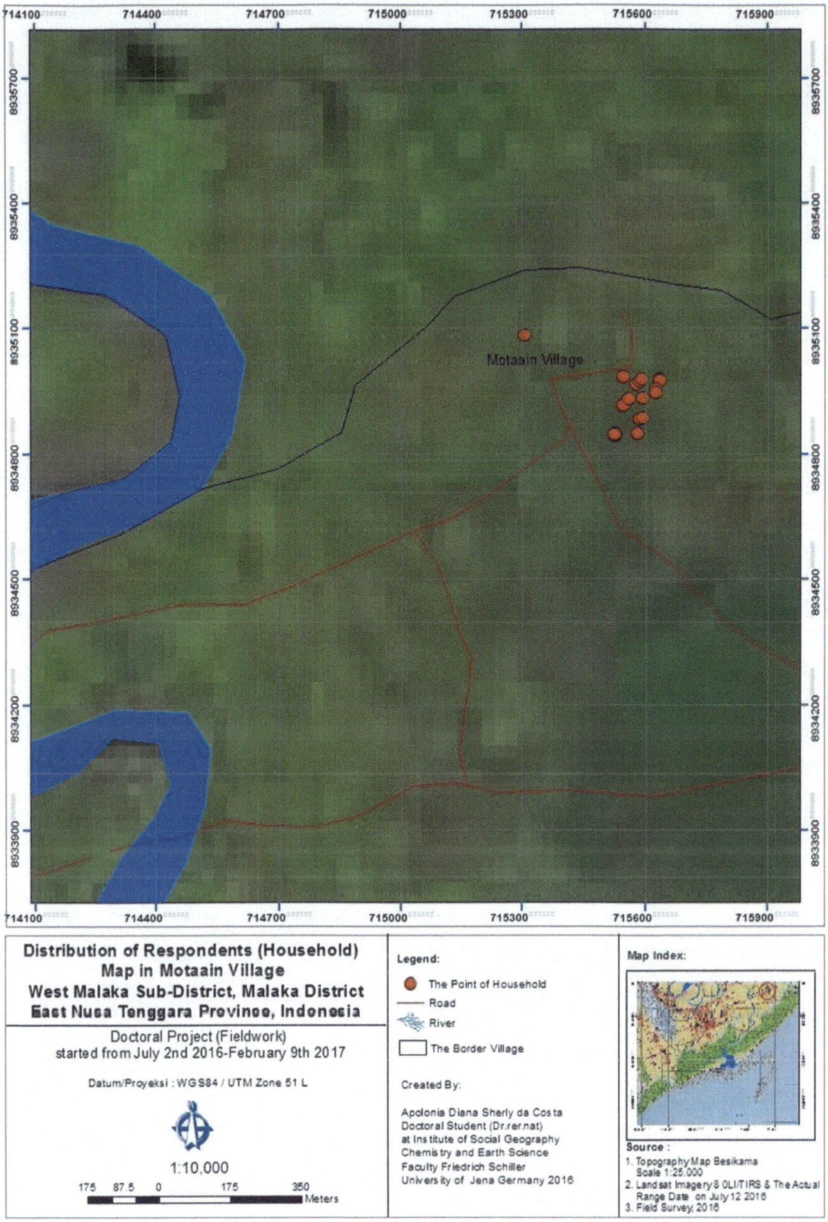

Fig. C.5 Distribution of respondents (Household) map in Motaain Village, West Malaka Subdistrict, Malaka District, East Nusa Tenggara Province, Indonesia. *Source* Author's Research Fieldwork, July 9, 2016, to February 9, 2017

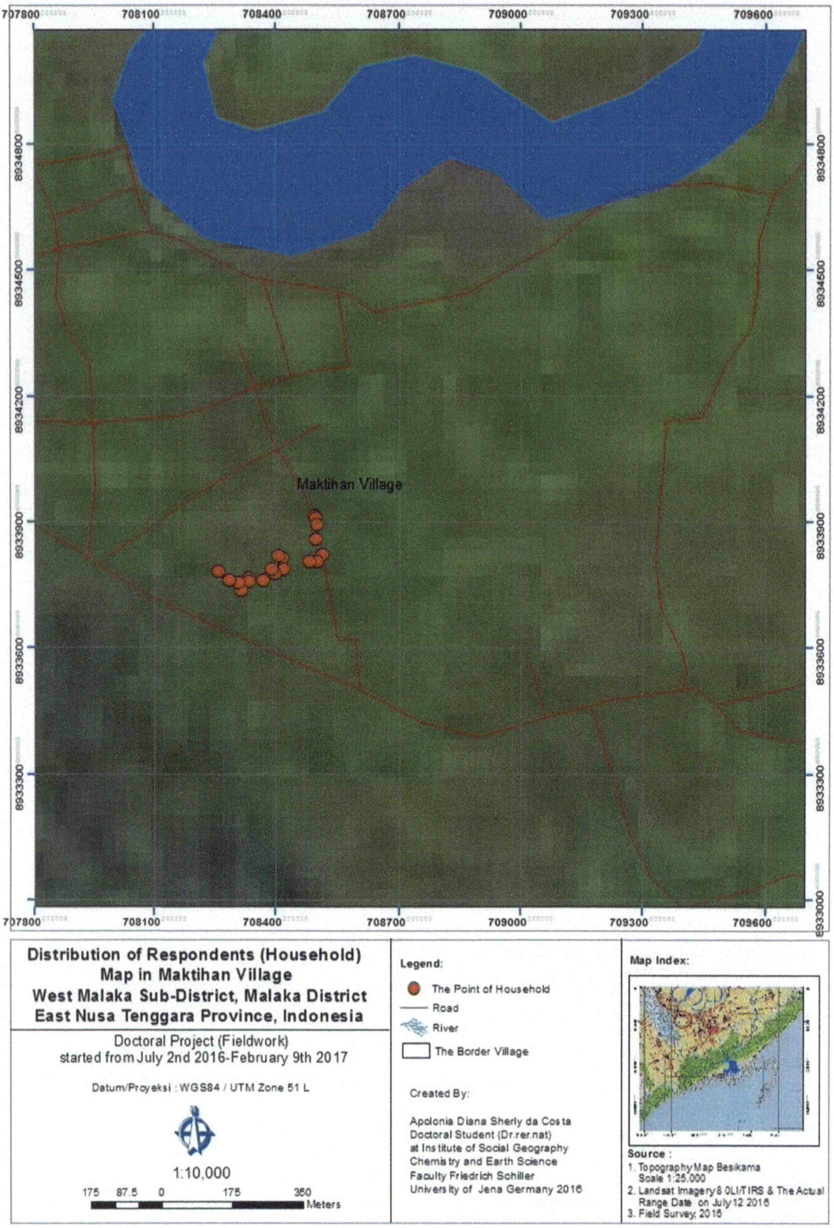

Fig. C.6 Distribution of respondents (Household) map in Maktihan Village, West Malaka Subdistrict, Malaka District, East Nusa Tenggara Province, Indonesia. *Source* Author's Research Fieldwork, July 9, 2016, to February 9, 2017

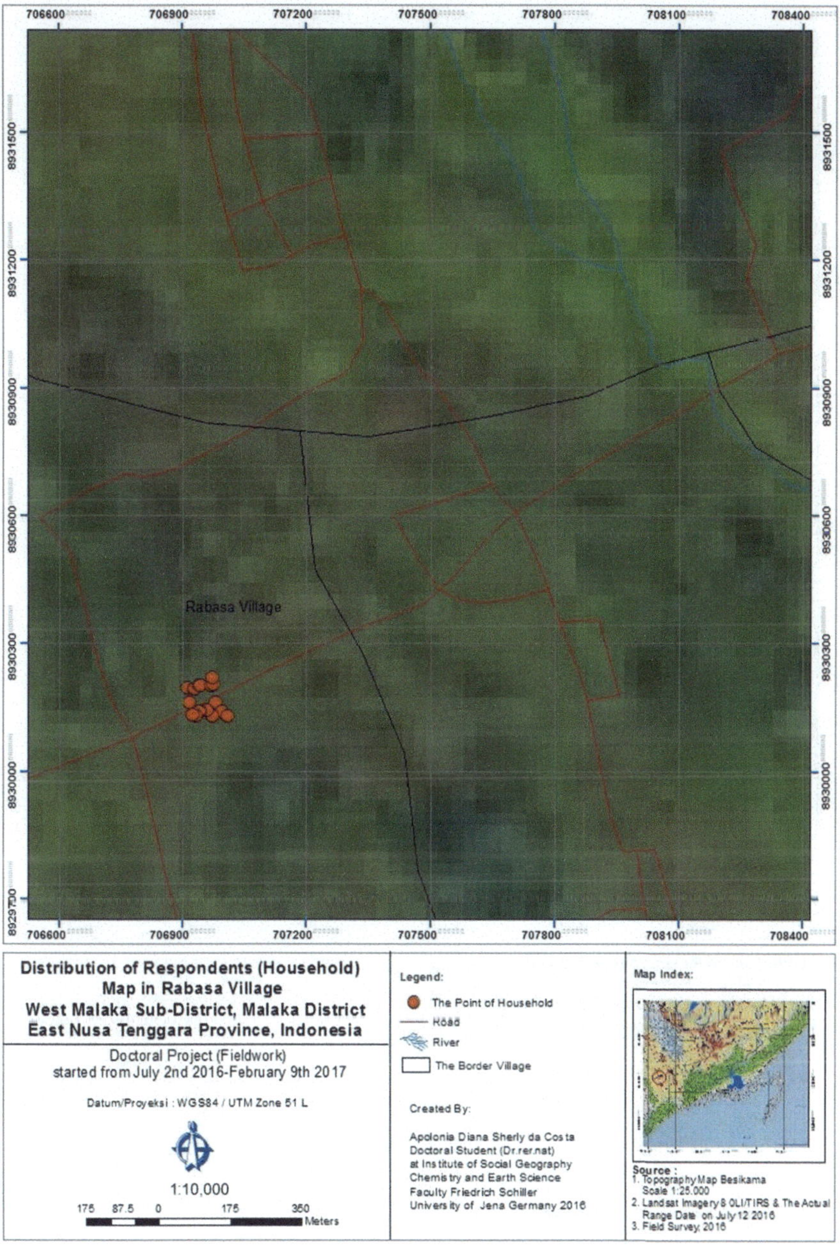

Fig. C.7 Distribution of respondents (Household) map in Rabasa Village, West Malaka Subdistrict, Malaka District, East Nusa Tenggara Province, Indonesia. *Source* Author's Research Fieldwork, July 9, 2016, to February 9, 2017

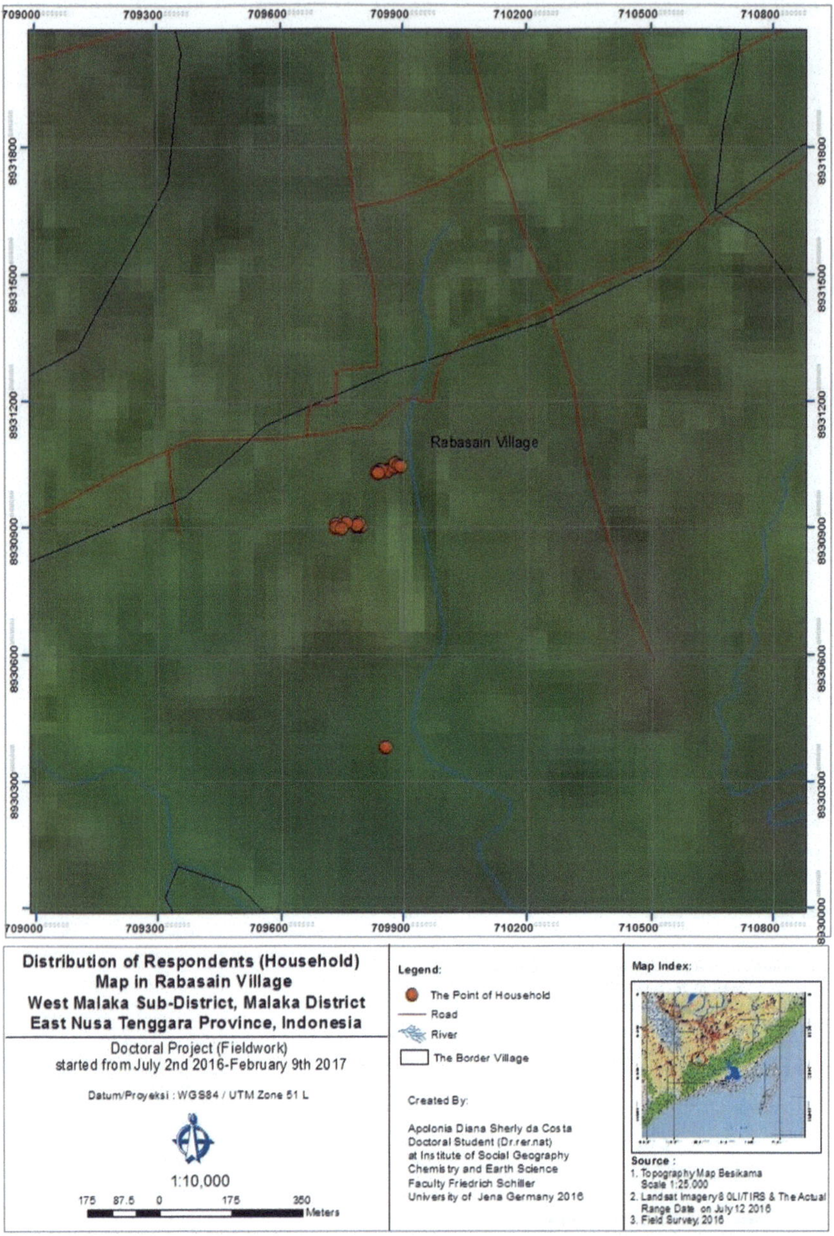

Fig. C.8 Distribution of respondents (Household) map in Rabasahain Village, West Malaka Subdistrict, Malaka District, East Nusa Tenggara Province, Indonesia. *Source* Author's Research Fieldwork, July 9, 2016, to February 9, 2017

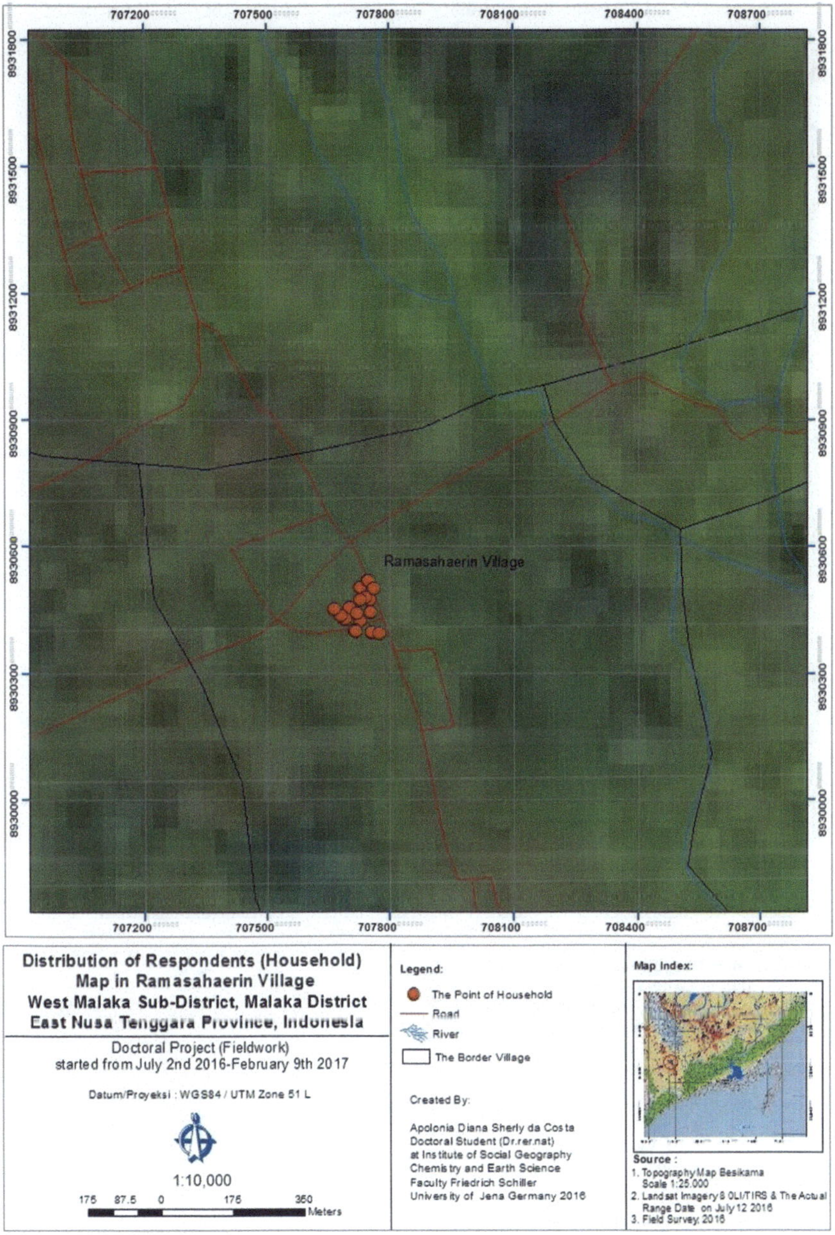

Fig. C.9 Distribution of respondents (Household) map in Rabasahaerain Village, West Malaka Subdistrict, Malaka District, East Nusa Tenggara Province, Indonesia. *Source* Author's Research Fieldwork, July 9, 2016, to February 9, 2017

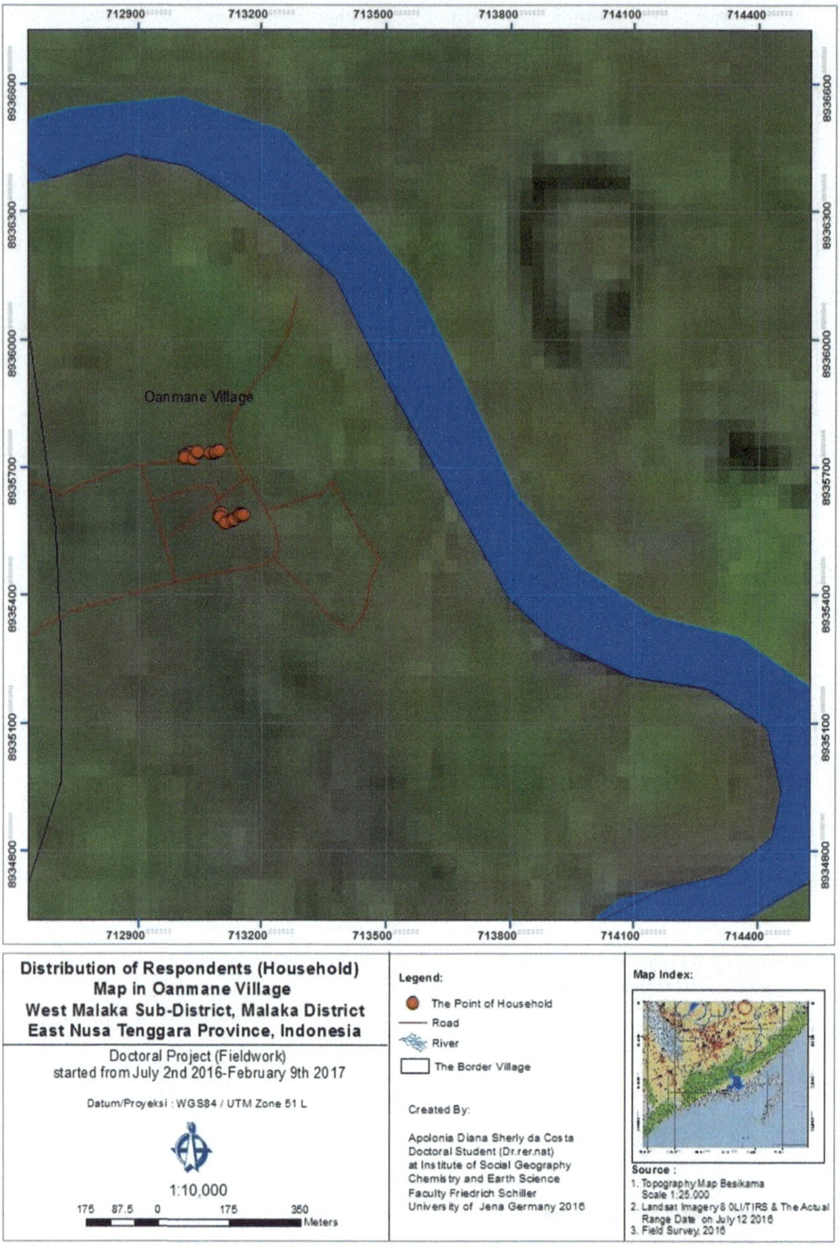

Fig. C.10 Distribution of respondents (Household) map in Oanmane Village, West Malaka Subdistrict, Malaka District, East Nusa Tenggara Province, Indonesia. *Source* Author's Research Fieldwork, July 9, 2016, to February 9, 2017

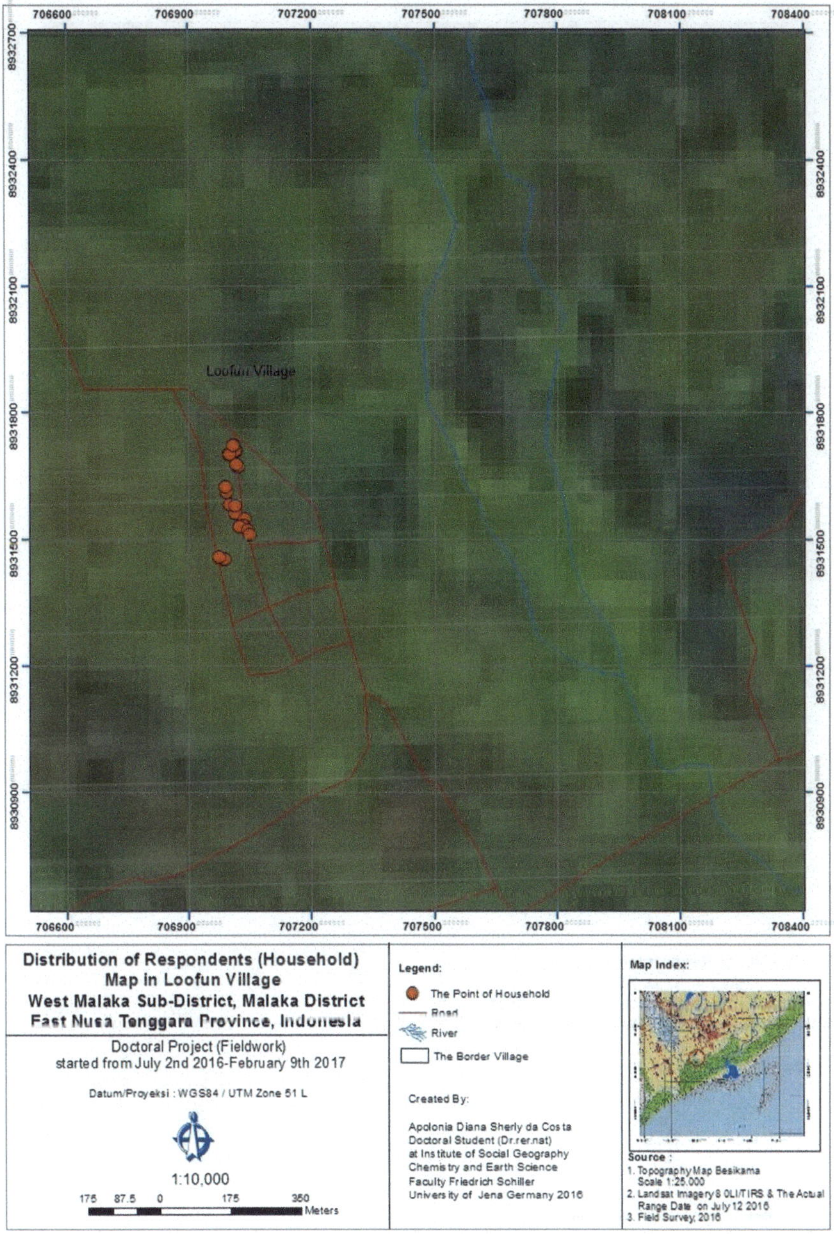

Fig. C.11 Distribution of respondents (Household) map in Lofoun Village, West Malaka Subdistrict, Malaka District, East Nusa Tenggara Province, Indonesia. *Source* Author's Research Fieldwork, July 9, 2016, to February 9, 2017

Fig. C.12 Distribution of respondents (Household) map in Naas Village, West Malaka Subdistrict, Malaka District, East Nusa Tenggara Province, Indonesia. *Source* Author's Research Fieldwork, July 9, 2016, to February 9, 2017

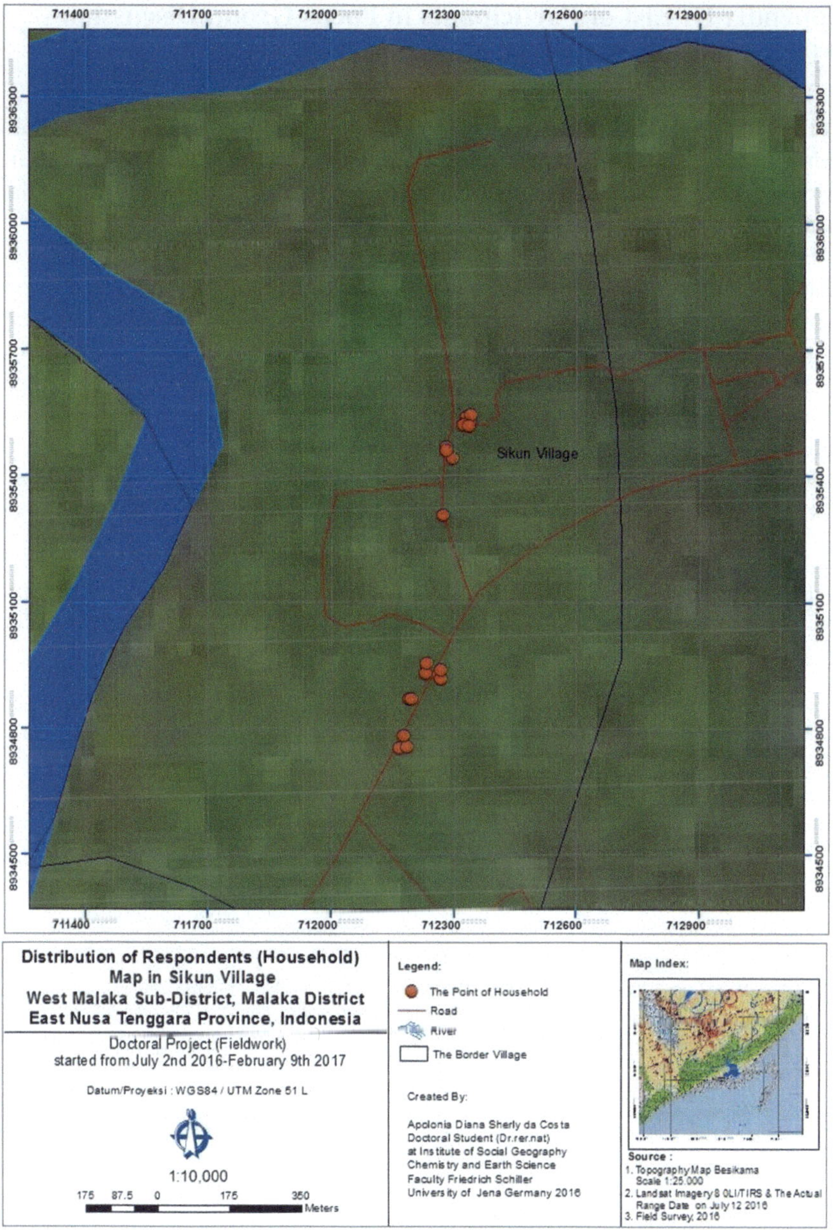

Fig. C.13 Distribution of respondents (Household) map in Sikun Village, West Malaka Subdistrict, Malaka District, East Nusa Tenggara Province, Indonesia. *Source* Author's Research Fieldwork, July 9, 2016, to February 9, 2017

Appendix D: List of Participants in Focus Group Discussion (FGD) and Stakeholder Meetings (SMEs)

At the West Malaka Subdistrict's Office, Besikama Village in the West Malaka Subdistrict of Malaka District, in the West Timor Island, East Nusa Tenggara Province, Indonesia, on November 19, 2016, at 10 a.m. to 15.00 p.m. (Indonesian Middle-Zone am/pm)

Leader for An Introduction Section: Ephi Bria Seran, M.Phil (Head of the West Malaka Subdistrict) Notulist: Rony Seran (Former Local Staff of The United Nations (UN) in the West Malaka Subdistrict and Former Researcher at Community Association for Disaster Management (CADM), The NGO)) Leader for each group in the FGD and SMEs: Patrisius Seran Klau (Local Researcher at Community Association for Disaster Management (CADM), The NGO)) and Leader of Disaster Group Volunteer in the West Malaka Subdistrict)

Researcher and Leader in the FGD and SMEs: I, Apolonia Diana Sherly da Costa (Table D.1)

Table D.1 List of participants in Focus Group Discussion (FGD) and Stakeholder Meetings (SMEs) at the West Malaka Subdistrict's Office, Besikama Village in the West Malaka Subdistrict of Malaka District, in the West Timor Island, East Nusa Tenggara Province, Indonesia

No.	Name	Female/ Male (F/M)	Occupation	Role in the community	Village	Group of FGD and SMEs (I/ II/III)
1	Yuliana Bano	F	Civil servant and Farmer	Head of Umatoos Village	Umatoos	I
2	Yuliana Luruk	F	Householdwife	Customary Leader in Umatoos Village	Umatoos	I
3	Yosefina Hoar Kwaik	F	Householdwife	Villager of Fafoe Village	Fafoe	I
4	Lambertus Klau Nahak	M	Civil servant and Farmer	Head of Maktihan Village	Maktihan	I
5	Wilem Leki Manek	M	Farmer and member of Disaster Group Volunteer	Customary Leader in Maktihan Village	Maktihan	I

(continued)

Table D.1 (continued)

No.	Name	Female/ Male (F/M)	Occupation	Role in the community	Village	Group of FGD and SMEs (I/ II/III)
6	Stanislaus Bria	M	Farmer and member of Disaster Group Volunteer	Customary Leader in Umaloor Village	Umaloor	I
8	Benekditus Seran Taek	M	Civil Servant and Farmer	Head of Rabasahain Village and Representative of the Catholic Church	Rabasahain	I
9	Zakarias Kalu	M	Farmer and member of Disaster Group Volunteer	Customary Leader in Rabasahain Village	Rabasahain	I
10	Melianus Lae	M	Civil Servant and Farmer	Head of Besikama Village	Besikama	I
11	Uma, Lana'in	F	Farmer and member of Disaster Group Volunteer	Customary Leader in Besikama Village	Besikama	I
12	Metri Teik	F	Civil Servant and Farmer	Secretary of Lasaen Village and Representative of the Catholic Church	Lasaen	I
13	Daniel Dahu	M	Farmer and member of Disaster Group Volunteer	Customary Leader in Lasaen Village	Lasaen	I
14	Hendrikus Nahak	M	Farmer and member of Disaster Group Volunteer	Secretary Rabasa Village and Representative of the Catholic Church	Rabasa	I
15	Yohanes Bria	M	Farmer and member of Disaster Group Volunteer	Customary leader in Rabasa Village	Rabasa	I

(continued)

Table D.1 (continued)

No.	Name	Female/ Male (F/M)	Occupation	Role in the community	Village	Group of FGD and SMEs (I/ II/III)
16	Hilarius Bere	M	Farmer and member of Disaster Group Volunteer	–	Sikun	I
17	Aprilia S. Fahik	F	Representative of local bank sector	One of Customary Leader in Besikama Village	Besikama	I
18	Arnoldus Seran	M	Civil Servant and Farmer	Head of Naas Village and Representative of Executive and Legislative councils	Naas	II
19	Bertolomeus Tahu	M	Farmer and member of Disaster Group Volunteer	Customary Leader in Naas Village	Naas	II
20	Kristina Bere Bau	F	Farmer and member of Disaster Group Volunteer	One of Customary Leaders and a stakeholder in Naas Village	Naas	II
21	Hendrikus B. Luan	M	Civil Servant and Farmer	Head of Raimataus Village	Raimataus	II
22	Petrus B. Seran	M	Farmer and member of Disaster Group Volunteer	Customary leader in Raimataus Village	Raimataus	II
23	Maria Anita Koi	F	Farmer and member of Disaster Group Volunteer	One of Customary Leaders and a stakeholder in Raimataus Village	Raimataus	II
24	Yovita Hoar	F	Farmer and member of Disaster Group Volunteer	Stakeholder in Raimataus Village	Raimataus	II

(continued)

Table D.1 (continued)

No.	Name	Female/ Male (F/M)	Occupation	Role in the community	Village	Group of FGD and SMEs (I/ II/III)
25	Yasinta Seuk	F	Farmer and member of Disaster Group Volunteer	Stakeholder in Motaulun Village	Motaulun	II
26	Yulius Bria Koe	M	Farmer and member of Disaster Group Volunteer	Secretary of Oanmane Village	Oanmane	III
27	Kosmas Koe	M	Farmer and member of Disaster Group Volunteer	Customary Leader in Oanmane Village	Oanmane	III
28	Benekditus Toe	M	Farmer and member of Disaster Group Volunteer	Stakeholder in Loofoun Village	Loofoun	III
29	Yohanes Bria	M	Farmer and member of Disaster Group Volunteer	Customary Leader in Fafoe Village	Fafoe	III
30	Maria Hoar Bria	F	Farmer and member of Disaster Group Volunteer	Stakeholder in Motaain Village	Motaain	III
31	Edmundus Bere Klau	M	Farmer and member of Disaster Group Volunteer	Head of Rabasa Village and Representative of the Catholic Church	Rabasa	III
32	Gregorius Nahak	M	Farmer and member of Disaster Group Volunteer	Stakeholder in Rabasahaerain Village	Rabasahaerain	III
33	Maria Nelde Luruk	F	Farmer and member of Disaster Group Volunteer	Stakeholder in Rabasa Village	Rabasa	III
34	Yustinus Telik	M	Farmer and member of Disaster Group Volunteer	Stakeholder in Umatoos Village	Umatoos	III

(continued)

Table D.1 (continued)

No.	Name	Female/ Male (F/M)	Occupation	Role in the community	Village	Group of FGD and SMEs (I/ II/III)
35	Petrus Nahak	M	Farmer and member of Disaster Group Volunteer	One of the Customary Leaders and a Stakeholder in Umatoos Village	Umatoos	III
36	Antonius Seran	M	Civil Servant and Farmer	Head of Oanmane Village and Representative of the local UNDP	Oanmane	III
37	Simor Seran	M	Civil Servant and Farmer	Head of Lasaen Village	Lasaen	III

Source Author's Fieldwork: FGD and SMEs, November 19, 2016. The West Malaka Subdistrict's Office, Besikama Village

Appendix E1: English Version

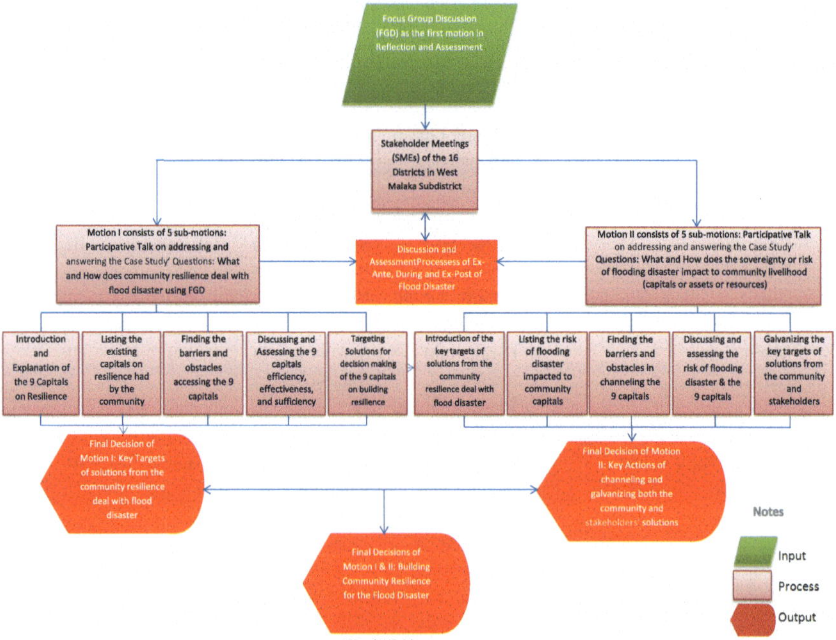

FGD and SMEs Scheme

Appendix E2: Indonesian Version

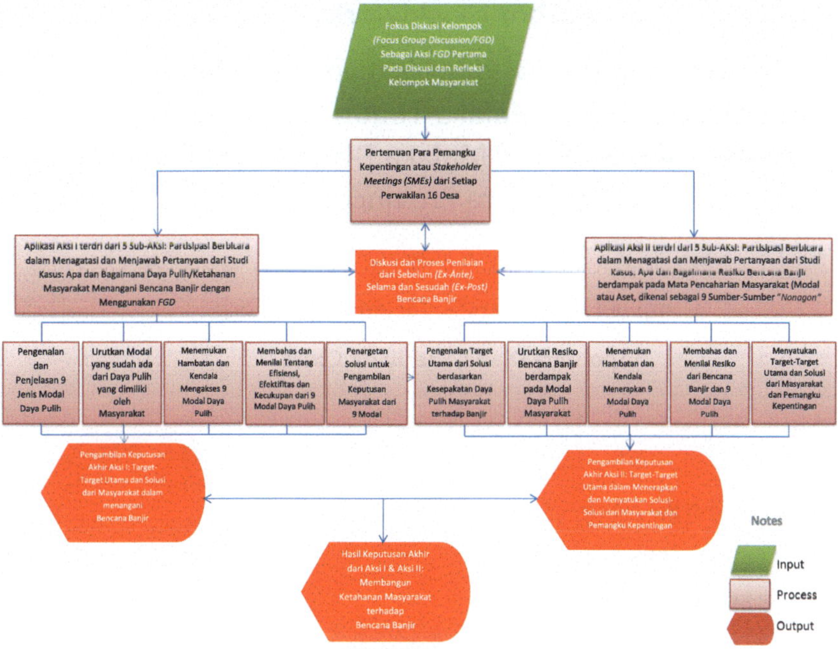

Skema FGD and SMEs

Appendix E3: English Version

Author's Fieldwork:

Focus Group Discussion (FGD) and Stakeholder Meetings (SMEs) in Malaka-Timor, Indonesia (West Malaka Subdistrict of Malaka District, East Nusa Tenggara Province)

Compiled by Researcher/Scientist (Author):

Apolonia Diana Sherly da Costa

Place (Venue) of the Activity of FGD dan SMEs : ____________________

Date of FGD and SMEs : ____________________2016

Duration of FGD and SMEs : ____________________

Name of the Participants in the FGD and SMEs:

1. Head of Village : ____________________
2. Customary Leader : ____________________

Village origin of the Village Head

and the Customary Leader : ____________________

or

Other Participants (Stakeholders) : ____________________

I. Answering the research questions of the first research objective: To analyze how community resilience is built for facing the flood disaster and how does community build their resiliency for facing the flood disaster.

1. Discussion: What does mean by *Community Resilience and Building Community Resilience?*

   (a) *Does Community Resilience have a meaning as: (Please cross (x) in the following (a,b,c, d, or e) which is correct according to your opinion))*

   a. An independent ability to construct human resources in order to deal with floods which this means as a smart community.

b. Both individual and social abilities to deal with floods.
c. An ability to prepare and to change a risky situation becomes a resiliency situation or create an asset despite floods.
d. A system or a process to filter or to process such a disturbance event and do efforts to manage it.
e. *A process in order to reach assets like in the previous situation (before flooding or ex-ante) which means a process with ignoring all of the risk things due to floods.* It denies a flood ever happens.

(b) *Does Building Community Resilience have a meaning as: (Please cross (x) in the following (a,b,c, d, or e) which is correct according to your opinion))*
a. A similar system like community resilience in order to strengthen the capacity and resiliency system to deal with floods.
b. *A different system in assessing (valuing a process)* that the past experiences (i.e., floods) could be able to improve or strengthen the existing assets or to create the new assets of resilience. It can be understood also as building community resilience which contains long-term strategies.

2. Listing community resilience's assets (i.e., existing and new assets of resilience).
3. Finding obstacles in utilizing the community resilience's assets.
4. Reflection and assessment.

Discussion Questions 2, 3, and 4: Participative Assessment of the Sustainable Use of Resilience that owned and applied by the Community Affected by a Flood Disaster

Participative assessment based on community perception for their own resilience (sustainable livelihood resources)	Phase of flood event (Before flooding or ex-ante, during flooding, and after flooding or ex-post)			Participative assessment of the sustainable use of resilience which is assessed by the community from (i.e., effective, efficiency, and sufficiency) Please give a mark (√), if saying Yes		
	Before flooding or *Ex-Ante*	During flooding	After flooding or *Ex-Post*	Effective	Efficiency	Sufficiency
Human Experience/Skill Education Health						
Natural Access to Natural Resources: Land, Water, Forests, Wildlife, Orchards, Tree, Animals, Forest , etc						

(continued)

(continued)

Participative assessment based on community perception for their own resilience (sustainable livelihood resources)	Phase of flood event (Before flooding or ex-ante, during flooding, and after flooding or ex-post)			Participative assessment of the sustainable use of resilience which is assessed by the community from (i.e., effective, efficiency, and sufficiency) Please give a mark (√), if saying Yes		
	Before flooding or *Ex-Ante*	During flooding	After flooding or *Ex-Post*	Effective	Efficiency	Sufficiency
Physical Housing, Access to clean water and sanitation (i.e., drinking water), access to infrastructure (i.e., road, bridge, school, hospital, irrigation/canal, dike/ embankment, vehicles, public transportation, local/traditional transportation, and livestock						
Economic Micro-credit, coping mechanism (i.e., savings), borrowing money from the neighbors/another households, friends, and relatives/families; selling natural assets such as paddy, corn/ maize, vegetables or bartering/ exchanging them; insurance and saving in the local bank						

(continued)

(continued)

Participative assessment based on community perception for their own resilience (sustainable livelihood resources)	Phase of flood event (Before flooding or ex-ante, during flooding, and after flooding or ex-post)			Participative assessment of the sustainable use of resilience which is assessed by the community from (i.e., effective, efficiency, and sufficiency) Please give a mark (√), if saying Yes		
	Before flooding or *Ex-Ante*	During flooding	After flooding or *Ex-Post*	Effective	Efficiency	Sufficiency
Social Society-Group, Relatives/ friendship Networks, Social Trust, Social Norms and Social capabilities which binding, bridging and linking to other sources of social capabilities						
Cultural A strong role of customary leader in each village for dealing with floods through Providing a spiritual motivation; rituals, praying to the ancestors and God with the way of "betel nut" as a wisdom/cultural symbol in order to strengthen community (per individual and household) to face flood						

(continued)

(continued)

Participative assessment based on community perception for their own resilience (sustainable livelihood resources)	Phase of flood event (Before flooding or ex-ante, during flooding, and after flooding or ex-post)			Participative assessment of the sustainable use of resilience which is assessed by the community from (i.e., effective, efficiency, and sufficiency) Please give a mark (√), if saying Yes		
	Before flooding or *Ex-Ante*	During flooding	After flooding or *Ex-Post*	Effective	Efficiency	Sufficiency
Religion A strong connection from the Priests, brothers, and nuns in providing pastoral (hospitality), planning for preparedness with customary leaders to strengthen community spiritual viability; spiritual training; providing direct communication; influencer, quick responder, and as a facilitator for floods and other onset crises						

(continued)

(continued)

Participative assessment based on community perception for their own resilience (sustainable livelihood resources)	Phase of flood event (Before flooding or ex-ante, during flooding, and after flooding or ex-post)			Participative assessment of the sustainable use of resilience which is assessed by the community from (i.e., effective, efficiency, and sufficiency) Please give a mark (√), if saying Yes		
	Before flooding or *Ex-Ante*	During flooding	After flooding or *Ex-Post*	Effective	Efficiency	Sufficiency
Law or Legal An integrated management for protecting natural environment stressors (i.e., floods), social environment (i.e., resiliency systems) from the executive councils (i.e., head subvillage and village, subdistrict, and district, etc.), the legislative council (i.e., local house representative and people's Consultative Assembly) to manifest a sustainable development						

(continued)

(continued)

Participative assessment based on community perception for their own resilience (sustainable livelihood resources)	Phase of flood event (Before flooding or ex-ante, during flooding, and after flooding or ex-post)			Participative assessment of the sustainable use of resilience which is assessed by the community from (i.e., effective, efficiency, and sufficiency) Please give a mark (√), if saying Yes		
	Before flooding or *Ex-Ante*	During flooding	After flooding or *Ex-Post*	Effective	Efficiency	Sufficiency
Political or Policy The Implementation of Law or Legal asset;access and participation in decision making, an integrated management and solution for the community affected by floods, gender relationship; management of the structural authority which influences to decision-making process with individuals, households, community, and access to a system of environmental security, the judiciary, and other controllers' sectors						

5. Targeting Solutions through Priority Scales of the Sustainable Assets of the Community Resilience

Community	Maj	Minor (less	Scoring			
Resilience's assets	or (More prioritized) Please, mark (√) if saying Yes	prioritized) Please, mark (√) if saying Yes	(1–10) Adjustment process for a resilience community that is " Sustainable Development" (Resilience Adjustment on Community Sustainable Actions)			
			Scoring 1 = Long-Term All of Assets and Subassets owned and applied (10)	Scoring 2 = Medium or Moderate-Term Some of Assets and Subassets owned and applied (5)	Scoring 3 = Nearly Medium or Moderate-Term Few of Assets and Subassets	Scoring 4 = Short-Term None of Assets and Subassets owned and applied (0)
Human Experience/Skill Education Health						
Natural Access to Natural Resources: Land, Water, Forests, Wildlife, Orchards, Tree, Animals, Forest, etc						
Physical Housing, Access to clean water and sanitation (i.e., drinking water), access to infrastructure (i.e., road, bridge, school, hospital, irrigation/ canal, dike/embankment, vehicles, public transportation, local/traditional transportation, and livestock						

(continued)

(continued)

Community	Maj	Minor (less	Scoring			
Resilience's assets	or (More prioritized) Please, mark (√) if saying Yes	prioritized) Please, mark (√) if saying Yes	(1–10) Adjustment process for a resilience community that is " Sustainable Development" (Resilience Adjustment on Community Sustainable Actions)			
			Scoring 1 = Long-Term All of Assets and Subassets owned and applied (10)	Scoring 2 = Medium or Moderate-Term Some of Assets and Subassets owned and applied (5)	Scoring 3 = Nearly Medium or Moderate-Term Few of Assets and Subassets	Scoring 4 = Short-Term None of Assets and Subassets owned and applied (0)
Economic Micro-credit, coping mechanism (i.e., savings), borrowing money from the neighbors/another households, friends, and relatives/families; selling natural assets such as paddy, corn/maize, vegetables or bartering/exchanging them; insurance and saving in the local bank						
Social Society-Group, Relatives/friendship Networks, Social Trust, Social Norms and Social capabilities which binding, bridging, and linking to other sources of social capabilities						

(continued)

(continued)

Community	Maj	Minor (less	Scoring			
Resilience's assets	or (More prioritized) Please, mark (√) if saying Yes	prioritized) Please, mark (√) if saying Yes	(1–10) Adjustment process for a resilience community that is " Sustainable Development" (Resilience Adjustment on Community Sustainable Actions)			
			Scoring 1 = Long-Term All of Assets and Subassets owned and applied (10)	Scoring 2 = Medium or Moderate-Term Some of Assets and Subassets owned and applied (5)	Scoring 3 = Nearly Medium or Moderate-Term Few of Assets and Subassets	Scoring 4 = Short-Term None of Assets and Subassets owned and applied (0)
Cultural A strong role of customary leader in each village for dealing with floods through Providing a spiritual motivation; rituals, praying to the ancestors and God with the way of "betel nut" as a wisdom/cultural symbol in order to strengthen community (per individual and household) to face flood						

(continued)

(continued)

Community	Maj	Minor (less	Scoring			
Resilience's assets	or (More prioritized) Please, mark (√) if saying Yes	prioritized) Please, mark (√) if saying Yes	(1–10) Adjustment process for a resilience community that is " Sustainable Development" (Resilience Adjustment on Community Sustainable Actions)			
			Scoring 1 = Long-Term All of Assets and Subassets owned and applied (10)	Scoring 2 = Medium or Moderate-Term Some of Assets and Subassets owned and applied (5)	Scoring 3 = Nearly Medium or Moderate-Term Few of Assets and Subassets	Scoring 4 = Short-Term None of Assets and Subassets owned and applied (0)
Religion A strong connection from the Priests, brothers, and nuns in providing pastoral (hospitality), planning for preparedness with customary leaders to strengthen community spiritual viability; spiritual training; providing direct communication; influencer, quick responder, and as a facilitator for floods and other onset crises						

(continued)

(continued)

Community	Maj	Minor (less	Scoring			
Resilience's assets	or (More prioritized) Please, mark (√) if saying Yes	prioritized) Please, mark (√) if saying Yes	(1–10) Adjustment process for a resilience community that is " Sustainable Development" (Resilience Adjustment on Community Sustainable Actions)			
			Scoring 1 = Long-Term All of Assets and Subassets owned and applied (10)	Scoring 2 = Medium or Moderate-Term Some of Assets and Subassets owned and applied (5)	Scoring 3 = Nearly Medium or Moderate-Term Few of Assets and Subassets	Scoring 4 = Short-Term None of Assets and Subassets owned and applied (0)
Law or Legal An integrated management for protecting natural environment stressors (i.e., floods), social environment (i.e., resiliency systems) from the executive councils (i.e., head subvillage and village, subdistrict, and district), the legislative council (i.e., local house representative and people's Consultative Assembly) to manifest a sustainable development						

(continued)

(continued)

Community	Maj	Minor (less	Scoring			
Resilience's assets	or (More prioritized) Please, mark (√) if saying Yes	prioritized) Please, mark (√) if saying Yes	(1–10) Adjustment process for a resilience community that is " Sustainable Development" (Resilience Adjustment on Community Sustainable Actions)			
			Scoring 1 = Long-Term All of Assets and Subassets owned and applied (10)	Scoring 2 = Medium or Moderate-Term Some of Assets and Subassets owned and applied (5)	Scoring 3 = Nearly Medium or Moderate-Term Few of Assets and Subassets	Scoring 4 = Short-Term None of Assets and Subassets owned and applied (0)
Political or Policy The Implementation of Law or Legal asset; access and participation in decision making, an integrated management and solution for the community affected by floods, gender relationship; management of the structural authority which influences to decision-making process with individuals, households, community, and access to a system of environmental security, the judiciary, and other controllers' sectors						

II. Answering the research questions from the second research objective: To analyze the risk of flood disaster impacted on community livelihood resources.

Research Questions for the second research objective:

1. How is the relationship between flood risk and community resilience?
2. How do the communities use their capitals or assets on resilience to deal with flood risk?
3. How can communities build or improve their capitals or assets on resilience in the study area?

The relationship between risk of flood disaster and community coping capacity in utilizing their own resilience's assets

Types of the flood risks that were experienced by the community (i.e., impacted to community) through the community resilience's assets (i.e., sustainable livelihood resources)	Flood Risk Management was assessed by the community coping capacity (community resilience's assets) Please, mention one, two, or several assets and how did flood risks impact the community livelihood resources. If all of them were impacted by floods, mark (√) If just some or none of them, please, mark (x)	Prevention	Protection	Preparedness	Recovery and Lesson Learned
Human Experience/Skill Education Health					
Natural Access to Natural Resources: Land, Water, Forests, Wildlife, Orchards, Tree, Animals, Forest, etc					

(continued)

(continued)

Types of the flood risks that were experienced by the community (i.e., impacted to community) through the community resilience's assets (i.e., sustainable livelihood resources)	Flood Risk Management was assessed by the community coping capacity (community resilience's assets) Please, mention one, two, or several assets and how did flood risks impact the community livelihood resources. If all of them were impacted by floods, mark (√) If just some or none of them, please, mark (x)	Prevention	Protection	Preparedness	Recovery and Lesson Learned
Physical Housing, Access to clean water and sanitation (i.e., drinking water), access to infrastructure (i.e., road, bridge, school, hospital, irrigation/canal, dike/ embankment, vehicles, public transportation, local/traditional transportation, and livestock					

(continued)

(continued)

Types of the flood risks that were experienced by the community (i.e., impacted to community) through the community resilience's assets (i.e., sustainable livelihood resources)	Flood Risk Management was assessed by the community coping capacity (community resilience's assets) Please, mention one, two, or several assets and how did flood risks impact the community livelihood resources. If all of them were impacted by floods, mark (√) If just some or none of them, please, mark (x)	Prevention	Protection	Preparedness	Recovery and Lesson Learned
Economic Micro-credit, coping mechanism (i.e., savings), borrowing money from the neighbors/another households, friend and relatives/families; selling natural assets such as paddy, corn/ maize, vegetables, or bartering/exchanging them; insurance and saving in the local bank					
Social Society-Group, Relatives/friendship Networks, Social Trust, Social Norms and Social capabilities which binding, bridging and linking to other sources of social capabilities					

(continued)

(continued)

Types of the flood risks that were experienced by the community (i.e., impacted to community) through the community resilience's assets (i.e., sustainable livelihood resources)	Flood Risk Management was assessed by the community coping capacity (community resilience's assets) Please, mention one, two, or several assets and how did flood risks impact the community livelihood resources. If all of them were impacted by floods, mark (✓) If just some or none of them, please, mark (x)	Prevention	Protection	Preparedness	Recovery and Lesson Learned
Cultural A strong role of customary leader in each village for dealing with floods through Providing a spiritual motivation; rituals, praying to the ancestors and God with the way of "betel nut" as a wisdom/ cultural symbol in order to strengthen community (per individual and household) to face flood					

(continued)

(continued)

Types of the flood risks that were experienced by the community (i.e., impacted to community) through the community resilience's assets (i.e., sustainable livelihood resources)	Flood Risk Management was assessed by the community coping capacity (community resilience's assets) Please, mention one, two, or several assets and how did flood risks impact the community livelihood resources. If all of them were impacted by floods, mark (√) If just some or none of them, please, mark (x)	Prevention	Protection	Preparedness	Recovery and Lesson Learned
Religion A strong connection from the Priests, brothers, and nuns in providing pastoral (hospitality), planning for preparedness with customary leaders to strengthen community spiritual viability, spiritual training; providing direct communication; influencer, quick responder, and as a facilitator for floods and other onset crises					

(continued)

(continued)

Types of the flood risks that were experienced by the community (i.e., impacted to community) through the community resilience's assets (i.e., sustainable livelihood resources)	Flood Risk Management was assessed by the community coping capacity (community resilience's assets) Please, mention one, two, or several assets and how did flood risks impact the community livelihood resources. If all of them were impacted by floods, mark (√) If just some or none of them, please, mark (x)	Prevention	Protection	Preparedness	Recovery and Lesson Learned
Law or Legal An integrated management for protecting natural environment stressors (i.e., floods), social environment (i.e., resiliency systems) from the executive councils (i.e., head subvillage and village, subdistrict, and district), the legislative council (i.e., local house representative and people's					

(continued)

(continued)

Types of the flood risks that were experienced by the community (i.e., impacted to community) through the community resilience's assets (i.e., sustainable livelihood resources)	Flood Risk Management was assessed by the community coping capacity (community resilience's assets) Please, mention one, two, or several assets and how did flood risks impact the community livelihood resources. If all of them were impacted by floods, mark (√) If just some or none of them, please, mark (x)	Prevention	Protection	Preparedness	Recovery and Lesson Learned
Political or Policy The Implementation of Law or Legal asset; access and participation in decision making, an integrated management and solution for the community affected by floods, gender relationship; management of the structural authority which influences to decision-making process with individuals, households, community, and access to a system of environmental security, the judiciary, and other controllers' sectors					

Result, Conclusion, and Suggestion:
THANK YOU GOD BLESS YOU

Appendix E4: Indonesian Version

Draf Kuesioner Penelitian Empiris (Bahasa Indonesia)

Oleh Peneliti/Ilmuwan (Penulis):

Apolonia Diana Sherly da Costa

Diskusi *Focus Group Discussion* (FGD) /Fokus Diskusi Kelompok dan *Stakeholder Meetings* (SMEs) / Pertemuan para Pemangku Kepentingan di Kecamatan Malaka Barat, Kabupaten Malaka, Provinsi Nusa Tenggara Timur, Indonesia

Tempat Pelaksanaan Kegiatan FGD dan SMEs : ____________

Tanggal Pelaksanaan : ____________2016

Waktu / Durasi Pelaksanaan : ____________

Nama Peserta Diskusi (Kepala Desa) dan Nama Kepala Adat:

1. Nama Kepala Desa: ____________
2. Nama Kepala Adat: ____________

Asal Desa dari Kepala Desa dan Kepala Adat : ____________

atau

Peserta Lainnya (Pemangku Kepentingan) : ____________

Pertanyaan Penelitian Tujuan Pertama: 1. Menganalisa *Community Resilience* (Ketahanan/Daya Pulih Masyarakat) dan *Building Community Resilience* (Meningkatkan Ketahanan/Daya Pulih Masyarakat) dalam menghadapi Bencana Banjir

I. Pertanyaan Penelitian dari Tujuan Pertama:

1. Apa yang dimaksud dengan Community Resilience (Ketahanan/Daya Pulih Masyarakat) dan Building Community Resilience (Meningkatkan Ketahanan/Daya Pulih Masyarakat)?
2. Apa saja kemampuan/capital/asset yang sudah ada sejauh ini dan atau yang baru dalam konteks (Ketahanan/Daya Pulih Masyarakat) dan Building Community Resilience (Meningkatkan Ketahanan/Daya Pulih Masyarakat dalam menghadapi bencana banjir?
3. Bagaimana Community Resilience (Ketahanan/Daya Pulih Masyarakat) dan Building Community Resilience (Meningkatkan Ketahanan/Daya Pulih

Masyarakat dapat dinilai dalam keberlanjutan jangka panjang (keberlanjutan pembangunan) ?

Pertanyaan Penelitian Tujuan Kedua: 2. Menganalisa resiko bencana banjir yang bedampak pada mata pencharian masyarakat (Kemampuan/capital/asset yang disebut sebagai 5 Sumber Daya (5 Pentagon Resources berdasarkan pendapat pakar resilience sebelumnya) yang didayakan oleh manusia terhadap lingkungan alam dan lingkungan social ekonomi nya). Di analisa dan dievaluasi dengan nano-resources (9 Sumber Daya), hasil pengamatan dan penemuan peneliti sendiri.

II. Pertanyaan Penelitian dari Tujuan Kedua:

1. Bagaimana hubungan resiko bencana banjir dan kemampuan masyarakat?
2. Bagaimana Masyarakat memanfaatkan kemampuan/capital/asset pada daya pulih/ketahanan mereka dalam penanganan resiko resiko bencana banjir?
3. Bagaimana masyarakat membangun atau meningkatkan kemampuan/ capital/asset (sumber daya) mereka dalam membangun ketahanan/ Daya Pulih mereka?

I Menjawab Pertanyaan Penelitian dari Tujuan Pertama: Menganalisa *Community Resilience* (Ketahanan/Daya Pulih Masyarakat) dan *Building Community Resilience* (Meningkatkan Ketahanan/Daya Pulih Masyarakat) dalam menghadapi Bencana Banjir

1. Pembahasan tentang Apa itu *Community Resilience and Building Community Resilience?*

   (a) *Apakah Community Resilience adalah Definisi dari: (Memberi tanda silang (x) pada pilihan yang dipilih))*
      a. Kemampuan Mandiri untuk merekonstruksi sumber daya saat ini untuk menangani Banjir Bencana (Dapat Dikatakan Masyarakat Cerdas).
      b. Kemampuan diri dan Kemampuan Sosial untuk mengatasi Banjir Bencana.
      c. Kemampuan untuk mempersiapkan dan untuk mengubah atau membuat keadaan sebelumnya dari risiko akibat Banjir Bencana menjadi sebuah aset.
      d. Sistem dan Proses untuk menyerap gangguan dari risiko bencana banjir dan menanggulanginya.
      e. Proses untuk kembali atau menolak segala risiko bencana banjir seperti pada keadaan sebelumnya, atau dengan kata lain Bahwasanya Tidak ada bencana banjir.

   (b) *Apakah Building Community Resilience adalah Definisi dari: (Memberi tanda silang (x) pada pilihan yang dipilih))*
      a. Sistem yang sama dari ketahanan masyarakat untuk memperkuat kapasitas dan sistem pemulihan untuk mengatasi bencana banjir.
      b. Sistem yang berbeda dalam menilai (memberikan penghargaan) pada proses (belajar dari pengalaman masa lalu) untuk meningkatkan atau memperkuat aset atau modal yang dimiliki oleh ketahanan masyarakat.

Membangun ketahanan masyarakat berisi strategi jangka panjang pada ketahanan masyarakat yang sudah terbentuk, yang berurusan dengan penanganan bencana banjir dimana ketahanan masyarakat saja tidak memilikinya.

2. Mengurutkan asset/kemampuan resilience yang sudah dan atau baru dimiliki oleh masyarakat
3. Menemukan Kendala atau tantangan dalam mendayagunakan asset/kemampuan resilience yang dimiliki oleh masyarakat?
4. Diskusi dan Penilaian

Pertanyaan 2, 3, dan 4: Penilaian Partisipatif tentang Keberlanjutan terhadap Bencana Banjir dengan Ketahanan Masyarakat melalui Penilaian Resilience Masyarakat

Penilaian Partisipatif Masyarakat melalui Penilaian Resilience (Ketahanan Masyarakat)	Fase Kejadian Banjir (Ex-Ante/ Sebelum, Selama/Berlangsung dan Ex-Post/Sesudah)			Penilaian Keberlanjutan Sistem Resilience Masyarakat (Ketahanan Masyarakat) Dinilai dari Efektifitas, Efisiensi dan Sufficiency (Memadai) Tandai (√) jika Ya		
	Ex-Ante/ Sebelum	Selama/ Berlangsung	Ex-Post/ Sesudah	Efektifitas	Efisiensi	Memadai *(Sufficiency)*
Manusia Pengalaman/ Ketrampilan Pendidikan Kesehatan						
Alam Akses Sumber Daya Alam: Tanah Air Margasatwa Tumbuhan Hutan, dll						
Fisik Perumahan Akses Air Bersih dan Sanitasi (PAM) Akses Infrastruktur (Jalan, Jembatan, Sekolah, Rumah Sakit Irigasi, Kendaraan, Kendaraan Umum, Ternak						
Ekonomi Kredit-Mikro Pola Penyesuaian melalui Tabungan, Meminjam Tetangga, Teman dan keluarga, Meminjam dari Kreditur, menjual/ perdagangan/menukar, penghasilan tetap, akses jaringan dan asuransi						

(continued)

(continued)

Penilaian Partisipatif Masyarakat melalui Penilaian Resilience (Ketahanan Masyarakat)	Fase Kejadian Banjir (Ex-Ante/ Sebelum, Selama/Berlangsung dan Ex-Post/Sesudah)			Penilaian Keberlanjutan Sistem Resilience Masyarakat (Ketahanan Masyarakat) Dinilai dari Efektifitas, Efisiensi dan Sufficiency (Memadai) Tandai (√) jika Ya		
	Ex-Ante/ Sebelum	Selama/ Berlangsung	Ex-Post/ Sesudah	Efektifitas	Efisiensi	Memadai (*Sufficiency*)
Sosial Grup sosial, Jaringan Kerabat, Kepercayaan Sosial, Norma sosial dan kemampuan-kemampuan sosial yang (Mengikat, menjembatani dan menghubungkan)						
Budaya Peran Pemimpin Adat pada kasus Bencana Banjir Melalui Penguatan Spiritual; ritual dan penguatan individu, kelompok, masyarakat melalui peranan aktif secara langsung, symbol-simbol kekuatan budaya, kearifan local masyarakat						
Agama Memberikan pelayanan pastoral (hospitality), perencanaan kesiapsiagaan bencana individu dan keluarga, pelatihan, layanan spiritual, menyediakan komunikasi langsung untuk mengingatkan sistem dan influencer, responden dan fasilitator untuk krisis—krisis mendadak lainnya						

(continued)

(continued)

Penilaian Partisipatif Masyarakat melalui Penilaian Resilience (Ketahanan Masyarakat)	Fase Kejadian Banjir (Ex-Ante/ Sebelum, Selama/Berlangsung dan Ex-Post/Sesudah)			Penilaian Keberlanjutan Sistem Resilience Masyarakat (Ketahanan Masyarakat) Dinilai dari Efektifitas, Efisiensi dan Sufficiency (Memadai) Tandai (√) jika Ya		
	Ex-Ante/ Sebelum	Selama/ Berlangsung	Ex-Post/ Sesudah	Efektifitas	Efisiensi	Memadai *(Sufficiency)*
Hukum Manajemen Terpadu perlindungan lingkungan alam dan Sosial sesuai regulasi naskah hukum; Manajemen Terpadu melindungi lingkungan Kemampuan Alam dan Sosial bagi (Ketahanan Manusia) ditelaah dari kacamata hukum, peran trial-politik; Legislatif, Eksekutif dan Yudikatif						
Politik (Implementasi Regulasi dari Kemampuan Hukum); Akses dan Partisipasi dalam pengambilan Keputusan pola terpadu ketahanan masyarakat; Hubungan Gender; Struktur Kekuasaan mempengaruhi kontrol atas pengambilan keputusan dengan Individu, Rumah Tangga, Masyarakat dan Masyarakat Luas dan akses ke Sistem Peradilan dan Lembaga Sektor Keamanan. (Individu, Kelompok dan Lembaga)						

5. Penargetan Solusi Melalui Skala Prioritas Pendayagunaan Aset/Kemampuan Resilience Masyarakat Malaka Barat

Aset/Kemampuan Resilience Masyarakat	Kurang diprioritaskan) Dengan tanda (x) jika ya	Major (Yang Utama diprioritaskan) Dengan tanda (√) jika ya	Skoring (*Scoring*) (1–10) *Peyesuaian Ketahanan pada Aksi Masyarakat yang Berkelanjutan (Resilience Adjustment on Community Sustainable Actions) Indikator Penilaian Jangka Panjang*			
			Skor 1 = Jangka Panjang Semua asset dan sub-aset di dayagunakan	Skor 2 = Medium Jangka Panjang Beberapa asset dan sub-aset di dayagunakan	Skor 3 = Menuju Jangka Pendek Sedikit asset dan sub-aset di	Skor 4 = Jangka Pendek Tidak sama sekali asset dan sub-aset di dayagunakan Masyarakat
Manusia Pengalaman/Ketrampilan Pendidikan Kesehatan						
Alam Akses Sumber Daya Alam: Tanah Air Margasatwa Tumbuhan Hutan, dll						
Fisik Perumahan Akses Air Bersih dan Sanitasi (PAM) Akses Infrastruktur (Jalan, Jembatan, Sekolah, Rumah Sakit Irigasi, Kendaraan, Kendaraan Umum, Ternak						
Ekonomi Kredit-Mikro Pola Penyesuaian melalui Tabungan, Meminjam Tetangga, Teman dan keluarga, Meminjam dari Kreditur, menjual / perdagangan / menukar, penghasilan tetap, akses jaringan dan asuransi						

(continued)

(continued)

Aset/Kemampuan Resilience Masyarakat	Kurang diprioritaskan) Dengan tanda (x) jika ya	Major (Yang Utama diprioritaskan) Dengan tanda (√) jika ya	Skoring (*Scoring*) (1–10) *Peyesuaian Ketahanan pada Aksi Masyarakat yang Berkelanjutan (Resilience Adjustment on Community Sustainable Actions) Indikator Penilaian Jangka Panjang*			
			Skor 1 = Jangka Panjang Semua asset dan sub-aset di dayagunakan	Skor 2 = Medium Jangka Panjang Beberapa asset dan sub-aset di dayagunakan	Skor 3 = Menuju Jangka Pendek Sedikit asset dan sub-aset di	Skor 4 = Jangka Pendek Tidak sama sekali asset dan sub-aset di dayagunakan Masyarakat
Sosial Grup sosial, Jaringan Kerabat, Kepercayaan Sosial, Norma sosial dan kemampuan-kemampuan sosial yang (Mengikat, menjembatani dan menghubungkan)						
Budaya Peran Pemimpin Adat pada kasus Bencana Banjir Melalui Penguatan Spiritual; ritual dan penguatan individu, kelompok, masyarakat melalui peranan aktif secara langsung, symbol-simbol kekuatan budaya, kearifan local masyarakat						

(continued)

(continued)

Aset/Kemampuan Resilience Masyarakat	Kurang diprioritaskan) Dengan tanda (x) jika ya	Major (Yang Utama diprioritaskan) Dengan tanda (√) jika ya	Skoring (*Scoring*) (1–10) *Peyesuaian Ketahanan pada Aksi Masyarakat yang Berkelanjutan (Resilience Adjustment on Community Sustainable Actions) Indikator Penilaian Jangka Panjang*			
			Skor 1 = Jangka Panjang Semua asset dan sub-aset di dayagunakan	Skor 2 = Medium Jangka Panjang Beberapa asset dan sub-aset di dayagunakan	Skor 3 = Menuju Jangka Pendek Sedikit asset dan sub-aset di	Skor 4 = Jangka Pendek Tidak sama sekali asset dan sub-aset di dayagunakan Masyarakat
Agama Memberikan pelayanan pastoral (hospitality), perencanaan kesiapsiagaan bencana individu dan keluarga, pelatihan, layanan spiritual, menyediakan komunikasi langsung untuk mengingatkan sistem dan influencer, responden dan fasilitator untuk krisis –krisis mendadak lainnya						
Hukum Manajemen Terpadu perlindungan lingkungan alam dan Sosial sesuai regulasi naskah hukum; Manajemen Terpadu melindungi lingkungan Kemampuan Alam dan Sosial bagi (Ketahanan Manusia) ditelaah dari kacamata hukum, peran trial-politik; Legislatif, Eksekutif dan Yudikatif						

(continued)

(continued)

Aset/Kemampuan Resilience Masyarakat	Kurang diprioritaskan) Dengan tanda (x) jika ya	Major (Yang Utama diprioritaskan) Dengan tanda (√) jika ya	Skoring (*Scoring*) (1–10) *Peyesuaian Ketahanan pada Aksi Masyarakat yang Berkelanjutan (Resilience Adjustment on Community Sustainable Actions) Indikator Penilaian Jangka Panjang*			
			Skor 1 = Jangka Panjang Semua asset dan sub-aset di dayagunakan	Skor 2 = Medium Jangka Panjang Beberapa asset dan sub-aset di dayagunakan	Skor 3 = Menuju Jangka Pendek Sedikit asset dan sub-aset di	Skor 4 = Jangka Pendek Tidak sama sekali asset dan sub-aset di dayagunakan Masyarakat
Politik (Implementasi Regulasi dari Kemampuan Hukum); Akses dan Partisipasi dalam pengambilan Keputusan pola terpadu ketahanan masyarakat; Hubungan Gender; Struktur Kekuasaan mempengaruhi kontrol atas pengambilan keputusan dengan Individu, Rumah Tangga, Masyarakat dan Masyarakat Luas dan akses ke Sistem Peradilan dan Lembaga Sektor Keamanan. (Individu, Kelompok dan Lembaga)						

II. Menjawab Pertanyaan Penelitian Tujuan ke 2:. Menganalisa resiko bencana banjir yang bedampak pada mata pencharian masyarakat (Kemampuan/capital/asset yang disebut sebagai 5 Sumber Daya (5 Pentagon Resources berdasarkan pendapat pakar resilience sebelumnya) yang didayakan oleh manusia terhadap lingkungan alam dan lingkungan social ekonomi nya). Di analisa dan dievaluasi dengan nano-resources (9 Sumber Daya), hasil pengamatan dan penemuan peneliti sendiri.

Pertanyaan Penelitian dari Tujuan Kedua:

1. Bagaimana hubungan resiko bencana banjir dan kemampuan masyarakat?
2. Bagaimana Masyarakat memanfaatkan kemampuan/capital/asset pada daya pulih/ketahanan mereka dalam penanganan resiko resiko bencana banjir?
3. Bagaimana masyarakat membangun atau meningkatkan kemampuan/capital/asset (sumber daya) mereka dalam membangun ketahanan/ Daya Pulih mereka?

Hubungan resiko bencana banjir dan kemampuan masyarakat dalam memanfaatkan kemampuan/capital/asset pada daya pulih/ketahanan mereka dan meningkatkan kemampuan/capital/asset ketahanan/ Daya Pulih mereka

Jenis Jenis Resiko dari bencana Banjir yang dialami dan dimiliki oleh Masyarakat (Berdampak pada Masyarakat)	Manajemen Resiko dilihat dari Sub-Parameter Kemampuan Resilience Masyarakat melalui Jika salah satu, dua atau beberapa Sebutkan resiko yang dialami dan dimiliki dari sub-parameter kemampuan resilience Masyarakat Jika semua sub-parameter tandai (√)Jika tidak semua sub-parameter tandai (x)	Pencegahan (Prevention)	Perlindungan (Protection)	Kesiapan (Preparedness)	Pemulihan dan Pelajaran (Recovery and Lesson Learned)
Manusia Pengalaman/ Ketrampilan Pendidikan Kesehatan					

(continued)

(continued)

Jenis Jenis Resiko dari bencana Banjir yang dialami dan dimiliki oleh Masyarakat (Berdampak pada Masyarakat)	Manajemen Resiko dilihat dari Sub-Parameter Kemampuan Resilience Masyarakat melalui Jika salah satu, dua atau beberapa Sebutkan resiko yang dialami dan dimiliki dari sub-parameter kemampuan resilience Masyarakat Jika semua sub-parameter tandai (√)Jika tidak semua sub-parameter tandai (x)	Pencegahan (Prevention)	Perlindungan (Protection)	Kesiapan (Preparedness)	Pemulihan dan Pelajaran (Recovery and Lesson Learned)
Alam Akses Sumber Daya Alam: Tanah Air Margasatwa Tumbuhan Hutan, dll					
Fisik Perumahan Akses Air Bersih dan Sanitasi (PAM) Akses Infrastruktur (Jalan, Jembatan, Sekolah, Rumah Sakit Irigasi, Kendaraan, Kendaraan Umum, Ternak					
Ekonomi Kredit-Mikro Pola Penyesuaian melalui Tabungan, Meminjam Tetangga, Teman dan keluarga, Meminjam dari Kreditur, menjual/ perdagangan/ menukar, penghasilan tetap, akses jaringan dan asuransi					

(continued)

(continued)

Jenis Jenis Resiko dari bencana Banjir yang dialami dan dimiliki oleh Masyarakat (Berdampak pada Masyarakat)	Manajemen Resiko dilihat dari Sub-Parameter Kemampuan Resilience Masyarakat melalui Jika salah satu, dua atau beberapa Sebutkan resiko yang dialami dan dimiliki dari sub-parameter kemampuan resilience Masyarakat Jika semua sub-parameter tandai (√)Jika tidak semua sub-parameter tandai (x)	Pencegahan (Prevention)	Perlindungan (Protection)	Kesiapan (Preparedness)	Pemulihan dan Pelajaran (Recovery and Lesson Learned)
Sosial Grup sosial, Jaringan Kerabat, Kepercayaan Sosial, Norma sosial dan kemampuan-kemampuan sosial yang (Mengikat, menjembatani dan menghubungkan)					
Budaya Peran Pemimpin Adat pada kasus Bencana Banjir Melalui Penguatan Spiritual; ritual dan penguatan individu, kelompok, masyarakat melalui peranan aktif secara langsung, symbol-simbol kekuatan budaya, kearifan local masyarakat					

(continued)

(continued)

Jenis Jenis Resiko dari bencana Banjir yang dialami dan dimiliki oleh Masyarakat (Berdampak pada Masyarakat)	Manajemen Resiko dilihat dari Sub-Parameter Kemampuan Resilience Masyarakat melalui Jika salah satu, dua atau beberapa Sebutkan resiko yang dialami dan dimiliki dari sub-parameter kemampuan resilience Masyarakat Jika semua sub-parameter tandai (√)Jika tidak semua sub-parameter tandai (x)	Pencegahan (Prevention)	Perlindungan (Protection)	Kesiapan (Preparedness)	Pemulihan dan Pelajaran (Recovery and Lesson Learned)
Agama Memberikan pelayanan pastoral (hospitality), perencanaan kesiapsiagaan bencana individu dan keluarga, pelatihan, layanan spiritual, menyediakan komunikasi langsung untuk mengingatkan sistem dan influencer, responden dan fasilitator untuk krisis –krisis mendadak lainnya					

(continued)

(continued)

Jenis Jenis Resiko dari bencana Banjir yang dialami dan dimiliki oleh Masyarakat (Berdampak pada Masyarakat)	Manajemen Resiko dilihat dari Sub-Parameter Kemampuan Resilience Masyarakat melalui Jika salah satu, dua atau beberapa Sebutkan resiko yang dialami dan dimiliki dari sub-parameter kemampuan resilience Masyarakat Jika semua sub-parameter tandai (√)Jika tidak semua sub-parameter tandai (x)	Pencegahan (Prevention)	Perlindungan (Protection)	Kesiapan (Preparedness)	Pemulihan dan Pelajaran (Recovery and Lesson Learned)
Hukum Manajemen Terpadu perlindungan lingkungan alam dan Sosial sesuai regulasi naskah hukum; Manajemen Terpadu melindungi lingkungan Kemampuan Alam dan Sosial bagi (Ketahanan Manusia) ditelaah dari kacamata hukum, peran trial-politik; Legislatif, Eksekutif dan Yudikatif					

(continued)

(continued)

Jenis Jenis Resiko dari bencana Banjir yang dialami dan dimiliki oleh Masyarakat (Berdampak pada Masyarakat)	Manajemen Resiko dilihat dari Sub-Parameter Kemampuan Resilience Masyarakat melalui Jika salah satu, dua atau beberapa Sebutkan resiko yang dialami dan dimiliki dari sub-parameter kemampuan resilience Masyarakat Jika semua sub-parameter tandai (√)Jika tidak semua sub-parameter tandai (x)	Pencegahan (Prevention)	Perlindungan (Protection)	Kesiapan (Preparedness)	Pemulihan dan Pelajaran (Recovery and Lesson Learned)
Politik (Implementasi Regulasi dari Kemampuan Hukum); Akses dan Partisipasi dalam pengambilan Keputusan pola terpadu ketahanan masyarakat; Hubungan Gender; Struktur Kekuasaan mempengaruhi kontrol atas pengambilan keputusan dengan Individu, Rumah Tangga, Masyarakat dan Masyarakat Luas dan akses ke Sistem Peradilan dan Lembaga Sektor Keamanan. (Individu, Kelompok dan Lembaga)					

Hasil, Kesimpulan dan Saran:
TERIMAKASIH TUHAN MEMBERKATI

The manufacturer's authorised representative in the EU is Springer Nature Customer Service Centre GmbH, Europaplatz 3, 69115 Heidelberg, Germany. If you have any concerns regarding our products, please contact ProductSafety@springernature.com

Printed and bound by CPI Group (UK) Ltd, Croydon, CR0 4YY
01/10/2026
02225857-0013